Methods in Molecular Biology

Series Editor
John M. Walker
School of Life and Medical Sciences
University of Hertfordshire
Hatfield, Hertfordshire, AL10 9AB, UK

For further volumes:
http://www.springer.com/series/7651

Erythropoiesis

Methods and Protocols

Edited by

Joyce A. Lloyd

Department of Human & Molecular Genetics, Virginia Commonwealth University, Richmond, VA, USA
Massey Cancer Center, Virginia Commonwealth University, Richmond, VA, USA

Humana Press

Editor
Joyce A. Lloyd
Department of Human & Molecular Genetics
Virginia Commonwealth University
Richmond, VA, USA

Massey Cancer Center
Virginia Commonwealth University
Richmond, VA, USA

ISSN 1064-3745 ISSN 1940-6029 (electronic)
Methods in Molecular Biology
ISBN 978-1-4939-7427-6 ISBN 978-1-4939-7428-3 (eBook)
DOI 10.1007/978-1-4939-7428-3

Library of Congress Control Number: 2017956114

Printed on acid-free paper

This Humana Press imprint is published by Springer Nature
The registered company is Springer Science+Business Media LLC
The registered company address is: 233 Spring Street, New York, NY 10013, U.S.A.

Preface

The aim of this volume is to provide a definitive source for molecular and cellular biology protocols to study erythropoiesis. The target audience is molecular biologists, biochemists, and geneticists who are studying erythropoiesis. The methods span the mouse, human, and zebrafish systems, because comparing model systems has historically provided a comprehensive approach to erythropoiesis.

Red blood cells are the stuff of life, so it is natural that erythropoiesis has been an active area of inquiry for centuries. Yet questions remain, and many of these have clinical relevance to anemia conditions. How does hemoglobin switching occur, and can the clock be turned back to produce fetal hemoglobin to ameliorate hemoglobinopathies? How can enough hematopoietic stem cells or red blood cell progenitors be produced *ex vivo* to treat blood disorders by autologous transplant? Is gene editing a safe option for correcting genetic diseases in stem cells? The detailed protocols in this book provide basic approaches to answer smaller questions within these larger domains of investigation.

The tools used to study erythropoiesis are continuously evolving, and there have been major developments in this arena in the twenty-first century. Mechanistic experiments have progressed from efforts in transformed cell lines to studies in genetically engineered mice, as well as the manipulation of genes and gene expression in primary human erythroid cells in culture. All of these approaches are still in use, and frequently are combined to answer the questions at hand.

This volume provides detailed protocols describing how to use genetics to study mouse and zebrafish erythropoiesis in whole animal models and for genetically manipulating cultured mouse and human erythroid cells. Protocols include strategies to analyze circulating red blood cell parameters, as well as progenitor cell status, stage of differentiation, and enucleation. Methods for the study of erythroid cell gene regulation using chromatin immunoprecipitation and chromatin conformation capture are detailed. Furthermore, protocols are provided to examine stress erythropoiesis, erythroblastic islands, and erythroblast structure. There are reviews on the available mouse models of erythropoiesis, as well as on future considerations for good manufacturing practice in order to translate the manufacture of erythrocytes to the clinic.

The coming years surely will bring new and exciting revelations in the field of erythropoiesis, using methods like those described in this volume. I sincerely thank the authors for their painstaking contributions to this book and their gracious cooperativeness throughout the writing process. I look forward to many innovative discoveries by these authors, and others using the approaches detailed here.

Richmond, VA, USA *Joyce A. Lloyd*

Contents

Contributors

XIULI AN • *Laboratory of Membrane Biology, New York Blood Center, New York, NY, USA; School of Life Science, Zhengzhou University, Zhengzhou, People's Republic of China*

DANIEL E. BAUER • *Division of Hematology/Oncology, Boston Children's Hospital, Boston, MA, USA; Dana-Farber Cancer Institute, Harvard Medical School, Harvard Stem Cell Institute, Boston, MA, USA*

LAURA F. BENNETT • *Department of Veterinary and Biomedical Sciences and Center for Molecular Immunology and Infectious Disease, Laura Bennett and Robert Paulson are Intercollege Graduate Program in Genetics, Robert Paulson and Chang Liao are Pathobiology Graduate Program, The Pennsylvania State University, University Park, PA, USA*

EMERY H. BRESNICK • *Department of Cell and Regenerative Biology, Carbone Cancer Center, University of Wisconsin School of Medicine and Public Health, Madison, WI, USA; UW-Madison Blood Research Program, University of Wisconsin, Madison, WI, USA*

JOHN D.M. CAMPBELL • *MRC Centre for Regenerative Medicine, The University of Edinburgh, Edinburgh, Scotland, UK; Advanced Therapeutics, Scottish National Blood Transfusion Service, National Science Laboratory, Edinburgh, Scotland, UK*

LIXIANG CHEN • *School of Life Science, Zhengzhou University, Zhengzhou, People's Republic of China*

MATTHEW P. COSGRIFF • *Discipline of Anatomy and Histology, School of Medical Sciences, Bosch Institute, University of Sydney, Camperdown, NSW, Australia*

MERLIN CROSSLEY • *School of Biotechnology and Biomolecular Sciences, University of New South Wales, Sydney, NSW, Australia*

ANN DEAN • *Laboratory of Cellular and Developmental Biology, National Institute of Diabetes, Digestive and Kidney Diseases, National Institutes of Health, Bethesda, MD, USA*

VELIA M. FOWLER • *Department of Molecular Medicine, The Scripps Research Institute, La Jolla, CA, USA*

STUART T. FRASER • *Discipline of Anatomy and Histology, School of Medical Sciences, Bosch Institute, University of Sydney, Camperdown, NSW, Australia; Discipline of Physiology, School of Medical Sciences, University of Sydney, Camperdown, NSW, Australia*

XIN GAO • *Department of Cell and Regenerative Biology, Carbone Cancer Center, University of Wisconsin School of Medicine and Public Health, Madison, WI, USA; UW-Madison Blood Research Program, University of Wisconsin, Madison, WI, USA*

GORDON D. GINDER • *Massey Cancer Center, Virginia Commonwealth University, Richmond, VA, USA; Department of Internal Medicine, Virginia Commonwealth University, Richmond, VA, USA*

FARRAH HERMES • *Department of Human and Molecular Genetics, Virginia Commonwealth University, Richmond, VA, USA*

KYLE J. HEWITT • *Department of Cell and Regenerative Biology, Carbone Cancer Center, University of Wisconsin School of Medicine and Public Health, Madison, WI, USA; UW-Madison Blood Research Program, University of Wisconsin, Madison, WI, USA*

MARTIN D. KAFINA • *Division of Hematology, Department of Medicine, Brigham & Women's Hospital, Boston, MA, USA*

THEODOSIA KALFA • *Cancer and Blood Diseases Institute, Cincinnati Children's Hospital Medical Center, Cincinnati, OH, USA; University of Cincinnati Medical School, Cincinnati, OH, USA*
ANNE KONISKI • *Department of Pediatrics, Center for Pediatric Biomedical Research, University of Rochester Medical Center, Rochester, NY, USA*
ANNA KOVILAKATH • *Department of Human and Molecular Genetics, Virginia Commonwealth University, Richmond, VA, USA*
IVAN KRIVEGA • *Laboratory of Cellular and Developmental Biology, National Institute of Diabetes, Digestive and Kidney Diseases, National Institutes of Health, Bethesda, MD, USA*
CHANG LIAO • *Department of Veterinary and Biomedical Sciences and Center for Molecular Immunology and Infectious Disease, Laura Bennett and Robert Paulson are Intercollege Graduate Program in Genetics, Robert Paulson and Chang Liao are Pathobiology Graduate Program, The Pennsylvania State University, University Park, PA, USA*
JOYCE A. LLOYD • *Department of Human and Molecular Genetics, Virginia Commonwealth University, Richmond, VA, USA; Massey Cancer Center, Virginia Commonwealth University, Richmond, VA, USA*
NEIL W.A. MCGOWAN • *MRC Centre for Regenerative Medicine, The University of Edinburgh, Edinburgh, Scotland, UK; Advanced Therapeutics, Scottish National Blood Transfusion Service, National Science Laboratory, Edinburgh, Scotland, UK*
KATHLEEN E. MCGRATH • *Department of Pediatrics, Center for Pediatric Biomedical Research, University of Rochester Medical Center, Rochester, NY, USA*
SKYE C. MCIVER • *Department of Cell and Regenerative Biology, Carbone Cancer Center, University of Wisconsin School of Medicine and Public Health, Madison, WI, USA; UW-Madison Blood Research Program, University of Wisconsin, Madison, WI, USA*
CHARU MEHTA • *Department of Cell and Regenerative Biology, Carbone Cancer Center, University of Wisconsin School of Medicine and Public Health, Madison, WI, USA; UW-Madison Blood Research Program, University of Wisconsin, Madison, WI, USA*
SAFA MOHAMAD • *Department of Human and Molecular Genetics, Virginia Commonwealth University, Richmond, VA, USA*
JOANNE C. MOUNTFORD • *Advanced Therapeutics, Scottish National Blood Transfusion Service, National Science Laboratory, Edinburgh, Scotland, UK*
ROBERTA B. NOWAK • *Department of Molecular Medicine, The Scripps Research Institute, La Jolla, CA, USA*
JAMES PALIS • *Department of Pediatrics, Center for Pediatric Biomedical Research, University of Rochester Medical Center, Rochester, NY, USA*
MATTHEW P. PARKER • *Department of Biochemistry and Molecular Biology, University of Kansas Medical Center, Kansas City, KS, USA*
ROBERT F. PAULSON • *Department of Veterinary and Biomedical Sciences and Center for Molecular Immunology and Infectious Disease, Laura Bennett and Robert Paulson are Intercollege Graduate Program in Genetics, Robert Paulson and Chang Liao are Pathobiology Graduate Program, The Pennsylvania State University, University Park, PA, USA*
BARRY H. PAW • *Division of Hematology, Department of Medicine, Brigham & Women's Hospital, Boston, MA, USA; Department of Medicine, Harvard Medical School, Boston, MA, USA; Department of Pediatrics, Harvard Medical School, Boston, MA, USA; Department of Pediatric Oncology, Dana-Farber Cancer Institute, Boston, MA, USA;*

Hematology-Oncology Division, Department of Medicine, Boston Children's Hospital, Boston, MA, USA; BWH Hematology, Harvard Institutes of Medicine, Boston, MA, USA

KENNETH R. PETERSON • *Department of Biochemistry and Molecular Biology, University of Kansas Medical Center, Kansas City, KS, USA; Department of Anatomy and Cell Biology, University of Kansas Medical Center, Kansas City, KS, USA*

KATE G.R. QUINLAN • *School of Biotechnology and Biomolecular Sciences, University of New South Wales, Sydney, NSW, Australia*

JOSEPHA-CLARA SEDZRO • *Molecular Genetics and Development, Institut de Recherches Cliniques de Montréal, Faculté de Médecine, Université de Montréal, Montreal, QC, Canada*

ALYSON S. SMITH • *Department of Molecular Medicine, The Scripps Research Institute, La Jolla, CA, USA*

MARIE TRUDEL • *Molecular Genetics and Development, Institut de Recherches Cliniques de Montréal, Faculté de Médecine, Université de Montréal, Montreal, QC, Canada*

DIVYA S. VINJAMUR • *Division of Hematology/Oncology, Boston Children's Hospital, Boston, MA, USA*

SHOU ZHEN WANG • *Massey Cancer Center, Virginia Commonwealth University, Richmond, VA, USA*

JIA HAO YEO • *Discipline of Anatomy and Histology, School of Medical Sciences, Bosch Institute, University of Sydney, Camperdown, NSW, Australia*

JINFEN J. YIK • *School of Biotechnology and Biomolecular Sciences, University of New South Wales, Sydney, NSW, Australia*

JING ZHANG • *UW-Madison Blood Research Program, University of Wisconsin, Madison, WI, USA; McArdle Laboratory for Cancer Research, University of Wisconsin, Madison, WI, USA*

Chapter 1

An Introduction to Erythropoiesis Approaches

Joyce A. Lloyd

Abstract

Many experimental models have been used to study erythropoiesis. Even prior to the advent of the genetic manipulation of animal models, erythropoiesis was examined in the mouse, chicken, sheep, goat, and rabbit, among other vertebrates. Erythroid cell lines derived from human blood cancers were also very useful, as they could be genetically manipulated more easily than whole animals. Genetic models in the mouse, zebrafish, and frog have provided a plethora of information advancing our understanding of erythropoiesis, and remain gold standards in the field for studies of hemoglobin switching, and experiments to study authentic blood cell development. Mouse and human embryonic stem (ES) and induced pluripotent (iPS) cells can be differentiated to erythroid cells in culture, though their use is somewhat limited by their propensity to express only the embryonic and fetal globin genes. Some very useful cell lines have been developed by manipulating ES or fetal liver erythroid progenitor cells from knockout mouse models. In recent years, our understanding of erythropoiesis has improved, due to the ability to knock down genes in native human hematopoietic stem and progenitor cells derived from umbilical cord blood or bone marrow, and differentiate them ex vivo to the erythroid lineage. These native cells, and cell lines derived from them, are now providing essential information about human erythropoiesis, which is complementary to that obtained from animal studies. This review provides some perspective about the cell and animal models used to study erythropoiesis over the years.

Key words Erythroid cell lines, Mouse models of erythropoiesis, Zebrafish models of erythropoiesis, Erythroid cell flow cytometry, Globin gene expression, Erythroid chromatin immunoprecipitation assays (ChIP), Chromosome conformation capture (3C)

1 Introduction

Erythropoiesis is the process by which red blood cells are generated from hematopoietic stem and progenitor cells. The study of erythropoiesis has been an active investigative field for many years. In mammals, the first erythroid cells are formed during a transient wave called primitive erythropoiesis, which occurs in the blood islands of the embryonic yolk sac, and by which nucleated cells enter the circulation. Later, definitive erythropoiesis occurs in the mammalian fetal liver, and then in the bone marrow. Definitive

Joyce A. Lloyd (ed.), *Erythropoiesis: Methods and Protocols*, Methods in Molecular Biology, vol. 1698,
DOI 10.1007/978-1-4939-7428-3_1, © Springer Science+Business Media LLC 2018

erythroid cells are enucleated prior to entry into the circulation, and are smaller than embryonic erythroid cells.

This review highlights historic and recently developed methods and approaches used to study erythropoiesis. It focuses mainly on those techniques that exploit genetic models, and particularly on the major systems that are in use today. This review is not meant to be comprehensive, particularly for those methods employed in the past but rarely used today.

Historically, erythropoiesis was studied in the mouse [1, 2], chicken [3], goat [4], sheep [5], and rabbit [6], among other vertebrates. This important work resulted in an understanding of the anatomic structure of the yolk sac blood islands, fetal liver, and bone marrow niches for erythropoiesis. In addition, the globin genes were cloned and sequenced from a variety of organisms. Interestingly, it became clear that multiple α-like and β-like globin genes are present in each organism, leading to pioneering studies of globin gene switching during development. In fact, the earliest attempts to manipulate globin switching as a model for treating human hemoglobinopathies, using pharmaceutical agents that dictate the re-expression of earlier globin genes, were carried out in sheep [7].

As the field has developed, many other systems have been used to study erythropoiesis. Mouse and human cell lines were developed that could be induced to differentiate along an erythroid lineage. Then, genetically modified animals became available for studying erythropoiesis, primarily using the mouse, zebrafish, and Xenopus systems. Genetically modified cell lines, derived from mouse fetal liver or mouse ES cells, have also provided important information. Most recently, it has become possible to genetically manipulate and differentiate native or immortalized hematopoietic stem and progenitor cells (HSPCs) ex vivo. These approaches are discussed in more detail below. This review provides some perspective about the cell and animal models used to study erythropoiesis over the years.

2 Immortalized Cell Lines to Study Erythropoiesis

Immortalized cell lines were useful in studying erythropoiesis because they could be genetically manipulated before this was possible in whole animals. The mouse erythroleukemia (MEL) cell line was derived by infection of mouse splenic cells with the Friend virus. MEL cells can be induced to differentiate into erythroid cells by treatment with dimethyl sulfoxide (DMSO) [8], and express adult hemoglobin.

Erythroid cell lines derived from human blood cancers have also been useful in studying erythropoiesis. The K562 cell line, derived from a patient with chronic myeloid leukemia, was described in 1976 [9]. HEL cells, a human erythroleukemia cell line, were first described in 1982 [10]. These two human cell lines can be chemically induced to differentiate to a more erythroid phenotype and to produce increased amounts of hemoglobin. The globin gene expression profile for these cell lines is similar to embryonic or fetal cells.

MEL, K562, and HEL cells can be transfected with various expression constructs, allowing for gene expression studies. More recently, methods have been developed to use CRISPR/Cas9 to edit genes in MEL and K562 cells, allowing for targeted mutations to be introduced into the cell lines (*see* Chapter 15). A limitation of using these models is that they do not have a normal genetic or chromosomal complement, and cannot be made to developmentally switch (or revert) from expression of one globin to another.

The human umbilical cord blood-derived erythroid progenitor (HUDEP-2) cell line was established by Kurita et al. by immortalizing umbilical cord blood CD34+ cells using lentiviral delivery of a tetracycline-inducible HPV16-E6/E7 transgene [11]. HUDEP-2 cells can be expanded in culture, and most closely resemble adult erythroid cells and express predominantly β-globin after growth in erythroid differentiation media. HUDEP-2 cells can be genetically manipulated by knocking down genes using transduced lentiviral-encoded shRNAs, or can be edited using CRISPR/Cas9, using methods detailed in Chapter 17. Because these cells have characteristics of adult erythroid cells, they can be used to identify genes and treatment agents that act to increase fetal hemoglobin, which is of interest for therapy of the β-hemoglobinopathies. Their limitation is that they represent a transformed cell line, and not native erythroid cells.

3 ES and iPS Cells

Mouse ES cells typically have a normal genetic complement, and they can be differentiated in culture to form erythroid cells. They have been useful in mechanistic work to study the roles of specific genes in erythropoiesis. Upon differentiation, ES cells have a propensity to express only the embryonic and fetal globin genes, limiting their usefulness as a model for adult erythropoiesis [12]. Interestingly, the same phenomenon occurs when human ES cells are differentiated in culture [13].

In recent years, human-induced pluripotent stem cells (iPS) have been manipulated to both correct genetic defects [14], and differentiate into erythroid cells. Although they have an attractive potential use in autologous cellular therapy, erythroid-differentiated iPS cells also typically produce ε- and γ-globin rather than β-globin [15]. Improvements to increase the amount of β-globin produced by these cells [16], and scale up the production of red blood cells in culture (*see* Chapter 18), are currently underway to increase therapeutic utility.

4 Animal Models for Erythropoiesis

The mouse, zebrafish, and Xenopus have been particularly useful systems for the study of erythropoiesis using genetically manipulated whole animal models. These models remain the gold standards in the field for studies of hemoglobin switching and investigation of authentic blood development. Although hemoglobin switching occurs in these organisms, there is only one embryonic to adult switch, not an embryonic to fetal followed by a fetal to adult switch, as in humans.

In the mid-1990s, Leonard Zon and colleagues initiated the use of the zebrafish as a genetic model to study erythropoiesis and globin gene regulation [17, 18]. This model system is highly useful because although vertebrates, zebrafish have a shorter generation time than mice, and afford the ability to examine many more progeny than with mice. *Xenopus* has also been a useful model because it is amenable to lineage tracing [19]. Many genes important in erythropoiesis have counterparts in zebrafish, Xenopus, and mammals. A benefit of the zebrafish system is the capacity to perform large genetic screens, and these have informed our current understanding of the genes required for erythropoiesis [20]. For zebrafish erythropoiesis protocols, *see* Chapter 2. Interestingly, although most erythropoietic processes are similar in mammals and zebrafish, there is no specific equivalent of the important mammalian erythroid transcription factor, Krüppel-like factor 1 or KLF1, in zebrafish.

Mice are an attractive model system due to their relatively short gestation period and because erythropoiesis in mice closely mirrors that in humans. In order to study regulation of the human globin genes in an in vivo system, the first transgenic mice carrying human globin genes were developed by Magram et al. [21] and Townes et al. [22]. The expression of the genes in these early models was not copy number-dependent, and was dependent on their position of integration in the mouse genome. With the advent of the discovery of the β-globin locus control region (LCR) [23, 24], it became possible to generate transgenic mice having the entire human β-globin locus. The first β-locus transgenic mice were developed using strategies to ligate two cosmid constructs [25]. Later, β-YAC transgenic mice with yeast artificial chromosome vectors (YACs) spanning the entire 150–250 kb human β-globin locus and flanking regulatory sequences were generated [26, 27]. These mice express the human globin genes in a tissue- and developmental-stage specific manner, and at levels similar to the mouse globin genes. The human ε- and γ-globin genes are expressed in yolk sac erythroid cells and the human β-globin gene is expressed in fetal liver-derived and bone marrow-derived erythroid cells. The β-YAC mice have been instrumental in studying numerous aspects of globin gene regulation in vivo. The limitation of this model is that human γ-globin is expressed embryonically in the mouse, rather than as a fetal gene, thus deviating from its expression pattern in humans.

Most erythroid-specific transcription factors, including GATA1 and KLF1, have counterparts in the mouse. Knockout models for erythroid specific transcription factors like GATA1 [28] and KLF1 [29, 30] have been extremely important in understanding mammalian erythropoiesis. In addition, several important human disease models, including mice with sickle cell disease, were developed in the early 1990s [31–35].

Mouse knockout and disease models are discussed in depth in the review in Chapter 3, and methods to analyze red cell parameters in mouse models are detailed in Chapter 6. It is often important to assess whether a genetic mutation or treatment condition affects the number or nature of the erythroid progenitor cell population. To assess this, erythroid progenitor cell assays in semisolid media can be employed. Chapter 7 provides a detailed protocol for analyzing erythroid progenitors using colony-forming assays.

Some mouse mutations affect the ability of the mouse to mount a stress erythropoiesis response, the number of erythroid progenitor cells that develop, or the integrity of the erythroblastic islands, where red blood cells develop in the bone marrow. Methods to analyze such mouse mutants are detailed in Chapters 5, 7, and 8, respectively. Chapter 12 describes high-resolution fluorescence microscope imaging methods to study erythroblast structure.

5 Cell Lines Derived from Mouse Embryonic Stem and Fetal Liver Erythroid Cells

Some very useful cell lines have been developed by transforming mouse embryonic stem cells or fetal liver erythroid progenitor cells, already having genetic modifications of transcription factors or other genes essential for erythropoiesis. For example, Weiss et al. generated an erythroid cell line from in vitro-differentiated GATA1 null murine embryonic stem (ES) cells, termed G1E [36]. These cells lack the erythroid transcription factor GATA1, until induced to express the GATA1 gene. B1.6 cells were derived from fetal liver cells from a mouse lacking KLF1, which were transduced with a J2 retrovirus to immortalize them, and infected with a retrovirus expressing an inducible KLF1 gene [37]. G1E and B1.6 cells have been used to examine the roles of GATA1 and KLF1, respectively, in erythroid differentiation. Blau et al. developed a cell line from β-YAC transgenic mice, chemical inducer of dimerization

(CID)-dependent cells, which can be used to test agents that could potentially increase fetal hemoglobin for therapeutic purposes [38].

6 Characterizing and Genetically Manipulating Native Erythroid Progenitor Cells

In recent years, our understanding of erythropoiesis has improved, due to the ability to knock down genes in mouse and human native erythroid progenitor cells. The use of native cells rather than cell lines has obvious advantages, and is often required for high impact publications in the current day. Methods for using mouse fetal liver cells to knock down and overexpress genes are discussed in Chapter 4. The approach involves enrichment of hematopoietic stem/progenitor cells, expansion ex vivo, infection with a retrovirus encoding an shRNA or cDNA, and ex vivo differentiation along the erythroid lineage.

Human hematopoietic stem and progenitor cells derived from umbilical cord blood and bone marrow can also be differentiated ex vivo to the erythroid lineage. These native cells are now providing essential information about human erythropoiesis, which is complementary to that obtained from mouse studies. In the last decade, researchers have refined methods of primary human cell culture to obtain larger numbers of erythroid cells from relatively few CD34+ hematopoietic precursor cells obtained from bone marrow, bone marrow mobilized cells in peripheral blood, or umbilical cord blood [39, 40]. Methods to isolate human CD34+ cells from umbilical cord blood, and expand and differentiate them in culture to obtain erythroid cells are detailed in Chapter 16. The hematopoietic stem and progenitor cells can be manipulated to knock down any gene of choice, using lentiviral particles expressing negative control (scramble) or gene-specific shRNAs [41–43].

7 Identifying and Characterizing Erythroid Cells

After erythroid differentiation in vivo or in culture, the majority of human erythroid cells have the cell surface markers, transferrin receptor and glycophorin A (GPA). Mouse erythroid cells express transferrin receptor and TER-119, a molecule associated with glycophorin A, in differing amounts at different stages of differentiation [44, 45]. Flow cytometry (FCM) or fluorescence-activated cell sorting (FACS) using antibodies for these cell surface markers is often employed to enrich for erythroid cells. Recently, a more precise method has been developed allowing for better isolation of distinct stages of erythroblasts from primary human bone marrow cells or from an in vitro CD34+ culture system. This method uses the surface markers GPA, band 3, and α4 integrin to examine terminal erythroid differentiation [46]. Methods for FCM and

FACS employing this newer separation protocol are included in Chapter 9. A more specialized protocol to assess the final stage of differentiation, erythroblast enucleation, using FCM is detailed in Chapter 11.

Imaging flow cytometry (IFC) is a recently developed technique, which provides bright field and fluorescent images allowing for the analysis of large numbers of cells, yet supplying morphological data for single cells. This technology can be used to examine erythroblast stages, as well as cell shape changes, heterochromatin, cell cycle, apoptosis, and enucleation of erythroid cells. Chapter 10 details methods for IFC analysis of erythroid intermediates and intracellular structures in primary erythroid cells.

8 Chromatin Immunoprecipitation (ChIP) and Chromosome Conformation Capture (3C)

Several important cell-specific and ubiquitous transcription factors are required to specify the erythroid lineage, and ensure the expression of globin and other crucial erythroid genes. Furthermore, the LCR enhancer, transcription factors, and chromatin modifications are important in maintaining a chromatin structure of the globin loci that is compatible with correct developmental gene expression. Therefore, an understanding of the landscape for transcription factor binding in normal erythroid cells, and in disease models, is essential to improve our understanding of erythropoiesis. This is typically accomplished through the use of chromatin immunoprecipitation (ChIP) assays (*see* Chapter 13). Also, changes in chromosome interactions throughout development, and in various disease states, have provided insights with regard to the architecture of the globin loci. Direct interactions between the various components of the locus, particularly between the LCR and the gene promoters, are required for globin gene expression. These interactions can be assessed through the use of chromatin conformation capture (3C) assays (*see* Chapter 14). Chapters 13 and 14 provide detailed ChIP and 3C protocols, specifically for use with erythroid samples.

9 Future Development of Erythropoiesis Methods

The mouse remains the most practical animal model in which to study authentic erythropoiesis. However, as discussed in Chapter 3, the model has its limitations. One of these is that in mice, the human γ-globin genes are primarily expressed in the yolk sac rather than the fetal liver, as is the case in man. Furthermore, the γ-globin gene is almost completely silenced in adult mice, rather than having a residual amount of expression as in man. Multiple transcription factors that are important in globin switching, such as KLF1 and BCL11A, significantly differ in their sequences or developmental expression patterns between mouse and man. Therefore, to improve the mouse model system it may be possible to develop humanized mice, not just with the human globin loci, but also with relevant human transcription factor genes replacing the mouse genes. It would also be important that the human transcription factor genes are engineered to have developmental expression patterns that match those in man. This approach may not be straightforward, but could potentially greatly improve the value of the mouse as a model system, particularly for testing treatments and drug therapies designed to elevate fetal hemoglobin.

Although this book is not primarily focused on clinical applications of erythropoiesis methods, it is clear that the field is moving closer to clinical applications for the treatment of genetic diseases. One attractive possibility is the future use CRISPR/Cas 9 gene editing to cure genetic mutations in resulting erythroid cells. The ability to correct genes is already available, but improving and assessing the safety of gene editing, and the ability to obtain and expand hematopoietic stem cells is still under development. There is promise in correcting mutations in iPS cells, because this would allow for autologous transplants, but much work remains to be done to make this a feasible system. The current promise and challenges of scaling up the production of red blood cells for therapeutic uses are reviewed in Chapter 18.

Acknowledgments

Anna Kovilakath provided thoughtful suggestions and critical evaluation of this manuscript.

References

1. Haar JL, Ackerman GA (1971) A phase and electron microscopic study of vasculogenesis and erythropoiesis in the yolk sac of the mouse. Anat Rec 170:199–223
2. Haar JL, Ackerman GA (1971) Ultrastructural changes in mouse yolk sac associated with the initiation of vitelline circulation. Anat Rec 170:437–455
3. Ginder GD, Wood WI, Felsenfeld G (1979) Isolation and characterization of recombinant clones containing the chicken adult beta-globin gene. J Biol Chem 254:8099–8102
4. Haynes JR, Rosteck P Jr, Schon EA, Gallagher PM, Burks DJ, Smith K, Lingrel JB (1980) The isolation of the beta A-, beta C-, and gamma-globin genes and a presumptive embryonic globin gene from a goat DNA recombinant library. J Biol Chem 255:6355–6367
5. Garner KJ, Lingrel JB (1988) Structural organization of the beta-globin locus of B-haplotype sheep. Mol Biol Evol 5:134–140
6. Evans MJ, Lingrel JB (1969) Hemoglobin messenger ribonucleic acid. Synthesis of 9S and ribosomal ribonucleic acid during erythroid cell development. Biochemistry 8:3000–3005
7. Perrine SP, Swerdlow P, Faller DV, Qin G, Rudolph AM, Reczek J, Kan YW (1989) Butyric acid modulates developmental globin

gene switching in man and sheep. Adv Exp Med Biol 271:177–183

8. Scher W, Parkes J, Friend C (1977) Increased carbonic anhydrase activity in Friend erythroleukemia cells during DMSO-stimulated erythroid differentiation and its inhibition by BrdU. Cell Differ 6:285–296
9. Klein E, Ben-Bassat H, Neumann H, Ralph P, Zeuthen J, Polliack A, Vanky F (1976) Properties of the K562 cell line, derived from a patient with chronic myeloid leukemia. Int J Cancer 18:421–431
10. Martin P, Papayannopoulou T (1982) HEL cells: a new human erythroleukemia cell line with spontaneous and induced globin expression. Science 216:1233–1235
11. Kurita R, Suda N, Sudo K, Miharada K, Hiroyama T, Miyoshi H, Tani K, Nakamura Y (2013) Establishment of immortalized human erythroid progenitor cell lines able to produce enucleated red blood cells. PLoS One 8: e59890
12. Lindenbaum MH, Grosveld F (1990) An in vitro globin gene switching model based on differentiated embryonic stem cells. Genes Dev 4:2075–2085
13. Chang KH, Nelson AM, Cao H, Wang L, Nakamoto B, Ware CB, Papayannopoulou T (2006) Definitive-like erythroid cells derived from human embryonic stem cells coexpress high levels of embryonic and fetal globins with little or no adult globin. Blood 108:1515–1523
14. Chang CJ, Bouhassira EE (2012) Zinc-finger nuclease-mediated correction of alpha-thalassemia in iPS cells. Blood 120:3906–3914
15. Bouhassira EE (2013) Therapeutic potential of hematopoietic cells derived from pluripotent stem cells. Expert Opin Biol Ther 13:1099–1102
16. Uchida N, Haro-Mora JJ, Fujita A, Lee DY, Winkler T, Hsieh MM, Tisdale JF (2017) Efficient generation of beta-globin-expressing erythroid cells using stromal cell-derived induced pluripotent stem cells from patients with sickle cell disease. Stem Cells 35:586–596
17. Chan FY, Robinson J, Brownlie A, Shivdasani RA, Donovan A, Brugnara C, Kim J, Lau BC, Witkowska HE, Zon LI (1997) Characterization of adult alpha- and beta-globin genes in the zebrafish. Blood 89:688–700
18. Stainier DY, Weinstein BM, Detrich HW III, Zon LI, Fishman MC (1995) Cloche, an early acting zebrafish gene, is required by both the endothelial and hematopoietic lineages. Development 121:3141–3150
19. Ciau-Uitz A, Monteiro R, Kirmizitas A, Patient R (2014) Developmental hematopoiesis: ontogeny, genetic programming and conservation. Exp Hematol 42:669–683
20. Orkin SH, Zon LI (1997) Genetics of erythropoiesis: induced mutations in mice and zebrafish. Annu Rev Genet 31:33–60
21. Magram J, Chada K, Costantini F (1985) Developmental regulation of a cloned adult beta-globin gene in transgenic mice. Nature 315:338–340
22. Townes TM, Lingrel JB, Chen HY, Brinster RL, Palmiter RD (1985) Erythroid-specific expression of human beta-globin genes in transgenic mice. EMBO J 4:1715–1723
23. Forrester WC, Takegawa S, Papayannopoulou T, Stamatoyannopoulos G, Groudine M (1987) Evidence for a locus activation region: the formation of developmentally stable hypersensitive sites in globin-expressing hybrids. Nucleic Acids Res 15:10159–10177
24. Grosveld F, van Assendelft GB, Greaves DR, Kollias G (1987) Position-independent, high-level expression of the human beta-globin gene in transgenic mice. Cell 51:975–985
25. Strouboulis J, Dillon N, Grosveld F (1992) Developmental regulation of a complete 70-kb human beta-globin locus in transgenic mice. Genes Dev 6:1857–1864
26. Porcu S, Kitamura M, Witkowska E, Zhang Z, Mutero A, Lin C, Chang J, Gaensler KM (1997) The human beta globin locus introduced by YAC transfer exhibits a specific and reproducible pattern of developmental regulation in transgenic mice. Blood 90:4602–4609
27. Peterson KR, Clegg CH, Navas PA, Norton EJ, Kimbrough TG, Stamatoyannopoulos G (1996) Effect of deletion of 5′HS3 or 5′HS2 of the human beta-globin locus control region on the developmental regulation of globin gene expression in beta-globin locus yeast artificial chromosome transgenic mice. Proc Natl Acad Sci U S A 93:6605–6609
28. Fujiwara Y, Browne CP, Cunniff K, Goff SC, Orkin SH (1996) Arrested development of embryonic red cell precursors in mouse embryos lacking transcription factor GATA-1. Proc Natl Acad Sci U S A 93:12355–12358
29. Perkins AC, Sharpe AH, Orkin SH (1995) Lethal beta-thalassaemia in mice lacking the erythroid CACCC-transcription factor EKLF. Nature 375:318–322
30. Nuez B, Michalovich D, Bygrave A, Ploemacher R, Grosveld F (1995) Defective haematopoiesis in fetal liver resulting from inactivation of the EKLF gene. Nature 375:316–318

31. Fabry ME, Costantini F, Pachnis A, Suzuka SM, Bank N, Aynedjian HS, Factor SM, Nagel RL (1992) High expression of human beta S- and alpha-globins in transgenic mice: erythrocyte abnormalities, organ damage, and the effect of hypoxia. Proc Natl Acad Sci U S A 89:12155–12159
32. Trudel M, Saadane N, Garel MC, Bardakdjian-Michau J, Blouquit Y, Guerquin-Kern JL, Rouyer-Fessard P, Vidaud D, Pachnis A, Romeo PH (1991) Towards a transgenic mouse model of sickle cell disease: hemoglobin SAD. EMBO J 10:3157–3165
33. Ryan TM, Townes TM, Reilly MP, Asakura T, Palmiter RD, Brinster RL, Behringer RR (1990) Human sickle hemoglobin in transgenic mice. Science 247:566–568
34. Greaves DR, Fraser P, Vidal MA, Hedges MJ, Ropers D, Luzzatto L, Grosveld F (1990) A transgenic mouse model of sickle cell disorder. Nature 343:183–185
35. Rubin EM, Witkowska HE, Spangler E, Curtin P, Lubin BH, Mohandas N, Clift SM (1991) Hypoxia-induced in vivo sickling of transgenic mouse red cells. J Clin Invest 87:639–647
36. Weiss MJ, Yu C, Orkin SH (1997) Erythroid-cell-specific properties of transcription factor GATA-1 revealed by phenotypic rescue of a gene-targeted cell line. Mol Cell Biol 17:1642–1651
37. Coghill E, Eccleston S, Fox V, Cerruti L, Brown C, Cunningham J, Jane S, Perkins A (2001) Erythroid Krüppel-like factor (EKLF) coordinates erythroid cell proliferation and hemoglobinization in cell lines derived from EKLF null mice. Blood 97:1861–1868
38. Blau CA, Barbas CF III, Bomhoff AL, Neades R, Yan J, Navas PA, Peterson KR (2005) {gamma}-Globin gene expression in chemical inducer of dimerization (CID)-dependent multipotential cells established from human {beta}-globin locus yeast artificial chromosome ({beta}-YAC) transgenic mice. J Biol Chem 280:36642–36647
39. Fibach E, Manor D, Oppenheim A, Rachmilewitz EA (1989) Proliferation and maturation of human erythroid progenitors in liquid culture. Blood 73:100–103
40. Migliaccio G, Di Pietro R, di Giacomo V, Di Baldassarre A, Migliaccio AR, Maccioni L, Galanello R, Papayannopoulou T (2002) In vitro mass production of human erythroid cells from the blood of normal donors and of thalassemic patients. Blood Cells Mol Dis 28:169–180
41. Gnanapragasam MN, Scarsdale JN, Amaya ML, Webb HD, Desai MA, Walavalkar NM, Wang SZ, Zu ZS, Ginder GD, Williams DC Jr (2011) p66Alpha-MBD2 coiled-coil interaction and recruitment of Mi-2 are critical for globin gene silencing by the MBD2-NuRD complex. Proc Natl Acad Sci U S A 108:7487–7492
42. Amaya M, Desai M, Gnanapragasam MN, Wang SZ, Zu ZS, Williams DC Jr, Ginder GD (2013) Mi2beta-mediated silencing of the fetal gamma-globin gene in adult erythroid cells. Blood 121:3493–3501
43. Vinjamur DS, Alhashem YN, Mohamad SF, Amin P, Williams DC Jr, Lloyd JA (2016) Krüppel-like transcription factor KLF1 is required for optimal gamma- and beta-globin expression in human fetal erythroblasts. PLoS One 11:e0146802
44. Zhang J, Socolovsky M, Gross AW, Lodish HF (2003) Role of Ras signaling in erythroid differentiation of mouse fetal liver cells: functional analysis by a flow cytometry-based novel culture system. Blood 102:3938–3946
45. Socolovsky M, Nam H, Fleming MD, Haase VH, Brugnara C, Lodish HF (2001) Ineffective erythropoiesis in Stat5a(-/-)5b(-/-) mice due to decreased survival of early erythroblasts. Blood 98:3261–3273
46. Hu J, Liu J, Xue F, Halverson G, Reid M, Guo A, Chen L, Raza A, Galili N, Jaffray J, Lane J, Chasis JA, Taylor N, Mohandas N, An X (2013) Isolation and functional characterization of human erythroblasts at distinct stages: implications for understanding of normal and disordered erythropoiesis in vivo. Blood 121:3246–3253

Chapter 2

Using the Zebrafish as an Approach to Examine the Mechanisms of Vertebrate Erythropoiesis

Martin D. Kafina and Barry H. Paw

Abstract

The zebrafish, *Danio rerio*, is a powerful model for the study of erythropoiesis and defining the genetic basis of hematological diseases. The mechanisms of erythroid differentiation are highly conserved in the zebrafish, permitting translational research studies and the modeling of erythropoiesis in higher vertebrates. An advantage of the system is the ability to manipulate gene expression and observe the effect on erythroid development in vivo, with relative ease and rapidity. The production of optically transparent embryos also makes it an attractive tool for visual analysis of circulating erythrocytes that can be used to study erythropoiesis. Through large-scale chemical mutagenesis screens, a variety of zebrafish blood mutants have been identified that are used for gene discoveries and the recapitulation of human diseases. Experimental techniques including in situ hybridization, *o*-dianisidine staining, flow cytometry, and microinjection are now commonly employed to study red blood cell biochemistry and erythropoiesis in the zebrafish. These techniques have been applied for identifying novel genes required for the hemoglobin synthesis, isolating blood cell lineages, visualizing genetic expression within erythroid tissues, and characterizing the phenotype of blood disorders. The applications of zebrafish methodology to the study of erythropoiesis and optimized step-by-step protocols are discussed in this chapter.

Key words Microinjection, Flow cytometry, In situ hybridization, *o*-dianisidine staining, Hemoglobin, *Danio rerio*

1 Introduction

The zebrafish is one of the most important systems for modeling erythropoiesis of humans and higher vertebrates. The power of using the zebrafish, a teleost, as an elegant model for developmental hematology dates back to the early 1960s [1]. At that time, researchers described the anlage of blood production in zebrafish, a structure known as the intermediate cell mass (ICM). This intraembryonic structure, analogous to the blood islands within the yolk sac of other species, is the site of primitive erythropoiesis. The primitive wave of erythropoiesis is marked by the production of essential hematopoietic transcription factors including *gata-1* and embryonic *globins* [2]. Embryonic erythroid cells begin entering

Joyce A. Lloyd (ed.), *Erythropoiesis: Methods and Protocols*, Methods in Molecular Biology, vol. 1698,
DOI 10.1007/978-1-4939-7428-3_2, © Springer Science+Business Media LLC 2018

circulation at 24 h post fertilization (hpf) and by 96 hpf hematopoietic cells can be detected in the zebrafish kidney, the counterpart to the mammalian hematopoietic bone marrow [3, 4]. Whole kidney marrow in the zebrafish contains all mature blood cell types including erythrocytes, myelomonocytes, and lymphocytes [5]. In the 1990s, the pioneers of zebrafish genetic screens identified more than 50 blood mutants with defective hematopoiesis by chemical-induced mutagenesis [6]. The genetic and molecular defects of these blood mutants enhanced our understanding of the pathology and molecular pathways associated with many human blood disorders [2]. Over the years, techniques including whole mount in situ hybridization (WISH), flow cytometry, *o*-dianisidine staining, and microinjection for gene knockdown or overexpression have been employed to pursue hematology research in zebrafish. These methods are often utilized in other vertebrate systems, including mice and in cell culture, for in vitro investigation. However, these systems lack many features found in zebrafish, which increase their experimental potential.

Zebrafish are a robust developmental and genetic system for studying erythropoiesis due to their rapid sexual maturity, high fecundity, and survivability in large numbers. Whereas mammalian embryogenesis takes place in utero, external fertilization and optical transparency of zebrafish embryos allows direct microscopic examination of organogenesis, visualization of blood development, and experimental manipulation of embryos ex utero. During zebrafish blood development, WISH allows the precise localization of gene expression including the patterning of blood transcription factors during early development. *o*-Dianisidine staining reveals sites of erythropoiesis and cellular hemoglobinization, which can be used to characterize the phenotype associated with a blood disorder. Flow cytometry and fluorescence-activated cell sorting (FACS) in transgenic zebrafish can be used to isolate cells for morphological analysis and to quantify specific blood lineages. Microinjection in zebrafish embryos enables researchers to manipulate gene expression by silencing or driving cRNA overexpression to reveal molecular function and recapitulate human diseases. These techniques have been demonstrated to be successful in the zebrafish, and instrumental for genetic dissection of the molecular mechanisms of erythropoiesis. They are introduced here and described in detail in Subheading 3 below.

WISH is valuable as a technique to localize a specific nucleic acid sequence for the study of gene expression. WISH can be used to detect particular transcripts with high spatial and temporal resolution by utilizing a labeled complementary probe to a target sequence. Hybridization of the probe to the target sequence reveals the organization of mRNA species within histologic sections or whole embryos. This protocol is especially useful for the analysis of gene expression during embryonic development and after experimental manipulation.

In the study of erythropoiesis, WISH allows tracing of erythropoietic genes and monitoring of the expression pattern during disease pathogenesis. The visualization of mRNA transcripts has been employed in studies of transcription factor expression patterns during early hematopoiesis, screening candidate genes required for heme production, and analyzing loss of function phenotypes.

The reproductive capability of the zebrafish coupled with WISH is a powerful method to study a targeted gene. Designing an experiment that utilizes WISH requires the selection of a target sequence and a corresponding readout [7]. First, a linearized template is prepared that contains the cDNA for the probe. After purifying the linearized DNA template, a labeled antisense RNA probe is synthesized for hybridization to the target mRNA species. Visualization and staining detection of the hybrid is achieved by immunohistochemistry using an alkaline phosphatase-conjugated antibody against digoxigenin and a chromogenic substrate [7].

An example where WISH was used to verify the expression of candidate erythropoietic genes is the detection of *gata1* in *cloche* and *dino* zebrafish mutants. The *cloche* mutant lacks endocardial cells, hematopoietic stems cells, and vascular progenitors due to deficient expression of *gata* transcription factors [8]. Genes involved in the production of blood cells are absent in the ICM of *cloche* mutants. In contrast, the *dino* zebrafish mutant, characterized by a ventralized phenotype [9] and an expanded ICM, contains upregulated erythropoietic markers. WISH in these zebrafish blood mutants provided convincing evidence for candidate erythropoietic genes by revealing their expression in erythroid tissue. Figure 1 (top panel) shows prominent expression of *gata-1* in the blood island ICM in zebrafish embryos at 24 hpf, whereas the *gata-1* WISH signal is absent in *cloche* mutants and expanded in *dino* mutants. This expression pattern is characteristic of erythroid transcription factors and provides a reference for screening candidate genes required for heme production and erythropoiesis. Chen and colleagues found a similar expression pattern for *sorting nexin 3* (*snx3*), suggesting a role of *snx3* in erythropoiesis (Fig. 1, bottom panel) [10].

Amigo and colleagues utilized WISH to elegantly illustrate the expression pattern of *friend-of-gata1* (*fog-1*) (Fig. 2a–e) in juxtaposition with *gata1 and -2* expression (Fig. 2f–o) during the early stages of hematopoiesis. The expression pattern of *fog-1/gata* genes can be traced to evaluate their interaction and developmental onset during early hematopoiesis. WISH of *fog-1* in the zebrafish mutants *cloche*, *vlad tepes*, and *frascati* reveals coordinated mechanisms of *fog-1* expression and highlights its expression in the ICM, heart, and areas of neural tissue (Fig. 2p–s) [11].

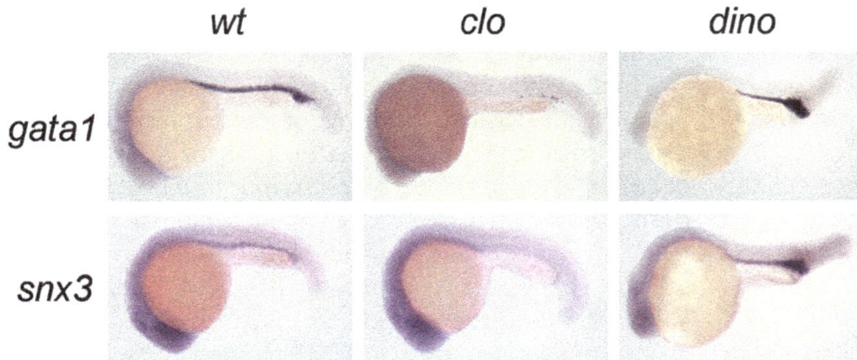

Fig. 1 Snx3 Is Required for Heme Production in zebrafish Embryos. Whole-mount in situ hybridization detects expression of *gata1* (*top panel*) and *snx3* (*bottom panel*) in the blood island or intermediate cell mass (ICM) in zebrafish embryos at 24 hpf. The ICM staining is absent in *cloche* mutants and is expanded in *dino* ventralized mutants. Reprinted with permission from Chen et al. [10] and *Cell Metabolism* (Elsevier, Inc)

Amigo and colleagues also discovered that knockdown of *fog-1* reduces ISH signal of erythroid marker *band-3* (Fig. 3a, b). *Band-3* is the major protein of the human erythrocyte membrane, and associates with multiple metabolic, ion transport, and cytoskeletal integrity (Fig. 3d–g). *Band-3* expression, a surrogate marker for erythroid differentiation, was rescued by coinjection of *fog-1* cRNA in the morphant embryos (Fig. 3c), suggesting a direct role of *fog-1* in erythroid differentiation [11].

These studies emphasize the flexibility and high resolution of WISH to reveal the dynamics of erythropoietic networks. WISH distinguishes the zebrafish as one of the most informative systems to detect and compare the expression patterns of different genes in the same specimen [12]. WISH in the zebrafish is fast, sensitive, and requires only a fragment of the gene of interest for three dimensional analyses of in vivo gene expression.

Histological *o*-dianisidine staining is used to visualize erythrocytes and detect hemoglobin. *o*-dianisidine staining is useful for a qualitative assessment of erythroid development and the study of anemia. O'Brien and colleagues demonstrated the identification of hemoglobin by its catalase reaction with peroxide and *o*-dianisidine [13]. Iuchi and Yamamoto adapted the detection method to fish by detecting *o*-dianisidine positive sites in rainbow trout embryos [14]. The method can also produce elegant staining in zebrafish embryos for the pronounced hemoglobin detection in the blood islands of the yolk sac. This method was applied during the discovery of the essential mitochondrial solute carrier for iron, mitoferrin1. Mitoferrin1 (*mfrn, slc25a37*) is the principle importer of iron into the mitochondria of vertebrate erythropoietic cells. A mutation in the *mfrn* gene is found in the zebrafish mutant *frascati* (*frs*) [15]. *Frs* mutants have a deficiency in mitochondrial iron uptake and consequently have defects in hemoprotein production that lead

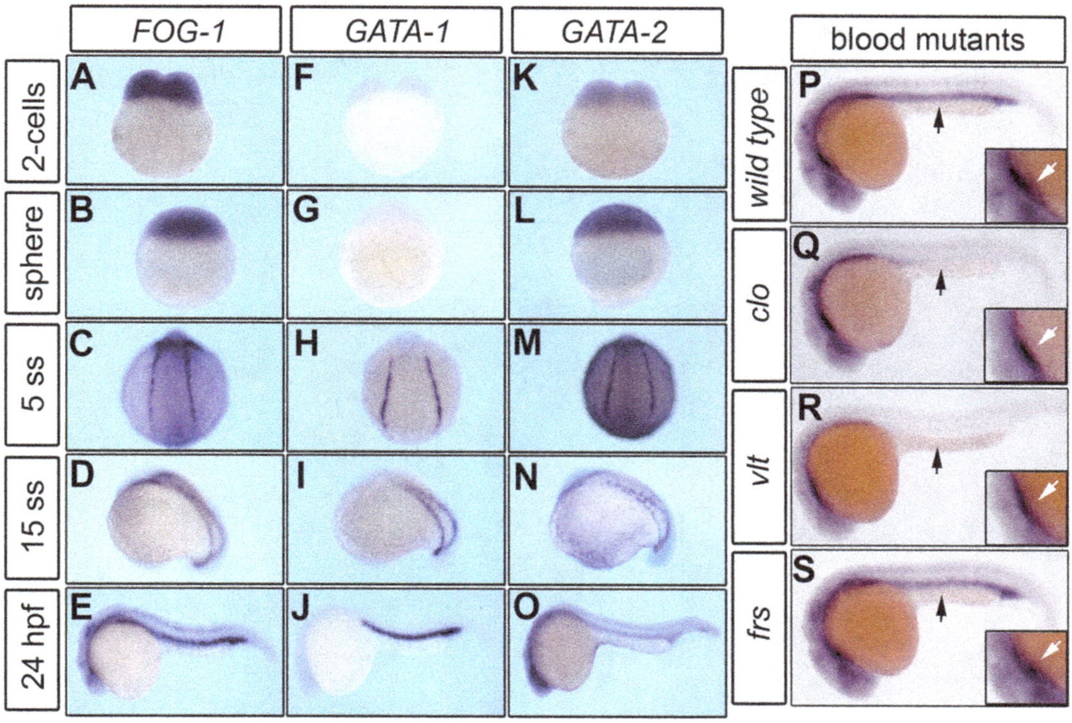

Fig. 2 *FOG-1* expression pattern during early hematopoiesis. The expression pattern of *FOG-1* and *GATA* factors was visualized by whole mount in situ hybridization (WISH) during zebrafish embryonic development (**a–o**). The maternal expression of *FOG-1* (**a**, **b**) and *GATA-2* (**k–l**) transcripts is initially noted at the 2-cell and sphere stages. The zygotic expression pattern of *FOG-1* in the blood island is coincident with the expression of *GATA-1* and *GATA-2* in the lateral plate mesoderm (LPM) (**c**, **h**, **m**) and the ICM (**d**, **e**, **i**, **j**, **n**, **o**). *FOG-1* is highly expressed in the ICM and neural tissues of a wild-type embryo at 24 hpf (**p**). Mutants involved in defective formation of hematopoietic (*cloche, clo*; **q**) or erythroid progenitors (*vlad tepes, vlt*; **r**) are deficient in the expression of *FOG-1* in the ICM (*black arrow*), whereas expression in the heart is preserved (*white arrow*, *inset* magnification). The expression of *FOG-1* in the ICM is normal in a mutant with a defect in late erythroid maturation (*frascati, frs*; **s**). Reprinted with permission from Amigo et al. [11] and *Blood* (The American Society of Hematology)

to profound hypochromic anemia. *frs* embryos are noticeably pale with an absence of hemoglobinization in the yolk sac and heart at 2–3 dpf, which is confirmed by *o*-dianisidine staining (Fig. 4, top panel). The defect in iron homeostasis continues into the juvenile stage of development, and is associated with cardiomegaly, resulting from elevated cardiac output due to the chronic anemia (Fig. 4, bottom panel).

Frs mutants also show that *o*-dianisidine staining is a powerful tool to verify that a candidate gene is disrupted in a particular blood mutant. Hemoglobin was measured by *o*-dianisidine staining in wild-type embryos (Figs. 5a and 6, panel 1), uninjected *frs* embryos (Figs. 5b and 6, panel 2), and mutant embryos injected with either zebrafish (Fig. 5a, panels 3–5) or human (Fig. 6, panels 3–5) *mfrn*

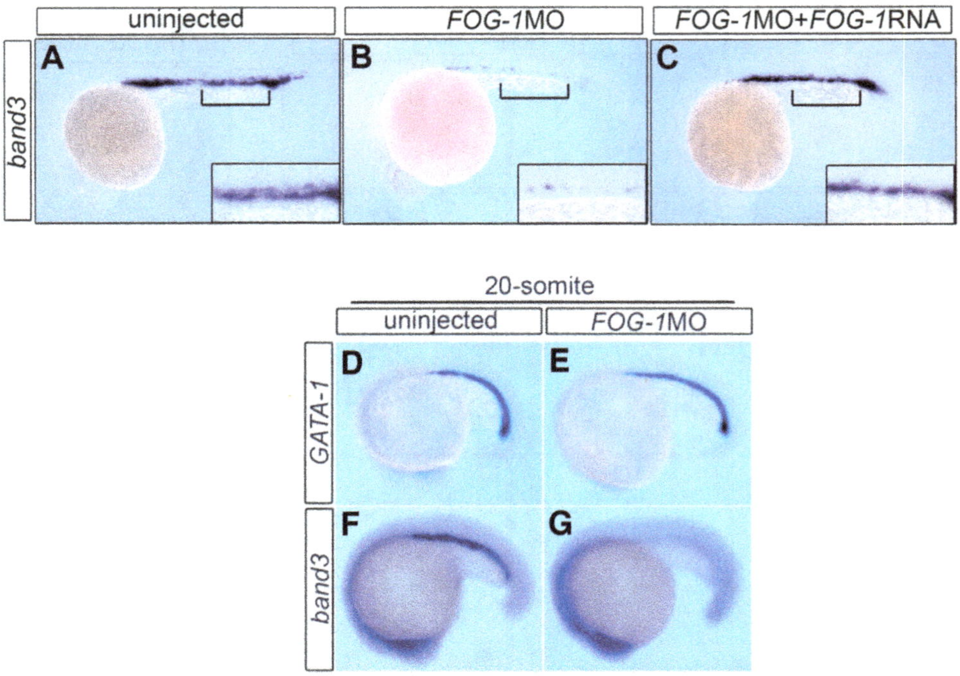

Fig. 3 *FOG-1* is necessary for erythroid/thrombocytic development and dispensable for hematopoietic progenitors specification in zebrafish. Uninjected control embryos (**a**), *FOG-1* MO-injected embryos (**b**), and *FOG-1* MO coinjected with *FOG-1* cRNA (**c**). Lateral views of 24-hpf embryos processed by WISH to reveal *band3* (*slc4a1*) expression (**a**–**c**). *FOG-1* morphant embryos display reduction or complete absence of *band3* expression in the ICM (**a**, **b** *black brackets*, *inset magnification*). The anemic phenotype is fully rescued by coinjection of *FOG-1* cRNA in the morphant embryos (**c**). Uninjected control embryos (**d**, **f**) and *FOG-1* MO injected embryos (**e**, **g**). Lateral views of fixed embryos at 20-somite stages were processed by WISH, revealing *GATA-1* (**d**, **e**) and *band3* (**f**, **g**) expression. At 20-somite stages, loss of FOG-1 function results in a drastic reduction of committed erythroid progenitors expressing band3 (**f**, **g**) suggesting that *FOG-1* is required for erythroid maturation. Reprinted with permission from Amigo et al. [11] and *Blood* (The American Society of Hematology)

cRNA. Hemoglobinization was restored in the injected *frs* embryos to wild-type levels. By using *o*-dianisidine staining as an indicator of phenotypic rescue, ectopic expression of *mfrn* could rescue anemia in *frs* embryos [15, 16]. Taken together, the injections robustly restored hemoglobinization and served as verification that *frs* zebrafish carry a mutation in the *mfrn* gene. This was confirmed by genotyping of the rescued animals (Fig. 5b). These experiments demonstrate that *o*-dianisidine staining is an experimental readout for confirming an anemic phenotype. A step-by-step protocol for performing *o*-dianisidine staining of zebrafish embryos according to Iuchi and Yamamoto is outlined below [14].

Flow cytometry is a quantitative method for the isolation of blood lineages and can be used to study defects in zebrafish blood mutants. Production of all major blood cell types in the adult

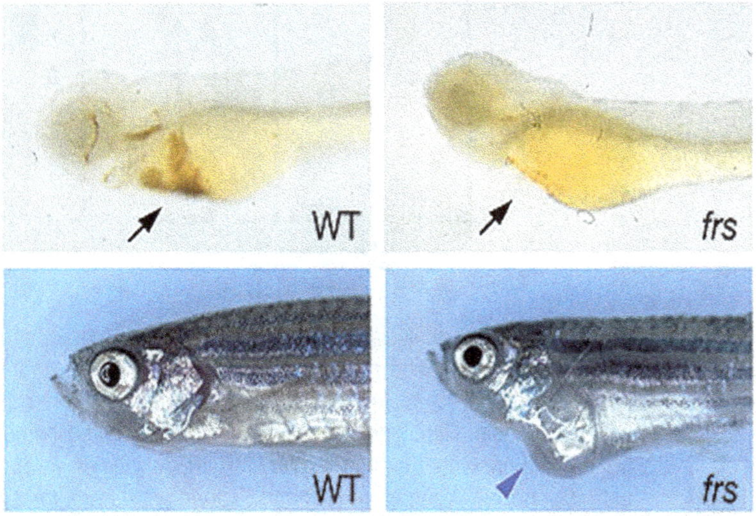

Fig. 4 Analysis of the zebrafish *frs* mutation. *frs* embryos and juvenile fish are profoundly anemic. Wild-type and *frs* embryos were stained to detect hemoglobinized cells with *o*-dianisidine (*top panel*). Juvenile *frs* fish exhibit cardiomegaly from chronic anemia (*bottom panel*). Reprinted with permission from Shaw et al. [15] and *Nature* (Macmillan Publishers, Ltd)

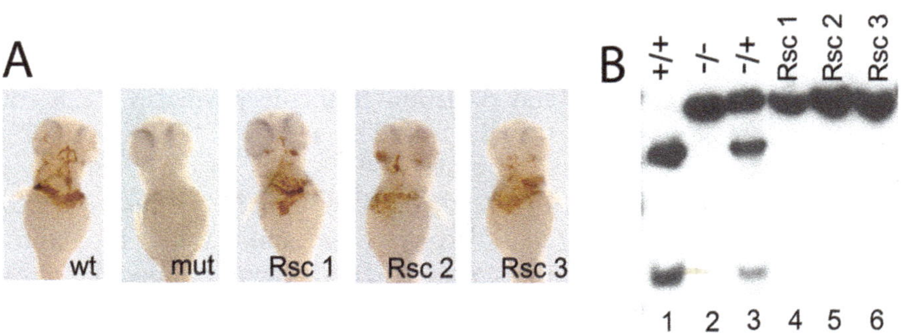

Fig. 5 Functional studies of the *mfrn* gene. (**a**) Ectopic expression of *mfrn (slc25a37)* cRNA rescues the anemia in *frascati (frs)* embryos. Control wild-type, *frs* mutant (mut), and rescued *frs* embryos (Rsc 1–3) were stained with *o*-dianisidine to detect hemoglobinized cells. (**b**) Control wild-type (*lane 1*), mutant (*lane 2*), heterozygote (*lane 3*), and rescued *frs* (*lanes 4–6*) embryos were genotyped by restriction enzyme digestion with *BsrI* for the frs^{tq223} mutation, confirming the phenotypically rescued embryos were indeed genotypic *frs* mutants. Reprinted with permission from Shaw et al. [15] and *Nature* (Macmillan Publishers, Ltd)

zebrafish occurs in the whole kidney marrow, which morphologically resembles the mammalian hematopoietic bone marrow and contains all mature blood cell types including erythrocytes, myelomonocytes, and lymphocytes [5]. Flow cytometric separation of these blood lineages is based on the combined light scattering profiles of forward scatter (FSC) and side scatter (SSC), which are directly proportional to cell size and cellular granularity, respectively. FSC-low subsets are composed entirely of mature

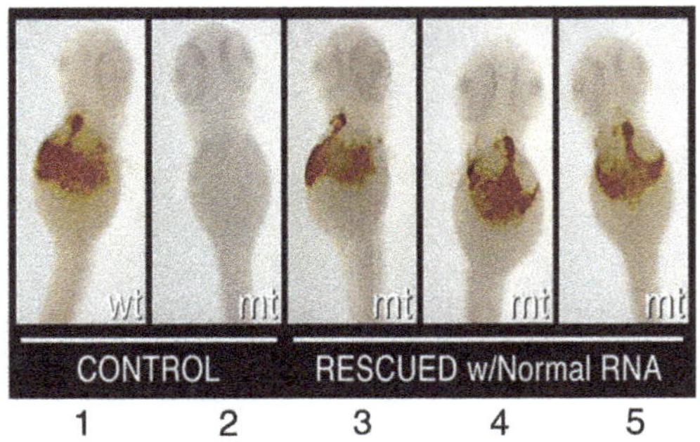

Fig. 6 Functional analysis of the normally and abnormally processed *MFRN1* cDNA. Complementation of anemia in *frascati* zebrafish embryos with normal human *MFRN1* cRNA. Wild-type (wt) and *frascati* (mt) embryos are shown in the first two panels. Genotypic mt embryos exhibit restored hemoglobinized cells after injection with the normal *MFRN1* cRNA (*panels 3–5*). Reprinted with permission from Wang et al. [16] and *Experimental Hematology* (Elsevier B.V.)

erythrocytes and FSC-high/SSC-high subsets are composed only of mature myelomonocytic cells [5]. Alterations of these populations including cell count deviation and morphological abnormalities can be quantitated by flow cytometry.

Transcription factors *gata-1* and *fog-1* are essential for erythrocyte and megakaryocyte development. Disruption of these transcriptional regulators leads to impaired differentiation of blood progenitors and deviation from the normal red cell count. The flow cytometric assay used in Fig. 7 is based on FSC and SSC light scattering profiles. GFP+ cells in the zebrafish strain *Tg (GATA-1:eGFP)* were purified by FACS (fluorescence-activated cell sorting) to isolate erythroid and myeloid progenitors of the ICM. Compared to the wild-type blood cell populations (Fig. 7a), knockdown of FOG-1 led to expansion of the myeloid lineage at the expense of the erythroid lineage (Fig. 7b). Additionally, FACS purification of the GFP+ cells revealed an underlying pathology associated with loss of *fog-1*. Histologic analysis of sorted GFP+ cells from the *fog-1* morphants revealed myelomonocytic cells with dyserythropoietic morphology (Fig. 7d). In contrast, the control embryos contained predominantly erythroblasts (Fig. 7c) [11]. These observations provide evidence for a regulatory role of *fog-1* in the commitment of hematopoietic progenitors to the erythroid: myeloid lineages. This study emphasizes the precision and sensitivity of flow cytometry-based profiling to isolate hematopoietic lineages and reveal deviations from their basal values.

Other strategies for sorting cell populations based on the gating of cell surface markers have emerged. Previous transplantation studies on irradiated zebrafish of the strain *Tg(CD41:eGFP)* identified CD41 GFP low cells and CD41 GFP high cells. The two populations were identified as hematopoietic stem cells (HSCs) and

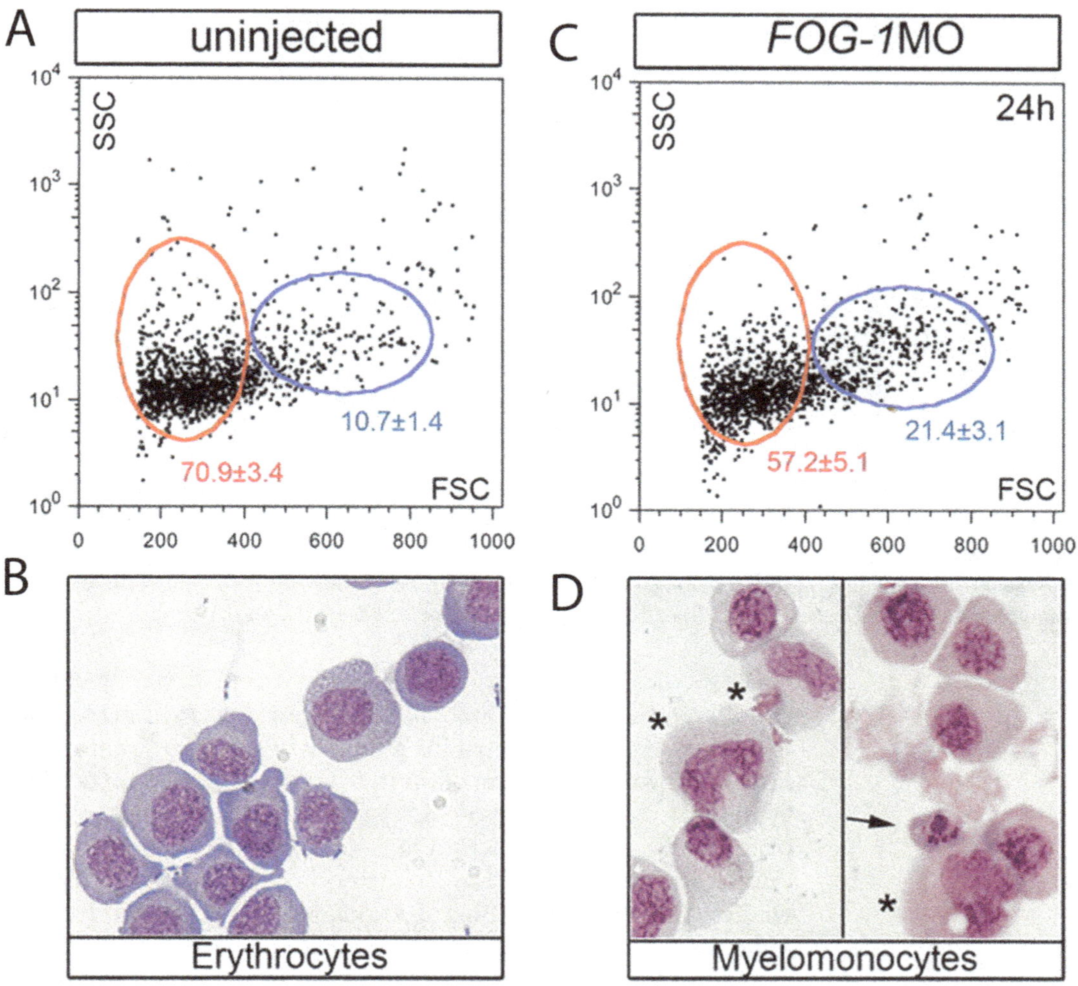

Fig. 7 Loss of *FOG-1* expands myelopoiesis at the expense of erythropoiesis. Gated erythroid (*red circle*) and myeloid cells (*blue circle*) are shown using forward (FSC) and side scatter (SSC) flow cytometry (**a** and **c**). The eGFP+ cells were purified by fluorescence-activated flow cytometry (FACS) (**b** and **d**). Histologic analysis of sorted eGFP+ cells from *FOG-1* morphants shows myelomonocytic (**d** *asterisks*) and dyserythropoietic morphology (**d** *arrows*); in comparison, the sorted cells from control embryos are predominantly erythroblasts (**b**). A representative from three independent experiments is shown. Populations of cells within the gate are enumerated as mean percentages of total cells ± SD, showing statistically significant differences (*$p < 0.05$; **a** and **c**). Reprinted with permission from Amigo et al. [11] and *Blood* (The American Society of Hematology)

thrombocytes respectively [17]. The identifications were based on cellular ability to reconstitute hematopoiesis and ultrastructural morphology [18]. These distinct populations of GFP-tagged CD41 cells were applied to a study involving a family of hematopoietic transcription factors associated with *growth factor independence* (*gfi*) genes. In mammals, *Gfi1a* and *Gfi1b* function as hematopoietic transcription factors with distinct and overlapping roles as regulators of HSCs and subsequent lineage commitment; namely, *Gfi1a* is crucial for myeloid and T-cell development, while

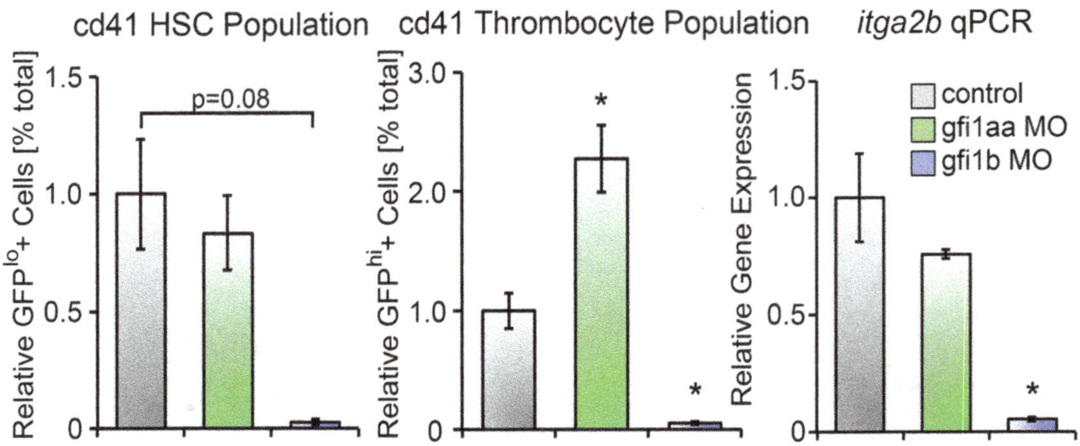

Fig. 8 Loss of *gfi1b* silences definitive HSC. FACS of *Tg(cd41:eGFP)* embryos at 96 hpf injected with MO targeting either *gfi1aa* or *gfi1b* reveals a marked reduction in the GFP^{lo} and GFP^{hi} populations in *gfi1b* morphants (mean ± SEM, *t*-test, $^*p < 0.05$, $n = 3$). qRT-PCR of *gfi* morphants shows significantly reduced expression of *itga2b* (*cd41*) in *gfi1b* morphants at 96 hpf (mean ± SEM, *t*-test, $^*p < 0.05$, $n = 3$). Reprinted with permission from Cooney et al. [19] and *Developmental Biology* (Elsevier B.V.)

Gfi1b is crucial for erythroid and megakaryocytic development. In the zebrafish, there are three *gfi* family genes: *gfi1ab, gfi1aa, and gfi1b* [19]. *gfi1aa* is essential for primitive hematopoiesis; however, little is known about its function in definitive hematopoiesis. Flow cytometric measurements of CD41 GFP+ cells in zebrafish suggest distinct roles for the transcription factor paralogs, *gfi1aa and gfi1b*, during primitive and definitive hematopoiesis and define a temporal window for their functioning. FACS analysis of CD41 GFP+ cells indicates a reduction in HSC (primitive) population from loss of *gfi1b* (Fig. 8, left panel) and an increase in thrombocyte (definitive) population after loss of *gfi1aa* (Fig. 8, right panel) [19]. The flow cytometry of transgenic zebrafish with gene knockdown studies of the *gfi* genes allows one to ascribe a hierarchical, epistatic relationship relative to regulators of primitive or definitive hematopoiesis.

The development of transgenic lines *Tg(GATA-1:eGFP)*, *Tg (CD41:GFP)*, and *Tg(lcr-globin:GFP)* that express fluorescently labeled markers as reporters for blood lineages has enabled studies of erythropoiesis via FACS in zebrafish. FACS can be performed in zebrafish blood mutants to determine cellular alterations and the presence of erythroid precursors in the strain *Tg(GATA-1:eGFP)*. Thrombocyte populations can be quantified by utilizing the surface marker CD41 [17] and hemoglobinized cell expression can be measured in the zebrafish strain *Tg(lcr-globin:eGFP)* [20]. Flow cytometry can identify underlying pathology, deviations in blood cell counts, and the functional role of genes in the development of specific blood lineages.

FACS isolation of GFP+ cells in the zebrafish strain *Tg(lcr-globin:eGFP)* [20] was performed in this study to measure the

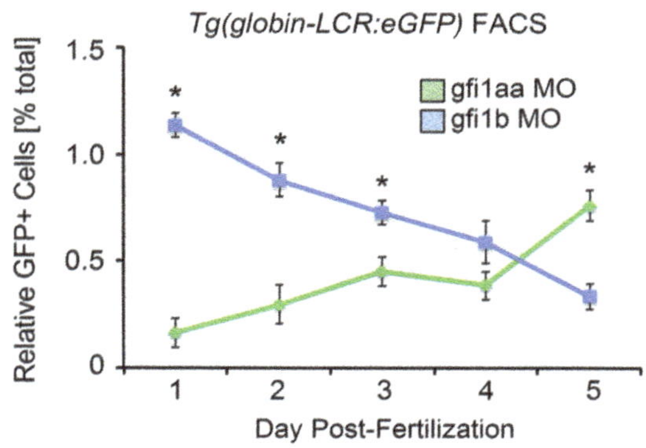

Fig. 9 *gfi1aa* and *gfi1b* have distinct roles at different stages in erythropoiesis. FACS of *Tg(lcr-globin:eGFP)* embryos injected with MO targeting either *gfi1aa* (*green*) or *gfi1b* (*blue*) quantifies the relative change in erythrocytes over time, showing that loss of either *gfi1aa* or *gfi1b* differentially impacts erythropoiesis at different stages of development (mean $\pm$ SEM, *t*-test, $*p < 0.05$, $n = 3$). Reprinted with permission from Cooney et al. [19] and *Developmental Biology* (Elsevier B.V.)

change in erythrocytes derived from primitive vs. definitive hematopoiesis over time after knockdown of both the *gfi1aa* and *gfi1b* genes. *Tg(lcr-globin:eGFP)* embryos, which contain erythrocytes tagged with GFP, are a crucial tool for quantifying erythropoiesis [20]. According to the FACS data for *gfi* morphants, the genetic paralogs function at precise stages of erythropoiesis. Loss of *gfi1aa* leads to a reduction of hemoglobinized cells that significantly recovers by 5 days post fertilization (dpf), when definitive erythropoiesis is underway. A reciprocal relationship can be observed in the *gfi1b* morphants that become progressively anemic in later stages of erythropoiesis (Fig. 9) [19]. These measurements provide an interesting loss of function analysis suggesting a temporal relationship between the *gfi* genes in their regulation of erythroid development and their ontogeny.

In vivo injection studies in the zebrafish can be used for the identification and validation of disease-carrying mutations, or as a system for recapitulating defects in erythropoiesis. Microinjection can be performed on fertilized zebrafish embryos at the unicellular stage of development. Microinjection can be used for introducing DNA or RNA during early stages of zebrafish development, allowing the study of protein expression. For example, injection of complementary RNA can facilitate the expression of proteins carrying mutations found in clinical hematology cases. Alternatively, delivery of an antisense oligomer, termed a morpholino (MO), can effectively silence genes to generate a transient null phenotype [21]. However, rigorous experimental controls need to be implemented for proper interpretation of the results from MO injections [22]. MOs provide tools for obstructing cellular processes

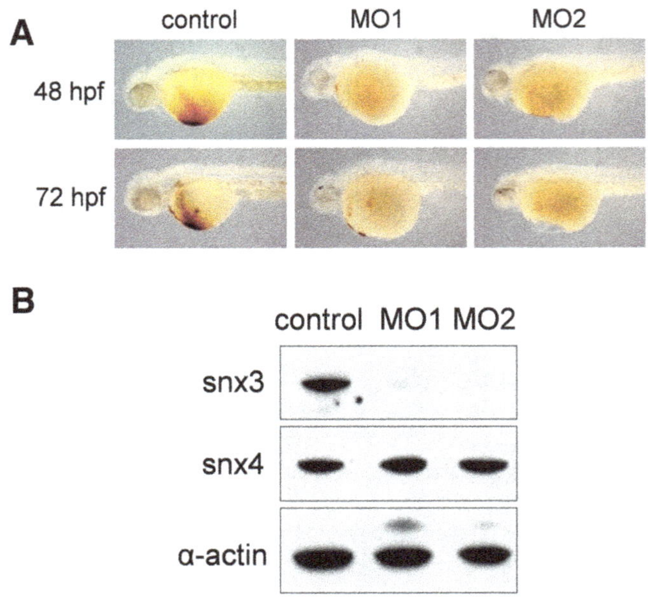

Fig. 10 Snx3 is required for heme production in zebrafish embryos. (**a**) Injections of MOs against snx3 result in a profound anemia in zebrafish embryos at both 48 and 72 hpf. (**b**) Western analysis confirms efficient silencing of snx3 by a translation-blocking morpholino (MO1) and a pre-mRNA splice-blocking morpholino (MO2). The expression of snx4 and α-actin control proteins was unaffected, ensuring the specificity in knockdown for snx3. Reprinted with permission from Chen et al. [10] and *Cell Metabolism* (Elsevier, Inc)

including translation initiation and pre-mRNA splicing. Translation blocking MOs work by steric inhibition of the translation initiation complex. Assays including western analysis are needed to confirm knockdown of protein expression. Figure 10 demonstrates MO knockdown of *Snx3*, which results in an anemic phenotype characterized by a reduction of *o*-dianisidine positive cells at 48 and 72 hpf (Fig. 10a) [10]. Western analysis confirms that the MO targets snx3, resulting in a disappearance of the snx3 protein band, while potential off targets including snx4 and actin remain intact. Thus, Fig. 10 demonstrates an approach to control for the specificity of MO injections that is critical when interpreting a loss of function phenotype (i.e., anemia). Another effective class of MO is the splice blocking MO, which blocks sites involved in the splicing of pre-mRNA. After controlling for normal RNA processing, quantification of steady-state mRNA by quantitative RT-PCR (qRT-PCR) can measure knockdown efficiency of the target mRNA species [23]. Please refer to http://www.gene-tools.com/morpholino_antisense_oligos for an in-depth explanation of MO oligomers.

When microinjection is coupled with flow cytometry and various staining techniques, pathologic mechanisms of hematologic diseases can be elucidated. Hildick-Smith and colleagues employed MO technology to transiently knock down *SFXN4* expression.

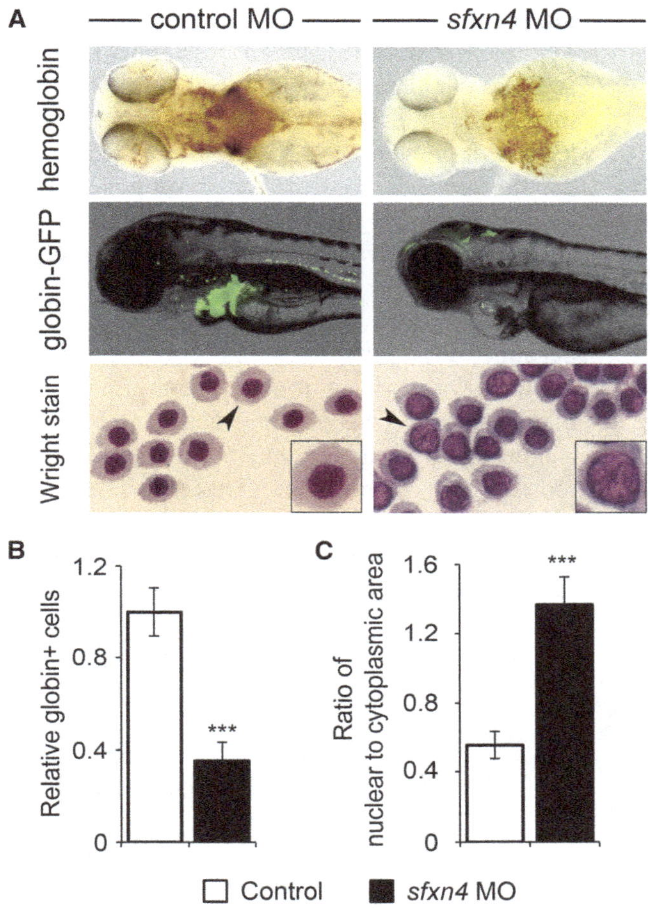

Fig. 11 *sfxn4* knockdown in zebrafish recapitulates the defects in erythropoiesis and mitochondrial respiration seen in individual one. Phenotypic characterization of control and zebrafish morphants for *sfxn4*. Zebrafish *sfxn4* morphants showed a defect in hemoglobinization (*brown color*) when stained with *o*-dianisidine (*upper*). The anemia in *sfxn4* morphants was evident by the reduction of GFP$^+$ erythroid cells in the *Tg(lcr-globin:eGFP)* transgenic zebrafish reporter line (*middle*). Cytospin analyses of flow-sorted erythroid cells from the control (*left*) and *sfxn4* morphant (*right*) revealed large nuclei with noncondensed chromatin in the latter (*lower*). Enlarged magnifications of individual cells (*arrow*) are shown in the *insets*. Reprinted with permission from Hildick-Smith et al. [21] and *American Journal of Human Genetics* (Elsevier, Inc.)

The MO used in this study was engineered to inhibit *SFXN4* RNA splicing. In embryos injected with *SFXN4* MO, termed "morphants," hemoglobinization and erythroid maturation were determined by *o*-dianisidine staining and globin-GFP levels by flow cytometry in *Tg(lcr-globin:GFP)* transgenic embryos. Compared to control embryos, morphants have reduced *o*-dianisidine staining (Fig. 11a, top panel) and globin-GFP expression (Fig. 11a, middle panel). Precise quantification and isolation of globin-GFP cells by FACS revealed significant anemia (Fig. 11b) and morphologic evaluation of the sorted cells revealed an increased ratio of erythroid nuclei to cytoplasmic area, which is characteristic of maturation arrest (Fig. 11a, bottom panel) [24]. This study is an example of the specificity and versatility of microinjection that can easily manipulate the genetics of zebrafish erythropoiesis. Experimental manipulation of early-stage zebrafish embryos by microinjection, followed by measurements of erythroid development, can produce essential information about gene function and the genetic basis of hematologic disease.

A distinct advantage of carrying out injection studies in zebrafish includes the functional and evolutionary conservation of genetic pathways between fish and humans. Genetic conservation permits translational research in the zebrafish making them an ideal platform for studying erythropoiesis and carrying out drug screening. In an injection rescue experiment by Hildick-Smith and colleagues, MO-mediated knockdown of *SFXN4* led to a significant reduction in hemoglobinized cells. The anemic phenotype was rescued by combining an equimolar concentration of SFXN4 morpholino and *SFXN4* cRNA from either *Danio rerio*Zebrafish or *Homo sapiens* (Fig. 12) [24]. This study demonstrates the

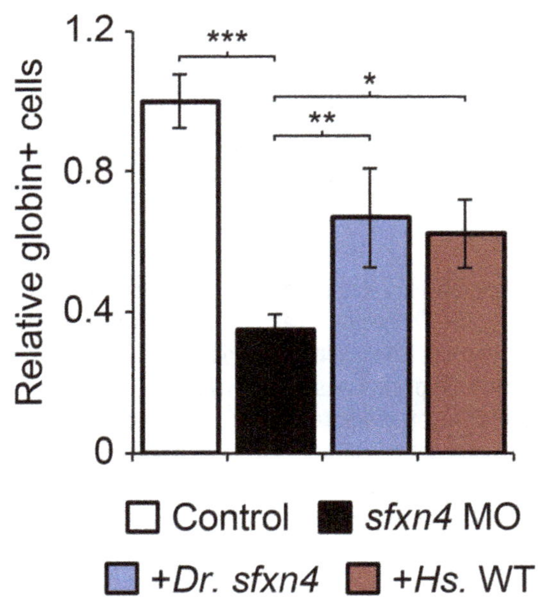

Fig. 12 *SFXN4* Is functionally conserved across vertebrate species in both erythropoiesis and mitochondrial respiration. Normal *SFXN4* cRNA from either zebrafish or human partially complemented the anemia in *sfxn4* morphant embryos (**$p < 0.005$, *$p < 0.05$). Zebrafish and human *FLAG-SFXN4* cDNA constructs in pXT7 were used for generating 5′ capped mRNA with the use of the mMessage mMachine T7 Kit (Ambion). The generated mRNA was mixed with MO at an equimolar concentration to the MO injection and injected at the 1-cell stage. The translational expression of the transgenic mRNA was confirmed by immunoblot analysis with FLAG antisera. Reprinted with permission from Hildick-Smith et al. [21] and *American Journal of Human Genetics* (Elsevier, Inc.)

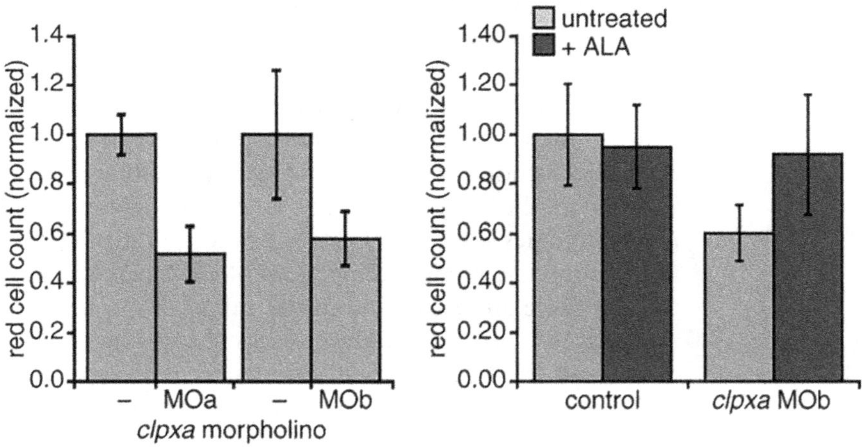

Fig. 13 mtClpX is important for vertebrate heme biosynthesis and erythropoiesis. (*Left*) Erythrocyte development at 72 hpf was quantified by flow cytometry, using dissociated cells from *Tg(lcr-globin:eGFP)* zebrafish. $p \leq 0.01$ for erythrocyte reduction by *clpxa* knockdown with either morpholino. (*Right*) Rescue of *clpxa* MOb-induced anemia by ALA supplementation. *Tg(lcr-globin:eGFP)* zebrafish embryos were supplemented with 2 mM ALA from 24 to 72 hpf, upon which GFP+ erythrocytes were quantified by flow cytometry. $p = 0.025$ for rescue of anemia in *clpxa* knockdown embryos by ALA supplementation. Reprinted with permission from Kardon et al. [22] and *Cell* (Elsevier, Inc.)

functional conservation of a mitochondrial protein across vertebrate species [16].

Phenotypic rescue experiments by chemical supplementation can be designed using microinjection. Following injections, zebrafish embryos are dechorionated at 24 hpf for direct absorbance of a chemical. Kardon and colleagues demonstrate that knockdown of a critical mitochondrial heme biosynthesis unfoldase, *ClpX*, leads to a reduced red cell count in zebrafish (Fig. 13, left). ClpX interacts with the initial enzyme in proto-heme biosynthesis, aminolevulinic acid synthase (ALAS), and activates its function. Knockdown of Clpx phenocopies a deficiency of ALAS. Supplementation of zebrafish Clpx or ALAS morphants with aminolevulinic acid (ALA) significantly rescues the anemic phenotype by-passing the biochemical block (Fig. 13, right) [25]. This observation suggests a direct role of *ClpX* in the synthesis of ALA. Hence, this application of chemical complementation using microinjection highlights the potential of using the zebrafish as a tool for drug identification, chemical biology studies, and a gateway for clinical investigation [26].

The ease of manipulating genetic expression in the zebrafish provides experimental scientists with a tool to analyze the erythropoietic gene network. Analysis of injections by flow cytometry, *o*-dianisidine staining, in vivo live imaging, and other readouts for erythroid maturation can produce important information about the molecular mechanisms of erythropoiesis. The zebrafish has emerged as an ideal model for the study of vertebrate erythropoiesis. Advantages of using the zebrafish for experimental hematology include optical transparency to visualize organogenesis in real time, rapid embryological development, ease of gene silencing to study loss-of-function phenotype in vivo by anti-sense MO, ectopic overexpression studies, chemical biology, and highly conserved hardwiring of the hematopoietic genetic program. The research methodologies of in situ hybridization, *o*-dianisidine staining, flow cytometry, and microinjections of MO and cRNA have proven to be reliable and successful in the zebrafish model system. These methods have constituted the basis for the tracing of erythropoietic genes and monitoring their expression patterns during disease pathogenesis, analyzing blood cell morphology, measuring degree of hemoglobinization, and manipulating the genetic pathways of erythropoiesis. New areas of technology to generate targeted mutations in genes using CRISPR/cas9 genome editing in the zebrafish in recent years will provide additional tools for this model organism [27–29]. The use of zebrafish has been applied to several areas of clinical hematology research, mainly in the domains of drug discovery, forward genetic screens, evaluating loss-of-function phenotypes, and recapitulating human disease states. Zebrafish have been demonstrated to be instrumental in the understanding of blood cell development and discovering therapies for disorders of erythropoiesis.

2 Materials

2.1 RNA Run-Off with Digoxigenin-UTP or Fluorescein-UTP

1. 10× DIG RNA Labeling Mix or 10× Fluorescein RNA Labeling Mix (Roche).
2. 20 U/μl (stock) T7, T3 or SP6 RNA polymerase.
3. 10× T7 RNA transcription buffer.
4. RNasin Plus.
5. RNase-free DNase.
6. G-50 Sephadex Quick Spin Columns.
7. Qiagen Qiaquick PCR purification kit.

2.2 Whole Mount In Situ Hybridization: Day 1 Solutions, Materials, and Equipment

1. Methanol (100%).
2. 4% Paraformaldehyde (PFA): 800 mL H_2O, 100 mL of 10× PBS, 400 μl of 10 M NaOH, 40 g PFA, H_2O to 1000 mL. Bring the solution up to ~60 °C (warm to the touch) while mixing with a mixing bar. When PFA is in solution (~5 min) let it cool to RT. Bring pH to ~7.4 using 15–20 drops of conc. HCl, checking with pH paper. Bring the final volume up to 1000 mL. Store in 50 mL aliquots at −20 °C.

3. PBST: 100 mL 10× PBS, 10 mL 10% Tween 20, 890 mL H_2O.
4. Proteinase K solution: 400 μl of 10 mg/ml Proteinase K, 200 mL of 1× PBS.
5. 20× SSC: 175.3 g NaCl, 88.2 g Na-citrate, H_2O to 1000 mL. Adjust pH to 7.0 with concentrated HCl.
6. Hyb^- solution: 100 mL formamide, 50 mL 20× SSC, 2 mL 10% Tween 20, add H_2O up to 200 mL.
7. Hyb^+ solution: 100 mL formamide, 50 mL 20× SSC, 2 mL 10% Tween 20, 1 g RNA from torula yeast type VI, 200 μl of 5 USP units/μl Heparin, add H_2O to 200 mL.
8. Corning Netwell carriers.
9. Hybridization oven.
10. Platform orbital shaker.
11. Rocking shaker.

2.3 Whole Mount In Situ Hybridization: Day 2 Solutions

1. 2× SSCT, 50% formamide: 50 mL 20× SSC, 250 mL formamide, 5 mL 10% Tween 20 H_2O to 500 mL.
2. 2× SSCT: 20 mL 20× SSC, 2 mL 10% Tween 20, H_2O to 200 mL.
3. 0.2× SSCT: 4 mL 20× SSC, 4 mL 10% Tween 20, H_2O to 400 mL.
4. MABT pH 7.4: 23.2 g Maleic acid, 17.53 g NaCl, 55 g Tris–HCl, 20 mL 10% Tween 20, H_2O to 2 L. Dissolve maleic acid, NaCl and Tris–HCl in water, adjust pH to 7.4 with concentrated HCl, then adjust volume to 2 L.
5. Block solution: 30 mL heat inactivated lamb serum, 270 mL MABT, 6 g blocking reagent (Roche 1096176). Warm the MABT and dissolve the blocking reagent in the MABT. Then add serum.
6. Antibody-block solution: 0.2 μl/ml of 0.75 U/μl Anti-Digoxigenin-AP, Fab fragments or 0.2 μl/ml of 0.75 U/μl Fluorescein-AP, Fab fragments.

2.4 Whole Mount In Situ Hybridization: Day 3 Solutions

1. MABT/10% lamb serum: MABT with 10% lamb serum.
2. Staining solution: 60 mL 1 M Tris–HCl pH 9.5, 30 mL 1 M $MgCl_2$, 12 mL 5 M NaCl, 6 mL 10% Tween 20, 144 mg Levimasole, H_2O to 600 mL.
3. 100 mM Glycine, pH 2.2: 3.8 g glycine, H_2O to 500 mL. Adjust pH before adding all of the water.
4. AP substrate for DIG-labeling: BM Purple AP Substrate, precipitating (Roche 1442074) with 5 mM Levimasole

hydrochloride. Make a 0.5 M Levimasole hydrochloride stock solution.

5. AP substrate for fluorescein-labeling: 7.5 μl INT/BCIP Stock Solution in 1 mL staining solution. INT/BCIP Stock Solution is 33 mg/ml INT (2-[4-iodophenyl]-3-[4-nitrophenyl]-5-phenyltetrazolium chloride) and 33 mg/ml BCIP (5-bromo-4-chloro-3-indolyl-phosphate), toluidine-salt in DMSO, commercially available through Roche.

2.5 o-Dianisidine Staining of Zebrafish Embryos

1. *o*-Dianisidine Stock Solution: 100 mg o-dianisidine powder (fast blue B, free base) in 70 mL 100% ethanol. Stock solution must be prepared a day ahead; it takes overnight to dissolve the powder. Store at 4 °C, in the dark by wrapping tube with aluminum foil (solution is photosensitive). The stock solution is generally good for 4 weeks. Discard stock solution if it turns color from clear to brown.
2. *o*-Dianisidine Working Solution: 2 mL *o*-Dianisidine Stock Solution, 500 μl 0.1 M NaOAc, pH 4.5, 2 mL sterile water, 100 μl H_2O_2 (30% stock solution). Wear gloves, H_2O_2 is caustic to skin.
3. Aluminum foil.
4. Pronase from *Streptomyces griseus*.
5. 4% Paraformaldehyde (PFA).
6. Bleaching solution: 400 μl 10% KOH, 150 μl H_2O_2, 5 mL ddH_2O.
7. Sterile Water.
8. Glass Vials.
9. Pasteur Pipets.
10. PBST.

2.6 Preparing Single-Cell Suspension for FACS

1. 50× E3 medium: 29 g NaCl, 1.3 g KCl, 4.8 g $CaCl_2 \cdot 2H_2O$, 8.15 g $MgCl_2 \cdot 6H_2O$ Dissolve ingredients in H_2O, adjust the pH to 7.2 with NaOH and total volume of 2 L. Autoclave.
2. 1× E3 medium: 20 mL of 50×, H_2O to 1 L. Add 100 μL of 1% methylene blue.
3. 1× Hanks balanced salt solution (HBSS) without Ca, Mg or phenol red.
4. 70 μl cell strainer.
5. 40 μl cell strainer.
6. 50 mL Conical Centrifuge Tubes.
7. 12 × 75 mm polystyrene round-bottom tubes.
8. Disposable Pestles.
9. RNase-free Microcentrifuge tubes.

10. 10× Hanks' Balanced Salt Solution (HBSS).
11. Refrigerated microcentrifuge.

2.7 Microinjection

1. E3 medium.
2. Agarose.
3. 100 mm × 15 mm Petri dish.
4. Plastic slot maker and plastic cover with slits [30] (can be purchased from IBI Scientific, Model Number FM-700).
5. Parafilm.
6. Inverted Microscope.
7. Micromanipulator.
8. Forceps.
9. Micropipette puller.
10. 1 mm outer diameter (od) glass capillaries.

3 Methods

3.1 RNA Probe Synthesis and Purification

1. Combine 2–5 μg linearized plasmid, 2 μl 10× labeled nucleotide mix (DIG or F), 2 μl 10× transcription buffer, 0.5 μl RNasin and 2 μl T7, T3 or SP6 RNA polymerase. Adjust volume to 20 μl with water.
2. Incubate for 2 h at 37 °C.
3. Add 1 μl of RNase-free DNAse.
4. Incubate for 30 min at 37 °C.
5. Add 80 μl of RNAse-free H_2O.
6. Add 10 μl of 3 M Na-acetate ($C_2H_3O_2$).
7. Add 250 μl 100% EtOH.
8. Mix well, incubate for 20 min at −20 °C for precipitation.
9. Pellet in microcentrifuge for 30 min at maximum speed at 4 °C.
10. Remove the supernatant and dissolve RNA in 50 μl RNase-free H_2O (*see* **Note 1**).
11. Check yield of RNA probe synthesis by running 2–5 μl on 2% agarose gel.
12. If unincorporated nucleotides are visible, purify probe on a G50 RNA Quick Spin Column (or similar). *See* **Note 2**.

3.2 Whole Mount In Situ Hybridization (WISH) in Zebrafish

This is a 3-day procedure, but embryos need to be removed from fixative the day before the process is initiated. *See* **Note 3**.

1. On Day −1, remove 4% PFA solution from fixed embryos and add 100% MeOH to dehydrate. Store at −20 °C until use (≥1 h). *See* **Note 4**.
2. On Day 1, thaw 4% PFA (50 ml/Netwell carrier). *See* **Note 5**.
3. Bring hyb$^-$ (50 ml/Netwell carrier) and hyb$^+$ (1 ml/sample) to RT.
4. Wash embryos for 5 min on an orbital/platform shaker (low speed) in Netwell carriers in the following series of solutions: 1× 50% MeOH, 1× PBST (50 ml/Netwell carrier), 1× 30% MeOH, 1× PBST (50 ml/Netwell carrier), 2× 1× PBST (100 ml/Netwell carrier). *See* **Note 6**.
5. *Fix* embryos with 4% PFA (fresh or reused) for 20–30 min at RT on orbital/shaker (shaking). Leave embryos ≤24 hpf in this solution until **step 8**. Keep PFA for fixation in **step 9**. If no proteinase K digestion is required, continue with **step 10**.
6. Wash embryos two times, 5 min each, RT, shaking in 1× PBST (100 ml/Netwell carrier).
7. Digest embryos >24 hpf in proteinase K. Add 4 μl of 10 mg/ml Proteinase K to 2 mL PBST in Netwell carrier, 12-well dish. The digestion time is increased for larger embryos, and is as follows: 10 min for 25–30 hpf, 15 min for 31–36 hpf, 20 min for 48 hpf, 35 min for 3 dpf, 45–60 min for 4–5 dpf.
8. Wash two times in 1× PBST (100 ml/Netwell carrier) for 5 min at RT, shaking.
9. Fix embryos with 4% PFA for 20–30 min at RT on an orbital shaker. Turn on hyb-oven, preset 65 °C.
10. Wash two times in 1× PBST (100 ml/Netwell carrier) for 5 min at RT, shaking.
11. Pre-hybridize in hyb$^-$ (50 ml/Netwell carrier) for 15 min at 65–70 °C in a hybridization oven (hyb-oven), with shaking. *See* **Note 7**.
12. Pre-hybridize in hyb$^+$. Place embryos in microcentrifuge tubes, remove hyb$^-$, add 1 ml/sample of hyb$^+$, close tubes tightly, and incubate for 4 h at 65–70 °C in a hyb-oven with shaking. Replace leaking tubes.
13. Hybridize with RNA probe, overnight at 65–70 °C. Remove hyb$^+$ from prehybridization step, add 500 μl RNA probe of interest in hyb$^+$. For double in situ hybridizations with differentially labeled probes, add a mix of both probes at this step.
14. On Day 2, pre-warm SSCT-based wash solutions in microwave prior to use.
15. Remove hyb$^+$ solution containing RNA probe and save for re-use (*see* **Note 8**).

16. Wash three times with 1 mL 2× SSCT, 50% formamide per microcentrifuge tube. For each wash, incubate for 15 min in a hyb-oven at 65 °C, shaking.
17. Transfer embryos into Netwell carriers after third wash step, and do the remaining washes in the indicated volume per Netwell carrier.
18. Wash in 50 ml 2× SSCT for 15 min at 65 °C, shaking.
19. Wash two times in 100 mL 0.2× SSCT for 15 min at 65 °C, shaking.
20. Wash two times in 100 mL MABT for 5 min at RT, shaking.
21. Block embryos in 50 ml block solution for ≥1 h at RT, shaking.
22. Fill each well of a 12-well plate with 2 mL antibody-block solution. Transfer embryos from Netwell carriers to 12-well plates, incubate overnight at 4 °C (cold room) on a rocking shaker at low speed. For double in situ hybridizations, first do the fluorescein and then do the DIG antibody (*see* **step 30**).
23. On Day 3, return samples to Netwell carriers for the following series of washes, and wash with the indicated volume per carrier. Wash with 50 mL MABT/10% lamb serum for 20 min at RT, shaking.
24. Wash three times with 150 mL MABT for 20 min at RT, shaking.
25. Wash three times with 150 mL staining solution for 5 min at RT, shaking. Save staining solution from last wash for **step 28**.
26. For color development, transfer samples into 12-well dishes that are filled with 1 mL of either the AP substrate for DIG-labeling or the AP substrate for fluorescein-labeling. Try to minimize the amount of staining solution transferred along with the embryos.
27. Wrap the dish in aluminum foil and place on an orbital shaker at RT. Rock gently by hand from time to time. Depending on the probe, staining can already be intense by 1 h (globin). Use qualitative inspection over the course of 1–12 h for color development in ICM without background color development in other parts of embryo.
28. Stop color development by removing substrate solution with a transfer pipet and immediately add staining solution from **step 25**.
29. Wash three times with staining solution, 5 min each at RT. Put the embryos back into 4% PFA for expression pattern analysis and storage.
30. The following optional steps can be performed for double in situ hybridizations. Fix embryos for 20 min in 4% PFA at RT, shaking.

31. Transfer to Netwell carrier and use indicated volume per carrier for next series of washes. Inactivate AP by washing two times with 100 mL of 100 mM glycine, pH 2.2 for 15 min at RT, shaking.
32. Wash two times with 100 ml MABT for 5 min at RT, shaking.
33. Block embryos in 50 mL block solution for ≥1 h at RT, shaking.
34. Fill each well of a 12-well plate with DIG antibody-block solution. Transfer embryos from Netwell carriers to the 12-well plates, incubate overnight at 4 °C (cold room) on a rocking shaker at low speed.
35. Continue with **steps 23–29**.

3.3 o-Dianisidine (3,3′-Dimethoxybenzidine) Staining of Zebrafish Embryos

1. Dechorionate embryos. If >24 hpf, dechorionate by treating with 0.2% (w/v) pronase in E3 media for 15 min incubation at room temperature, followed by washing ten times with 1× E3 media at room temperature.
2. Without fixing in PFA, directly place embryos in *o*-dianisidine working solution in glass vials. Optimal conditions for o-dianisidine staining occur at 72 hpf (*see* **Note 9**).
3. Incubate for 3–5 min at room temperature, out of direct light (inside an opaque box or wrapped in aluminum foil). Incubation does not have to be in the dark. Invert the vials every 30 s to ensure the embryos are well mixed and exposed to the coloring reagents.
4. Staining solution may turn brown color and bubbles may be liberated. Examine samples for blood cell staining by placing glass jars under a dissecting microscope. When stained, stop the color reaction by washing samples three times in dH_2O. Embryos tend to be sticky, so use a pasteur pipet, fine-tip, or pulled-glass pipet.
5. Fix the stained samples in 4% PFA and store long-term at 4 °C in the dark.
6. After benzidine staining and fixation, rinse two times in PBST.
7. To remove melanin pigmentation, incubate samples for 15′ at RT in bleaching solution. Observe carefully, as too long and the embryos start to disintegrate.
8. Rinse two times in PBST, then store in 4% PFA for archiving.

3.4 Zebrafish Embryonic Cell Analysis and Sorting by Flow Cytometry

1. Set temperature on microcentrifuge to 10 °C.
2. Distribute live embryos in E3 into 1.5 mL RNAse-free tubes (~200/tube).
3. Aspirate E3; replace with 1 mL 1× HBSS; aspirate HBSS until there is just enough to cover the embryos.

4. Place embryos and cells on ice from this point on.
5. Crush embryos thoroughly with pestle until the solution is homogeneous and contains no observable particulates.
6. Add 1 mL HBSS to homogenized solution; filter through a 70 μm cell strainer, draining into a 50 mL tube.
7. Pulse-spin the strainer/tubes at low speed (135 × *g*).
8. Pass filtrate through a 40 μl cell strainer draining into a fresh 50 mL tube.
9. Pulse-spin the strainer/tubes at 135 × *g*.
10. Transfer filtrate to a polystyrene round-bottom tube.
11. Spin the tube at 135 × *g* for 15 min at 10 °C; the cells should form a pellet.
12. Discard the supernatant, resuspend the pellet in 1–2 mL 1× HBSS by vortexing gently.
13. Perform flow cytometry with forward and side scatter gatings for eGFP.

3.5 Microinjection (see Note 10)

1. To make injection chambers [30], pour approximately 20 mL of hot 1.5% agarose in 1× E3 medium into a 100× 15 Petri dish on a level surface. Wait until completely solidified.
2. Add an additional 20 mL of the 1.5% agarose to the dish. Set the plastic slot maker (teeth down) into the liquid agarose overlay, tapping to eliminate any bubbles.
3. After the agarose sets, add a small amount of E3 medium, wrap the Petri dish in parafilm, and store in the 4 °C refrigerator.
4. Warm the Petri dish to the temperature you prefer for injection (embryos can tolerate 18 °C for about 1 h, and the slower division rate allows more time to inject at the 1 and 2 cell stage).
5. Remove the plastic slot maker.
6. Position the dish on a microscope stage and adjust the angle of the pipette so that you can aim directly into the trough.
7. Transfer embryos into the troughs and set the plastic cover with slits into the depression in the agarose.
8. Using forceps, align the embryos all in the same orientation. Add enough culture medium so that the level reaches the plastic edges of the slits. Gently tamp the embryos down into the trough.
9. Using controls on the micromanipulator, force the pipette through the chorion and the yolk cell, entering the embryonic cell(s) from the yolk cell. (One might prefer to enter the embryonic cell directly; in this case, you would choose a different orientation of the embryo in the trough.)

10. As the pipette goes through the yolk, move the Petri dish slightly, so the pipette is close to the plastic cover.
11. Inject the desired volume of solution and then withdraw the pipette with a slow steady motion. The embryo will catch on the plastic cover and drop back into the trough.
12. Transfer the embryos into a dish with 1× E3 Medium and maintain at 28.5 °C for further development.

4 Notes

1. If the probe is dissolved in hyb$^+$ at this stage, the probe cannot be purified if there are unincorporated nucleotides present.
2. Usually, 10 μl of synthesized probe (2 μl of it resulted in a readily visible band on a 2% agarose gel) is more than enough for many rounds of in situ hybridizations. Probes can be reused. If the probes start to decrease in intensity, add another 10 μl to the existing hyb$^+$/probe-mix or prepare a fresh mix.
3. For a tutorial on WISH, please visit http://www.jove.com/video/50644/high-resolution-whole-mount-situ-hybridization-within-zebrafish
4. Prepare a word document with a list of samples. Arrange samples according to age (proteinase K treatment for all embryos >24 hpf) and hybridization probe (i.e., put samples that will be treated with globin asRNA at the very end, and segregate samples that will be treated with asRNA probes that stain the same region of the embryo into different Netwell carriers to avoid cross hybridization).
5. Use fresh or up to 2× used 4% PFA for embryos <24 hpf (Keep PFA stocks in aliquots at −20 °C). Reuse 4% PFA up to 6× for older embryos.
6. The handle can easily fall off and release the Netwells. When grasping handle, always press it apart with your thumb and forefinger.
7. Hyb$^-$ can be reused up to 10×.
8. Always change transfer pipet or pipet tip to avoid cross-contamination of probes during wash steps.
9. An alternative to bleaching and to improve *o*-dianisidine signal detection, embryos can be treated with 1-phenyl 2-thiourea (PTU; 0.003% w/v) at 24 hpf, before the onset of melanogenesis. To do this, after pronase treatment, incubate 24 hpf embryos in PTU until *o*-dianisidine staining is performed at 3 dpf. Use gloves, personal protective equipment, and safe handling while using PTU. PTU is a teratogen and may cause birth defects during pregnancy.

10. For a tutorial on microinjection including how to pull injection pipettes visit: http://www.jove.com/video/1115/microinjection-of-zebrafish-embryos-to-analyze-gene-function

Acknowledgments

We thank our colleagues for critical feedback and comments on this chapter: Eva Buys, Aaron Kithcart, Ludmila Flores, Brian Dulmovits, Lionel Blanc, Alex Cintolo, Thomas Pedulla, Lisa van der Vorm, Bryce Klehm, and Isy Mekler. The in situ methods were from Gabriele E. Ackermann. This work was supported by grants from the National Institutes of Health (R01 DK070838 and P01 HL032262 to B.H.P.).

References

1. Colle-Vandevelde A (1963) Blood anlage in teleostei. Nature 198:1223
2. Shafizadeh E, Paw BP (2004) Zebrafish as a model of human hematologic disorders. Curr Opin Hematol 11:255–261
3. Orkin SH, Zon LI (2008) Hematopoiesis: an evolving paradigm for stem cell biology. Cell 132:631–644
4. Jagannathan-Bogdan M, Zon LI (2013) Hematopoiesis. Development 140:2463–2467
5. Traver D, Paw BH, Poss KD, Penberthy WT, Lin S et al (2003) Transplantation and in vivo imagine of multilineage engraftment in zebrafish bloodless mutants. Nat Immunol 4:1238–1246
6. Carradice D, Lieschke GJ (2008) Zebrafish in hematology: sushi or science? Blood 111:3331–3339
7. Jing L, Durand EM, Ezzio C, Pagliuca SM, Zon LI (2012) In situ hybridization assay-based small molecule screening in zebrafish. Curr. Protocol Chem Biol 4:143–160
8. Reischauer S, Stone OA, Villasenor A, Chi N, Jin SW et al (2015) Cloche is a bHLH-PAS transcription factor that drives haemato-vascular specification. Nature 535:294–298
9. Schulte-Merker S, Lee KJ, McMahon AP, Hammerschmidt M (1997) The zebrafish organizer requires chordino. Nature 387:862–863
10. Chen C, Garcia-Santos D, Ishikawa Y, Seguin A, Li L et al (2013) Snx3 regulates recycling of the transferrin receptor and iron assimilation. Cell Metab 17:343–352
11. Amigo JD, Ackerman GE, Cope JJ, Yu M, Cooney JD et al (2009) The role and regulation of friend of GATA-1 (FOG-1) during blood development in the zebrafish. Blood 114:4654–4663
12. Schulte-Merker S (2002) Looking at embryos. In: Nusslein-Volhard C, Dahm R (eds) Zebrafish: a practical approach. Oxford Univ Press, New York, NY, pp 41–43
13. O'Brien BA (1961) Identification of haemoglobin by its catalase reaction with peroxide and O-dianisidine. Biotech Histochem 36:57–61
14. Iuchi I, Yamamoto M (1983) Erythropoiesis in the developing rainbowtrout, Salmo gairdneri irideus: histochemical and immunochemical detection of erythropoietic organs. J Exp Zool 226:409–417
15. Shaw GS, Cope JJ, Li L, Corson K, Hersey C et al (2006) Mitoferrin is essential for erythroid iron assimilation. Nature 440:96–100
16. Wang Y, Langer NB, Shaw GC, Yang G, Li L et al (2011) Abnormal mitoferrin-1 expression in patients with erythropoietic protoporphyria. Exp Hematol 39:784–794
17. Lin HF, Traver D, Zhu H, Dooley K, Paw BH (2005) Analysis of thrombocyte development in CD41-GFP transgenic zebrafish. Blood 106:3803–3810
18. Ma D, Zhang J, Lin HF, Italiano J, Handin RI (2011) The identification and characterization of zebrafish hematopoietic stem cells. Blood 118:289–297
19. Cooney JD, Hildick-Smith GJ, Shafizadeh E, McBride PF, Carroll KJ et al (2013) Teleost growth factor independence (gfi) genes differentially regulate successive waves of hematopoiesis. Dev Biol 373:431–441

20. Ganis JJ, Hsia N, Trompouki E, de Jong JL, DiBiase A et al (2012) Zebrafish globin switching occurs in two developmental stages and is controlled by the LCR. Dev Biol 366:185–194
21. Morcos PA (2007) Achieving targeted and quantifiable alteration of mRNA splicing with Morpholino oligos. Biochem Biophys Res Commun 358:521–527
22. Eisen JS, Smith JC (2008) Controlling morpholino experiments: don't stop making antisense. Development 135:1735–1743
23. Lichtenstein DA, Crispin AW, Sendamarai AK, Campagna DR, Schmitz-Abe K et al (2016) A recurring mutation in the respiratory complex 1 protein NDUFB11 is responsible for a novel form of X-linked sideroblastic anemia. Blood 128:1913–1917
24. Hildick-Smith GJ, Cooney JD, Garone C, Kremer LS, Haack TB et al (2013) Macrocytic anemia and mitochondriopathy resulting from a defect in sideroflexin 4. Am J Hum Genet 93:906–914
25. Kardon JR, Yien YY, Huston NC, Branco DS, Hildick-Smith GJ et al (2015) Mitochondrial ClpX activates a key enzyme for heme biosynthesis and erythropoiesis. Cell 161:858–867
26. Dang M, Fogley R, Zon LI (2016) Identifying novel cancer therapies using chemical genetics and zebrafish. Adv Exp Med Biol 916:103–124
27. Dow LE (2015) Modeling disease in vivo with CRISPR/Cas9. Trends Mol Med 21:609–621
28. Irion U, Krauss J, Nüsslein-Volhard C (2014) Precise and efficient genome editing in zebrafish using the CRISPR/Cas9 system. Development 141:4827–4830
29. Albain J, Durand EM, Yang S, Zhou Y, Zon LI (2015) A CRISPR/Cas9 vector system for tissue-specific gene disruption in zebrafish. Dev Cell 32:756–764
30. Weinberg E (2007) A device to hold zebrafish embryos during microinjection, Chapter 5. In: The zebrafish book, 4th edn. University of Oregon Press, Eugene, OR. https://zfin.org/zf_info/zfbook/chapt5/5.1.html

Chapter 3

Mouse Models of Erythropoiesis and Associated Diseases

Matthew P. Parker and Kenneth R. Peterson

Abstract

Animal models of erythropoiesis have been, and will continue to be, important tools for understanding molecular mechanisms underlying the development of this cell lineage and the pathophysiology associated with various human erythropoietic diseases. In this regard, the mouse is probably the most valuable animal model available to investigators. The physiology and short gestational period of mice make them ideal for studying developmental processes and modeling human diseases. These attributes, coupled with cutting-edge genetic tools such as transgenesis, gene knockouts, conditional gene knockouts, and genome editing, provide a significant resource to the research community to test a plethora of hypotheses. This review summarizes the mouse models available for studying a wide variety of erythroid-related questions, as well as the properties inherent in each one.

Key words Mouse models, Transgenic mice, Knockout mice, Conditional knockout mice, Cre-*loxP*, Erythropoiesis, Globin gene switching, Hemoglobin

1 Introduction

Mouse models provide valuable resources to understand the molecular mechanisms underlying normal cellular processes of development, repair, and regeneration, as well as associated genetic perturbations resulting in disease. In addition, mouse models serve as experimental vehicles in which to test pharmaceutical and gene therapies. The common house mouse, *Mus musculus*, has been utilized in this capacity for over a century, beginning with inbred strains of mice as human disease models, leading to the mouse genome project, transgenics and genome editing capabilities of present day. Mice have a physiology similar to humans and their long history of use offers many advantages over other model systems. Early studies produced several thousand spontaneous and radiation-induced mutant strains, which are still available from a number of commercial vendors. These are useful because mutations in mouse genes homologous to their human counterparts often result in disease etiologies similar to humans. Later, inbred and congenic strains were created; these provide a uniform genetic

Joyce A. Lloyd (ed.), *Erythropoiesis: Methods and Protocols*, Methods in Molecular Biology, vol. 1698,
DOI 10.1007/978-1-4939-7428-3_3,

background on which to perform studies free from the effect of modifier genes and polymorphisms. More recently, the advent of transgenic and gene targeting methodologies makes it possible to probe the biochemical and molecular pathways of human gene expression, offering the possibility that understanding these processes might lead to points of intervention in disease etiology. Variants of these two procedures made mice the animal of choice for modeling erythropoiesis and red blood cell diseases. Transgenesis allows introduction of human gene sequences or entire loci into the mouse genome or "knock-in" of human sequences to replace their mouse counterparts, while gene targeting to create gene "knockouts" or conditional knockouts, permits genetic studies of the endogenous globin loci, which often are arranged and regulated to a large degree like the human loci. Finally, the 21-day parturition period in mice compacts developmental studies into a useful time frame.

2 Globin Gene Switching and Erythropoiesis

Much of the technology utilized in modern mouse models was developed and/or applied to studies of the hematopoietic system, largely to work on globin gene switching and erythropoiesis. Novel transgenesis technology created opportunities for studying human gene expression in vivo within an animal model. Previous work was limited to analysis of transgenes in established cell lines. Murine transgenesis was particularly attractive, since endogenous globin synthesis had been well characterized during embryogenesis [1]. Developmental studies on hematopoiesis and globin gene switching could be carried out on staged embryos and fetuses during pregnancy beginning with the onset of erythropoiesis at day seven post-conception through post-partum in adult mice. All of the hematopoietic tissues present at a particular stage of development could be assessed, including yolk sac, liver, bone marrow, spleen, and peripheral blood.

3 Early Transgenic Models

Transgenic mice were first produced in the early 1980s by a number of investigators using simple gene or cDNA sequences linked to a promoter of choice. Most of the transgenes integrated in these mice suffered from position effect variegation (PEV) and as the technique developed, various *cis*-acting regulatory elements were included in transgene constructs in an attempt to overcome PEV, including enhancers, introns, insulators, and poly-adenylation signals. Only after locus control regions (LCRs) were discovered and incorporated into transgenes was PEV ameliorated [2].

Initial experiments involved microinjection of simple transgenes encompassing the human adult β-globin gene or the fetal γ-globin gene (Fig. 1, left side) [3, 4]. Expression was affected by position-of-integration of the transgenes within the murine genome and copy number-dependence was not observed, hallmarks of PEV. In spite of variable levels of expression associated with PEV, when the transgenes were expressed, correct regulation regarding developmental stage- and tissue-specificity was observed,

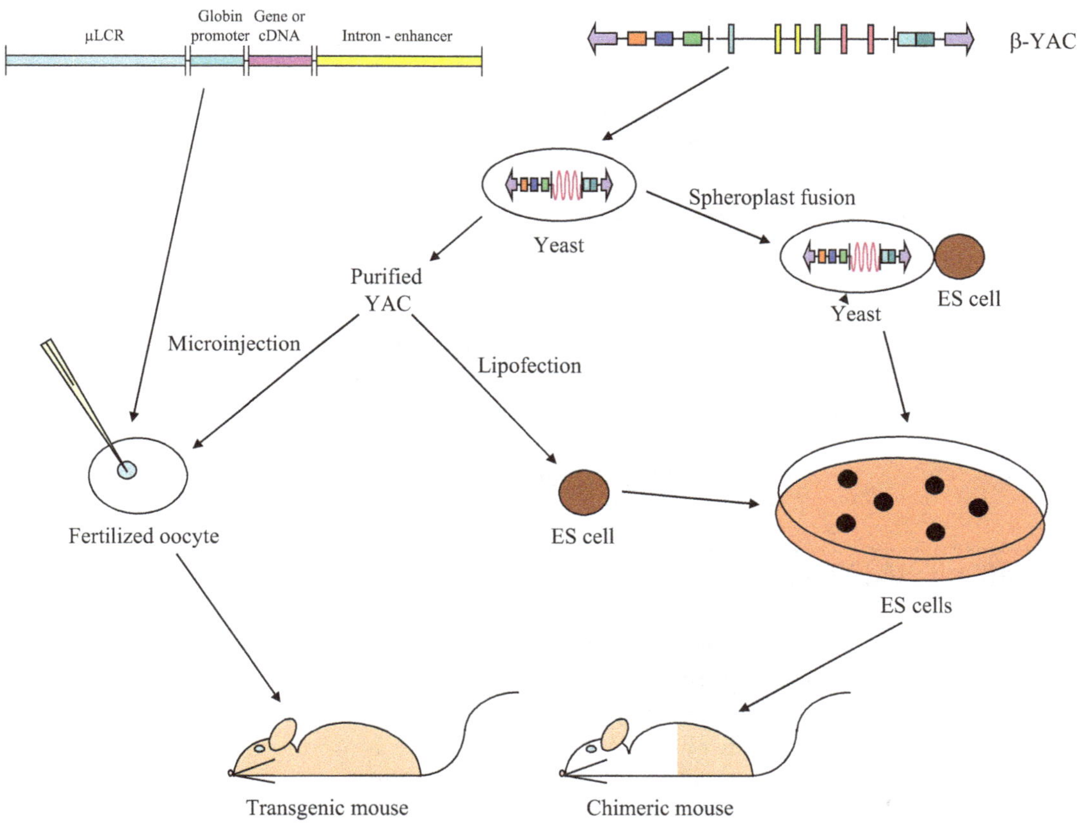

Fig. 1 Transgenesis with small recombinant or whole loci globin constructs. A prototypical small recombinant globin transgene is illustrated at the *upper left*. Most early transgenes consisted of LCR sequences (mini- or micro-LCRs, one or more 5′HSs, or the entire LCR), a globin promoter linked to its gene sequences, hybrid globin gene sequences, or a cDNA reporter (occasionally a non-globin promoter was linked to a globin gene or cDNA cassette), and a globin or heterologous intron and/or enhancer (or not). These purified transgenes were microinjected into the pronucleus of fertilized mouse oocytes to produce transgenic mice as shown schematically. Cosmid and bacterial artificial chromosomes (BACs) containing larger β-like globin locus inserts were similarly microinjected (not shown). Methods used to generate transgenic mice with yeast artificial chromosomes (YACs) are illustrated beginning with the human β-globin locus YAC (β-YAC) shown at the *upper right*. The β-YAC (155 or 215 kb) and other YACs ranging up to 650 kb in size may be purified from the yeast host and microinjected similar to small recombination DNA constructs or lipofected into embryonic stem (ES) cells, which are then utilized to produce chimeric mice and establish transgenic lines. Alternately, the YAC may be transferred to ES cells by yeast spheroplast-ES cell protoplast fusion. Lipofection or fusion must be employed when the YAC exceeds 650–800 kb in size

demonstrating that *cis*-regulatory elements controlling spatial and temporal expression were gene-proximal. Discovery of the β-globin locus LCR in 1987 provided a major breakthrough in globin gene switching [2]. The LCR consists of a collection of DNA regions upstream of the β-like globin gene cluster that are hypersensitive to DNaseI. The LCR opens up the globin domain and makes it available for transcription and is required for high-level expression of all the genes within the cluster. When this element was linked to either a globin gene or a heterologous gene, PEV disappeared; that is, site-of-integration-independent, copy number-dependent expression of the linked transgene was obtained [2]. Generally, all of these transgenes used fragments that encompassed one or two human β-like globin genes coupled to derivative LCR sequences, such as mini- or micro-LCR cassettes (μLCR), individual, or multiple DNaseI-hypersensitive sites (HSs). As transgenic technology advanced, and more DNA could be successfully microinjected into oocytes in an intact state, multiple β-like genes could be linked to μLCRs or the intact LCR could be linked to single-globin genes in cosmid constructs [5].

Analysis of α-globin transgenics identified the HS-40 upstream master regulatory region, and *cis*-regulatory elements controlling ζ-globin and α-globin gene regulation [6, 7]. Particularly notable were studies of the 3′ untranslated region (UTR) and its role in α-globin mRNA stability (reviewed in [8]).

4 YAC and BAC Transgenesis

The 1990s saw the advent of whole human loci transgenesis using first yeast artificial chromosomes (YACs), then bacteriophage P1 artificial chromosomes (PACs), and later bacterial artificial chromosomes (BACs). YACs and BACs allow the manipulation of very large stretches of DNA, permitting the analysis of entire gene clusters such as the β-globin locus. Transgenesis with these vectors resulted in murine models that more accurately mirrored human gene expression [9–11]. The first human β-globin locus yeast artificial chromosome (β-YAC) transgenic mice were produced in 1993 [12, 13], ushering in the era of whole locus transgene studies (Fig. 1, right side). More recently, the use of β-globin locus bacterial artificial chromosomes (β-BACs), coupled with "recombineering," has been employed [14]. Both types of transgenes allow the introduction of alterations within the context of the entire human β-locus using homologous recombination in either yeast or bacteria, respectively, without retention of exogenous DNA sequences. Thus, the effect of the mutation may be studied in an intact locus throughout development within the mouse. Initial studies demonstrated that the human globin gene expression pattern was, by and large, recapitulated during ontogeny in the mouse. Modified transgenes were utilized to produce hereditary persistence of fetal hemoglobin (HPFH) and sickle cell mouse models (reviewed in [15, 16]).

Many of these transgenics were mated with murine knockout mutant lines to produce the mouse models of sickle cell disease and thalassemia now utilized to study the pathophysiology of these diseases.

5 Knockout Models

Coincidental and in parallel to transgenic studies, were projects analyzing the regulation of the endogenous murine β-like and α-like globin genes. The strategy in this case utilized gene targeting and chimeric mice to produce knockout mutations of genes or globin regulatory sequences (Fig. 2). In general, the deficiencies

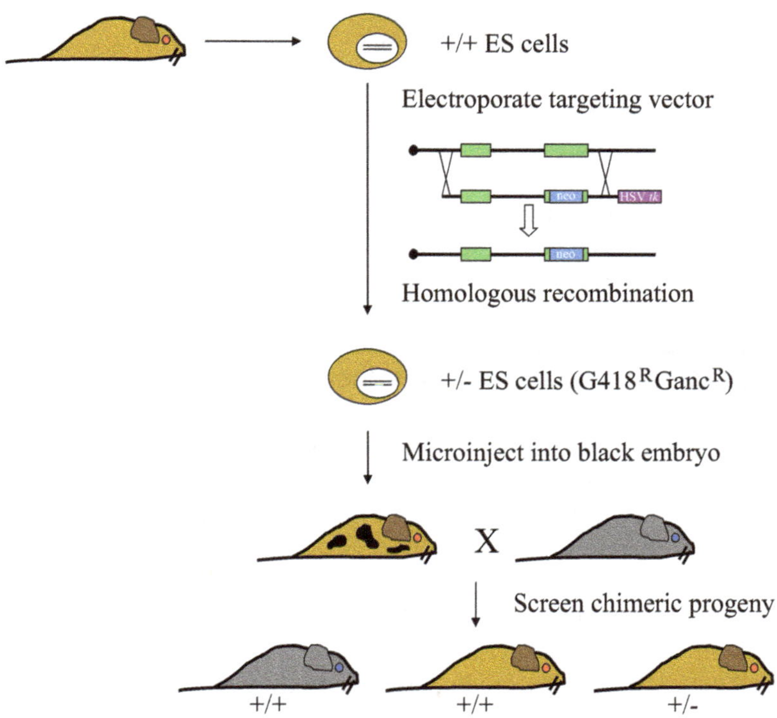

Fig. 2 Generation of targeted mutations in endogenous murine globin loci. ES cells usually derived from 129 mice (*brown coat color*) are electroporated with a gene-targeting construct, a generalized version of which is shown. Following drug selection and screening for proper integration of the targeting vector, the ES cells are microinjected into blastocysts normally derived from C57Bl6 mice (*black coat color*). Chimeric mice are identified initially based on coat color, which is a varying mixture of brown and black. If the ES cells contributed to the germ line, the targeted mutation will be inherited by the next generation of mice as indicated at the *bottom* of the figure. Following homologous recombination in the target ES cells, the neo cassette confers resistance to the drug G418; loss of the HSV*tk* cassette following homologous recombination results in ganciclovir (GANC) resistance

produced revealed similar mechanisms of action regarding globin gene switching in humans. One important caveat is that mice do not have a fetal globin analogous to the human γ-globin genes. A single switch from embryonic globin synthesis to definitive globin synthesis occurs, whereas two switches exist in humans, from primitive embryonic erythropoiesis to fetal definitive erythropoiesis and later from fetal definitive erythropoiesis to adult definitive erythropoiesis. However, the sites of hematopoiesis during development are conserved between mice and humans. Although γ-globin gene expression is confined to the fetal liver in humans, it is expressed mainly in the embryonic yolk sac in the mouse. Expression in the yolk sac can be attributed to the evolutionary homology of the human γ-globin and murine βh1-globin genes, both of which derive from an ancestral γ-globin gene. Mouse βh1-globin gene expression is limited to the embryonic yolk sac; the human γ-globin gene apparently "remembers" its evolutionary origin when in mice and is likewise expressed. However, unlike the murine βh1-globin gene, expression of the human γ-globin genes in the mouse continues during fetal liver definitive erythropoiesis and in a small population of adult definitive red cells [13]. Thus, studies on γ-globin gene expression can only be carried out in mice using the human transgene approach. Knockouts of adult α-globin and β-globin provided the mouse lines that were bred with human transgenic lines to produce mouse models of hemoglobinopathies, in which only human globin chains are incorporated into the hemoglobin molecule.

6 Sickle Cell, Thalassemia, and Cooley's Anemia Models

A number of mouse models for various hemoglobinopathies have been generated (Table 1). Using slightly different approaches for transgenesis and gene knockouts and knock-ins, alone or in combination, lines modeling sickle cell disease, β- or α-thalassemias, and Cooley's anemia have been produced [17]. In many instances, the human phenotype for each of these diseases is remarkably conserved in the mouse (reviewed in [17, 18]). These mice offer the possibility for understanding the pathophysiology of these diseases in more detail, may serve as vehicles to test emerging pharmaceutical compounds and gene therapy approaches, will be useful for the isolation and proliferation of hematopoietic stem cells (HSCs), embryonic stem cells (ESCs), or induced pluripotent stem cells (iPSCs), and for testing the efficacy of bone marrow transplantation and engraftment of gene-corrected stem cells.

Early models of sickle cell disease models (Table 1) include S Antilles and a modified version of this model [19, 20], S+S Antilles [21–23], and SAD [24, 25]. Although these models provided important information about sickle cell disease, they only mimicked

Table 1
Murine models of sickle cell disease, thalassemias and Cooley's anemia

	Genotype	Phenotype	Severity of disease
SCD mouse model			
S Antilles	$\alpha^{H}\ \beta^{S\ Antilles}$	Mild adult anemia Low irreversible sickling	Mild
S+S Antilles	$\alpha^{H}\ \beta^{S}\ \beta^{S}_{Antilles}$	Neonatal anemia Significant sickling	Mild—moderate
SAD	$\alpha^{H}\ \beta^{SAD}$	Neonatal anemia Low irreversible sickling Multi-organ pathology	Moderate
Berkeley model (murine knockout)	$\alpha^{H}\ \beta^{S}\ \gamma^{H}\ \delta^{H}$	Significant adult anemia High irreversible sickling Chronic multi-organ damage	Moderate—severe
Birmingham model (murine knockout)	$\alpha^{H}\ \beta^{S}\ \gamma^{H}$	Severe anemia High irreversible sickling Multi-organ pathology	Moderate—severe
San Francisco model (murine knockout)	$\alpha^{H}\ \beta^{S}\ \gamma^{H}\ \delta^{H}$ (YAC)	Anemia Irreversible sickling	Moderate—severe
Enhanced γ-globin expression models	$\alpha^{H}\ \beta^{S}\ (\gamma^{Ha}/\gamma^{Hb}/\gamma^{Hc})$ and $\alpha^{H}\ \beta^{SAD}\ \gamma^{H}$	Decreased severity of disease with increasing HbF levels	Moderate—severe
Thalassemia mouse model			
Carolina β^{0} (murine knockout)	Δb1/b2	Heterozygous mice dramatically decreased hematocrit, hemoglobin, red blood cell counts, MCV, MCH, and MCHC, dramatically increased reticulocyte counts, serum bilirubin concentrations, and red cell distribution widths, display tissue and organ damage, spontaneous iron overload in spleen, liver, kidneys Homozygous mice: die in utero	Severe
Birmingham β^{0} (murine knockout)	$\Delta\beta^{maj}/\beta^{min}$	Heterozygous mice: severely anemic, dramatically reduced hemoglobin levels, abnormal red cell morphology, splenomegaly, markedly increased reticulocyte counts Homozygous mice: die in utero	Severe
βIVS-2-654 (murine knockout/human knock-in)	$\Delta\beta^{maj}/\beta^{min}$/Hu βIVS-2-654	Heterozygous mice: reduced mouse β globin chains and no human β globin Homozygous mice: do not survive postnatally	Moderate

(continued)

Table 1 (continued)

	Genotype	Phenotype	Severity of disease
Cooley's anemia (β thalassemia major) (murine knockout/ human knock-in)	$\Delta\beta^{maj}/\beta^{min}$/Hu $\gamma\beta^{0}$ $\Delta\alpha1/\alpha2/$Hu $\Delta\alpha2/\alpha1$	Heterozygous $\gamma\beta^{0}$ knock-in mice: β thalassemia intermedia Homozygous mice: expire due to severe anemia upon completion of the HbF to adult switch after birth	Severe
Cooley's anemia (β thalassemia major) Preclinical model (murine knockout/human knock-in)	$\Delta\beta^{maj}/\beta^{min}$/Hu $\gamma^{HPFH}\delta\beta^{0}$ $\Delta\alpha1/\alpha2/$Hu $\Delta\alpha2/\alpha1$	Fully humanized mice survive postnatally by synthesizing predominantly HbF, some HbA2, completion of fetal to adult Hb switch after birth results in severe anemia marked by erythroid hyperplasia, ineffective erythropoiesis, hemolysis, and death	Severe
Berkeley α (murine knockout)	$\Delta\alpha1/\alpha2$	Homozygous mice: become hydropic, die late in gestation, model for human hydrops fetalis	Severe

α^{H}, γ^{H}, δ^{H}, human α-, γ- and δ-globin genes; β^{S}, human β-globin gene with HbS mutation; $\beta^{S\ Antilles}$, human β-globin gene with HbS Antilles mutation; β^{SAD}, human β-globin gene with HbS, HbS Antilles and HbD Punjab mutations; $\gamma^{Ha}/\gamma^{Hb}/\gamma^{Hc}$: human γ-globin genes expressing three different levels of HbF. See text for references

the sickle cell trait exhibited in human β^{S} heterozygous carriers, and did not present the severe hemolytic anemia that is a hallmark of homozygous sickle cell patients. The phenotypes associated with these mice were due to the inhibition of sickling by endogenous murine hemoglobin.

Newer SCD model mice (Table 1) express exclusively human globins and overcome the failure of earlier models to exhibit faithful sickle cell pathology. They include the Berkeley model, the Birmingham model, and the San Francisco model [33–36]. None of these models express murine α- and β-globin genes. Berkeley model mice express exclusively human α-, β^{S}-, and γ-globin genes [26]. The $^{G}\gamma$- and $^{A}\gamma$-globin genes were included because γ-globin has anti-sickling properties that decreased the likelihood of fetal death during gestation as the switch from primitive embryonic erythropoiesis to fetal definitive erythropoiesis occurred. This model also showed some characteristics of β-thalassemia, including an increased susceptibility to oxidative damage due to free-radical production. The Birmingham mouse was the first model produced that displayed the sickle trait [27]. Subsequent improvements led to the creation of an improved SCD mouse model containing human $^{A}\gamma$-, β^{S}-, and α_1-globin genes driven by the human β-globin LCR [28]. Although coinjection of human globin constructs in the Berkeley and Birmingham models led to the integration of all the genes at a single chromosomal site, normal gene order and spatial organization required for appropriate expression of the members of the β-like globin gene family was disrupted. The San Francisco model partially remedied this by creating novel sickle cell mice harboring a β^S-globin YAC containing the entire human β-globin gene cluster and the human α_2-globin gene linked to a mini-LCR [29].

One other group of SCD mouse models are the enhanced γ-globin expression models. Increasing the level of human γ-globin in transgenic mice to various levels in the SAD mouse improved life span, pathology, and hematological profile [30], thereby establishing the physiological range of HbF that would alleviate the SCD condition in mice. Another set of knockout mice expressing exclusively human globins and three levels of HbF (low, medium, and high) also showed a more faithful SCD pathology than the previous models [31]. In this model, murine knockout mice [26, 32] were mated with mice that expressed the co-integrated human mini-LCRα_2 and mini-LCRβ^S used to create the S+S-Antilles model. Lethality of these mice was rescued by breeding with transgenic mice expressing three different human γ-globin constructs (γL, γM, γH) with increasing postnatal HbF levels. These mice exhibited more balanced α- and β-globin chain synthesis and thus present a closer approximation of human SCD.

Thalassemia model mice were generated with deletions of adult murine β-like globin genes, β^{major} and β^{minor} [32, 33] and both adult α-globin genes [34]. The disease phenotypes of these mice were rescued by the introduction of human globin transgenes. Mouse models with deletions in β^{major}- and α-globin chains show varying severity of disease [33–35]. The βIVS-2-654 C→T mutation, which causes aberrant RNA splicing and leads to β^0 thalassemia, has also been modeled in mice [36]. Two models of β thalassemia major or Cooley's anemia were produced. Both are totally humanized; that is, they carry the human α2/α1 genes and a human $\gamma\beta^0$ gene cassette [37] or $\gamma^{HPFH}\delta\beta^0$ gene cassette [38]. In the homozygous state, both the mouse strains expire shortly after birth due to severe anemia following the completion of the γ-globin to β^0-globin switch. However, the later mouse model can be rescued from lethal anemia by regular blood transfusions.

7 Conditional Knockout Models

Manipulation of endogenous murine genes by gene targeting has been performed primarily in embryonic stem (ES) cells. Gene targeting uses homologous recombination to generate null alleles or gene "knockouts" or to substitute an altered transgene for the native chromosomal allele ("knock-ins"). The technique is a powerful tool for studying gene function in hematopoiesis. However, this classical method for producing null mutations results in the loss of gene expression in all tissues and at all developmental stages. The mutation may be lethal, or affect other tissues and developmental pathways not under analysis, consequently complicating interpretation of data. Thus, it is important that the inactivation of a particular gene occurs in a conditional manner, such as in a predefined tissue or at a certain stage of development. The gene-targeting approach to study gene function in mice has been refined with the advent of the bacteriophage P1 Cre-*loxP* site-specific recombination system. Using this technology, tissue- or developmental stage-specific gene knockouts may be produced. Genes are expressed normally in untargeted tissues and the effect of the null mutation is confined to the tissue of interest.

Cre is a 38 kDa protein that catalyzes reciprocal site-specific recombination between 34 bp *loxP* sequences on DNA (Fig. 3). If two *loxP*Cre/loxP sites are in the same orientation in a genomic locus, expression of Cre results in the deletion of *loxP*-flanked DNA sequence in both the prokaryotic and eukaryotic cells. Generally, the two *loxP* sites are introduced into the genome such that they flank an essential portion of the target gene without affecting its function. Following Cre recombinase expression the gene is conditionally inactivated, depending upon the cell type-specific or inducible promoter that defines the spatial and temporal synthesis of Cre protein.

A number of Cre drivers have been utilized to produce conditional knockouts in the hematopoietic/erythropoietic lineages. These are summarized in Table 2. The range of lineage specificity is wide; several are not strictly specific to erythroid, or more broadly, hematopoietic lineages. However even those with broad tissue/developmental range have been useful in the production of conditional knockouts, or activation of mutant versions of genes affecting erythropoiesis. Properties of these Cre transgenes, including name, tissue-specificity, developmental stage-specificity, and compound inducibility, are summarized in Table 2. The efficiency of recombination between *loxP* sites resulting in excision and production of null mutants varies between the available Cre drivers and in some instances efficiency is influenced by the *floxed* target. Many of the Cre driver mice are available commercially from vendors such as Jackson Laboratories. Wherever possible the reference describing the original synthesis of the Cre construct and generation of transgenic mice is included, so that an investigator desiring to obtain these mice will be able to contact the laboratory of the principal investigator where they were created.

Mice bearing *floxed* target genes important to erythropoiesis or some aspect of red blood cell biology have been culled from the literature and the list is summarized in Table 3. The name of each gene is listed, along with basic properties of the encoded protein, the major red cell property or process in which the gene product functions (and consequently that the null mutation affects),

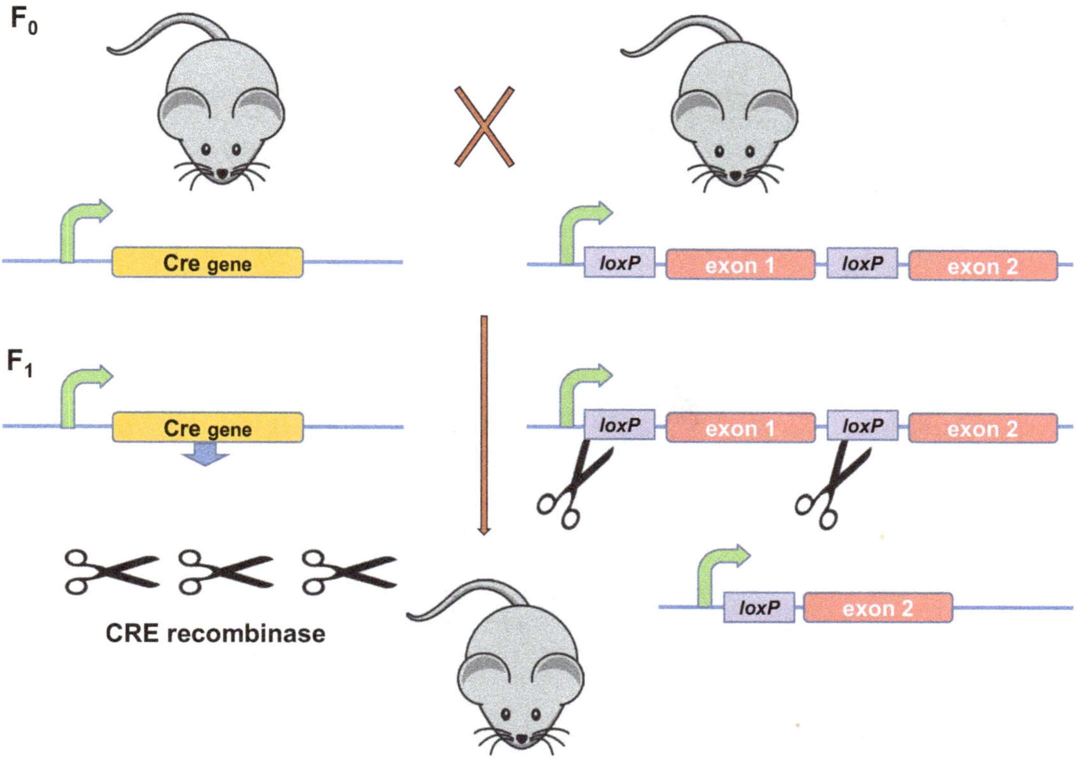

Fig. 3 Conditional gene knockouts in mice using the Cre recombinase-*loxP* recombination system. A conditional knockout allele is produced by gene targeting as shown in Fig. 2. In this instance, the targeting construct contains two 34 bp *loxP* sites inserted in intergenic DNA sequence, usually in introns flanking an exon, multiple exons, or an exon plus the promoter of the targeted gene (*right side* of figure). The *loxP* sites do not interfere with gene expression. The resultant mice are bred with Cre expression mice (*left side* of figure), in which Cre recombinase gene expression is driven by a tissue-specific, developmental stage-specific, or inducible promoter, or a combination of these. Cre recombinase catalyzes excision of the *loxP*-flanked (*floxed*) DNA sequences, resulting in a targeted deletion and leaving behind a single *loxP* site. Intergenic sequences are shown in blue; promoters are indicated as *curved green lines*

diseases modeled, and the earliest reference describing creation of the mice bearing the erythroid-specific null mutation (or activation of a mutated version of the gene in the erythroid compartment). Processes modeled by these Cre-*lox* mice include globin gene switching, hematopoietic stem cell function, erythropoiesis, and iron metabolism. Diseases modeled include paroxysmal nocturnal hemoglobinuria (PNH), polycythemia-like disease, erythrocytosis, chronic normocytic, normochromic anemia (preclinical model for the anemia of Epo deficiency), and human myelodysplastic syndromes (defects in cell cycle regulation and mitochondrial function).

Table 2
Cre drivers

Gene promoter controlling expression of Cre-recombinase	Tissues/cell types/stage-specificity	Reference
Erythropoietin receptor (EpoR) promoter (ErGFPcre)	Erythroid progenitors/erythroid-specific lineage. EpoR expression initiates at the erythroid burst-forming unit stage (BFU-E), and reaches maximal expression at the erythroid colony-forming unit (CFU-E) and/or proerythroblast stage, and decreases thereafter.	[42]
Vav promoter (VavCre)	Hematopoietic tissues and vascular endothelium.	[43]
Mx1 promoter (Mx-Cre or Mx1-Cre)	This promoter is silent unless activated via the administration of interferon α, interferon β, or a synthetic poly I:C double-stranded RNA. Expression is not tissue-specific or developmental stage-specific, and expression can be activated at any point during development.	[44]
GATA-1 promoter (GATA1-Cre)	GATA1 is a transcription factor necessary for erythroid-specific gene expression and is essential for both primitive and definitive erythropoiesis. Low level expression is achieved in $CD34^+$ hematopoietic cells and high level expression is observed in eosinophils, mast cells, megakaryocytes, and erythroblasts.	[45]
Rosa26 promoter fused to the Cre ORF and the ligand-binding domain of the human estrogen receptor (Rosa 26CreER)	Rosa26 is highly and ubiquitously expressed in most cells and tissues of adult mice. CreER recombinase is inactive unless stimulated by the administration of the synthetic estrogen receptor ligand, 4-hydroxytamoxifen (OHT), allowing for temporal regulation of Cre activity.	[46]
Receptor tyrosine kinase Tek (Tie2) promoter/enhancer (Tie2Cre)	The Tie2 promoter/enhancer is expressed consistently in endothelial cells during embryogenesis and adulthood. The Tie2Cre system can delete *floxed* targets in vascular endothelial cells.	[47]
Mouse mammary tumor virus (MMTV) long terminal repeat (LTR) promoter (MMTV-LTR Cre)	MMTV-LTR promoter drives high levels of expression in virgin and lactating mammary gland, salivary gland, seminal vesicle, skin, erythrocytes, B cells, and T cells.	[48]
μ-LCR cassette-β-globin promoter-Cre construct (HBB-cre)	μ-LCR cassette-β-globin promoter stimulates high levels of Cre recombinase in a erythroid/megakaryocytic cell lineage-specific manner.	[49]

(continued)

Table 2
(continued)

Gene promoter controlling expression of Cre-recombinase	Tissues/cell types/stage-specificity	Reference
EIIaCre	The adenovirus EIIa promoter targets Cre recombinase expression to early mouse embryos. Cre expression is mosaic, and occurs in a wide variety of tissues including germ cells.	[50]
iCre	Mammalian codon-optimized Cre recombinase that improves Cre translation/function and reduces the high CpG content of the prokaryotic coding sequences, thus diminishing the chance of epigenetic silencing in mammalian systems. Included here for informational purposes as newer Cre drivers will utilize this variant with or without tamoxifen inducibility.	[51]

A broad-coverage internet resource that may be used in conjunction with the mouse sources listed in this chapter is the International Mouse Strain Resource (IMSR, www.findmice.org), which offers users a catalog of worldwide mouse resources (live, cryopreserved, embryonic stem cells). A search engine is available on the home page that will perform a user-defined search of all or select resources. In addition, links are provided for direct access to the web sites of repositories, which describe holding resources of interest, descriptions of the holdings, and order forms. A link to the Mouse Genome Database (MGD, www.informatics.jax.org) is also provided for detailed information about the gene(s) involved, as well as the relationship of the mouse model(s) to human disease(s).

8 Reporter Mice

Mice with robust gene expression readouts in response to activation have utility in the screening or testing of potential compounds with therapeutic ramification in an in vivo setting. Reactivation of silenced γ-globin gene expression is of substantial clinical interest. A dual luciferase β-YAC transgenic mouse in which the $^{A}\gamma$-globin promoter was fused to firefly luciferase and the β-globin promoter was fused to Renilla luciferase was utilized to derive immortalized bone marrow cells that were subsequently used in a high-throughput screen (HTS) for γ-globin inducers [39]. Two other mouse models that serve a similar purpose is the so-called GG mouse, which carries a 183 kb human β-globin locus with eGFP

Table 3
Cre-*loxP*-mediated conditional knock-outs or knock-ins affecting erythropoiesis

Floxed allele	Gene product description	Process under investigation	Phenotype/process affected	Reference
Zbtb7a	Transcription factor: Zinc finger and BTB domain containing 7A	Hematopoietic Stem Cell (HSC) maintenance	Embryonic lethal, anemia	[52]
Myc	Transcription factor: Myc proto-oncogene protein	Primitive Hematopoiesis	Embryonic lethal, blocks HSC differentiation, absence of primitive erythrocytes	[53]
SCL/Tal-1	Transcription factor	Hematopoietic Stem Cells	Embryonic lethal, genesis of HSCs, differentiation into erythroid and megakaryocytic progenitors, anemia, thrombocytopenia	[54]
Asxl1 (also *Asxl1: nlacZ/ nGFP* knock-in)	Transcription factor: Additional sex combs like transcriptional regulator 1	Erythropoiesis	Embryonic lethal, myelodysplastic syndrome	[55]
*Meis*1	Transcription factor: Homeobox protein Meis1	Erythropoiesis	Primitive hematopoietic repopulating cells and megakaryocytic/ erythroid progenitor expansion, abnormal HSC function	[56]
Klf1	Transcription factor: Krüppel-like factor 1	Hemoglobin Switching	Embryonic lethal, β-like globin gene expression, anemia	[57]
Klf2	Transcription factor: Krüppel-like factor 2	Hemoglobin Switching	Embryonic lethal, β-like globin gene expression, anemia	[57]
Bcl11a	Transcription factor: C_2H_2 type zinc-finger	Hemoglobin Switching	γ-globin (HbF) repression	[58]
Mi2β	Chromodomain helicase DNA binding protein 4	Hemoglobin Switching	γ-globin (HbF) repression	[59]
Atg7	Interacting protein: Ubiquitin-activating enzyme E1-like protein	Erythropoiesis	Mitochondrial autophagy, splenomegaly, death via severe anemia	[60]
Med1	Mediator complex subunit 1: required for SP1 activation as part of CRSP complex	Erythropoiesis	Block in erythroid development	[61]

(continued)

Table 3 (continued)

Floxed allele	Gene product description	Process under investigation	Phenotype/process affected	Reference
Taf10	Interacting protein: TATA-box binding protein associated factor 10	Erythropoiesis	Embryonic lethal	[62]
Gfi1b	Transcription factor: Growth factor independent 1B transcription repressor	Erythropoiesis	Abnormal HSC function	[63]
Tsc1/Rptor	Interacting protein: Tuberous sclerosis 1 interacting protein/ regulatory associated protein of mTOR complex 1	Erythropoiesis	Role of mTORC1 in erythropoiesis, increased neonatal mortality, macrocytic/microcytic anemia	[64]
Ldb1	Interacting protein: LIM domain binding 1	Erythropoiesis	Embryonic lethal, abnormal erythroid gene expression, defective primitive erythropoiesis, continuously required for definitive erythropoiesis and megakaryopoiesis	[65]
Pit1	Interacting protein: Solute carrier family 20 (phosphate transporter) member 1	Erythropoiesis	Erythroid maturation, anemia, myelodysplastic syndromes	[66]
Sf3b1^{K700E} knock-in	Interacting protein: Splicing factor 3b subunit 1	Erythropoiesis	Impaired erythropoiesis and aberrant splicing, myelodysplastic syndromes	[67]
Miz-1	Transcription factor: Myc-interacting zinc finger protein 1	Erythropoiesis	Embryonic lethal, required for embryonic and stress-induced erythropoiesis, perturbed erythroid differentiation and development, severe anemia	[68]
Brg1	Transcription factor: Brahma-related gene 1	Erythropoiesis, Vascular Development	Embryonic lethal, altered globin expression, anemia, erythropoietic/vascular abnormalities	[69]
RhoA	Hub protein: Ras homolog family member A	Erythropoiesis	Embryonic lethal, cytokinesis in erythroblasts, failed definitive erythropoiesis, anemia	[70]

(continued)

Table 3
(continued)

Floxed allele	Gene product description	Process under investigation	Phenotype/process affected	Reference
Spry1	Interacting protein: Sprouty RTK signaling antagonist 1	Erythropoiesis	Perturbed erythroid development, reticulocytosis, heightened splenic erythropoiesis, regulator of erythropoiesis during anemia, transducer of EPOR signals, and candidate suppressor of Jak2 activity	[71]
Vegf	Growth factor: Vascular endothelial growth factor	Erythropoiesis	Erythropoietic lineage development, increased Gata1 levels, increased erythroid differentiation	[72]
Stat5	Transcription factor: Signal transducer and activator of transcription 5A	Erythropoiesis	Reduced transferrin receptor gene expression, microcytic, hypochromic anemia	[73]
cdc42	Rho GTPase Cdc42: Cell division control protein 42 homolog	Erythropoiesis	Balance between myelopoiesis and erythropoiesis, altered gene expression, fatal myeloproliferative disorder, anemia, splenomegaly	[74]
Tfr2	Membrane protein: Transferrin receptor 2	Iron Metabolism	Liver iron overload, inadequate hepcidin levels, iron-deficient anemia	[75]
Mfrn1	Mitochondrial iron transporter: Mitoferrin-1	Iron Metabolism	Embryonic lethal, reduced erythroblast formation, severe anemia	[76]
Piga	Enzyme: Phosphatidylinositol glycan anchor biosynthesis class A	Erythropoiesis	Embryonic lethal, paroxysmal nocturnal hemoglobinuria (PNH)	[45]
Jak2V617F knock-in	Non-receptor tyrosine kinase: Janus kinase 2 (JAK2)	Erythropoiesis	Polycythemia-like disease (all major features of human polycythemia vera)	[77]
Phd1, *Phd2*, *Phd3*	Enzyme: Prolyl hydroxylase domain protein 1, 2, or 3 (3 PHD isoforms)	Erythropoiesis	Erythrocytosis	[78]

(continued)

Table 3 (continued)

Floxed allele	Gene product description	Process under investigation	Phenotype/process affected	Reference
Epo	Growth factor: Erythropoietin (EPO)	Erythropoiesis	Embryonic lethal, chronic, normocytic, normochromic anemia, pre-clinical model for anemia of Epo deficiency	[79]
pRb	Tumor suppressor protein: Retinoblastoma protein	Erythropoiesis	Defects in cell cycle regulation and mitochondrial function, anemia, myelodysplastic syndromes	[80]
Trim33	Transcriptional corepressor: Tripartite motif-containing 33, also known as transcriptional intermediary factor 1 gamma (TIF1-γ)	Erythropoiesis	Embryonic lethal	[81]
Bcl-x	Anti-apoptotic factor: Bcl-2-like protein 1	Erythropoiesis	Required for the survival of erythroid cells at the end of maturation, which includes enucleated reticulocytes in circulation, hemolytic anemia, hyperplasia of immature erythroid cells, splenomegaly	[82]
Nix	Pro-apoptotic factor: BH3-only-like protein	Erythropoiesis	Reticulocytosis, thrombocytosis, splenomegaly, splenic and bone marrow erythroblastosis, reduced apoptosis during erythrocyte maturation	[83]
Tr2/Tr4	Transcription factor: Testicular receptor 2 (TR2), Testicular receptor 4 (TR4)	Hemoglobin Switching	Embryonic lethal, γ-globin gene repression	[84]
Zbp-89 (*Zfp148*)	Transcription factor: Zinc finger protein 148	Hematopoiesis	Stress erythropoiesis, erythroid lineage development, anemia, thrombocytopenia, myeloid and B lymphoid lineage anomalies, altered HSC gene expression	[85]

(continued)

Table 3 (continued)

Floxed allele	Gene product description	Process under investigation	Phenotype/process affected	Reference
Trim28	Transcriptional cofactor: Tripartite motif-containing protein 28, also known as transcriptional intermediary factor1 beta (TIF1β)	Erythropoiesis	Cell-autonomous development of immature erythroblasts in bone marrow, essential for erythroblasts differentiation, anemia, increased apoptosis, reduced erythroid transcription factor expression, reduced heme biosynthesis	[86]
Ccbe1	Lymphangiogenic factor: Collagen and calcium-binding epidermal growth factor domain-containing protein 1	Primitive Erythropoiesis	Secreted lymphangiogenic CCBE1 is essential for fetal, but not postnatal erythropoiesis, loss of CCBE1 signaling impairs erythroblastic island formation and function, not required in hematopoietic cells, severe anemia, increased apoptosis	[87]
Sur	Inhibitor of apoptosis: Survivin	Erythropoiesis	Essential for steady-state hematopoiesis and survival of adult, high level expression critical for proper erythroid differentiation, bone marrow depletion, pancytopenia, erythrocyte abnormalities, hypocellular spleen, hematopoietic cell death	[88]
Runx1	Transcription factor: Runt-related transcription factor 1	Hematopoiesis	Embryonic lethal, not required in adult hematopoietic compartment, abnormal HSCs, defective T- and B-cell maturation, inefficient platelet production, mild myeloproliferative phenotype	[89]

(continued)

Table 3 (continued)

Floxed allele	Gene product description	Process under investigation	Phenotype/process affected	Reference
α4	Transmembrane receptor: α4 integrin subunit	Hematopoietic Stem Cells	Progenitor cell influx into the peripheral blood, delayed erythroid and myeloid regeneration, hematopoietic abnormalities	[90]
Raf-1	Enzyme: RAF proto-oncogene serine/threonine-protein kinase	Erythropoiesis	Embryonic lethal, anemia, accelerated erythroid differentiation	[91]
Jak2	Non-receptor tyrosine kinase: Janus kinase 2	Hematopoiesis	Embryonic lethal, impaired hematopoiesis, splenic atrophy, severely diminished erythropoiesis and thrombopoiesis, modestly affected granulopoiesis and monocytopoiesis	[92]
Abcb7	Mitochondrial ATP-binding cassette sub-family B member 7	Hematopoiesis	X-linked sideroblastic anemia with ataxia (XLSA/A) due to inhibition of heme biosynthesis, damaged mitochondria, pancytopenia, hypocellular bone marrow	[93]
Ufbp1	Inhibitor of apoptosis in ER-stressed cells: Ufm1 binding protein 1	Hematopoiesis	Embryonic lethal, impaired hematopoiesis, defective erythroid development, pancytopenia, suppressed GATA-1 and KLF1 expression	[94]
Ezh2	Enzymatic component of Polycomb Repressive Complex 2: Enhancer of zeste homolog 2	Hematopoietic Stem Cells	Embryonic lethal, essential for fetal liver, but not bone marrow erythropoiesis, defective HSC expansion, compromised lymphopoiesis, reduced H3K27me3 levels suggesting epigenetic status of HSCs is developmentally regulated, anemia	[95]

(continued)

Table 3
(continued)

Floxed allele	Gene product description	Process under investigation	Phenotype/process affected	Reference
Rcor1	Transcriptional corepressor: REST (repressor element-1 silencing transcription factor) corepressor 1	Hematopoietic Stem Cells	Embryonic lethal, promotes erythropoiesis by repressing HSC and/or progenitor genes, as well as genes and signaling pathways that lead to myeloid cell fate, block in fetal erythropoiesis at the proerythroblast stage, anemia	[96]
Setd8	Histone Methyltransferase: monomethylates histone H4 lysine 20 (H4K20me1)	Erythropoiesis	Embryonic lethal, defect in primitive erythropoiesis, abnormalities in cell cycle progression, regulates erythroid maturation, represses Gata2 expression, severe anemia	[97]
Smad4	Signal transduction: Mothers against decapentaplegic homolog 4	Erythropoiesis	Not required for adult erythropoiesis, severe anemia due solely to blood loss, iron deficiency	[98]
Ulk1	Enzyme: Serine/threonine-protein kinase	Erythropoiesis	Autophagy machinery that leads to the elimination of organelles in erythroid cells	[99]
Ck2β	Enzyme: serine-threonine kinase composed of two catalytic (α) and two regulatory (β) subunits	Hematopoiesis	Embryonic lethal, regulates definitive hematopoiesis of all hematopoietic cell lineages	[100]
Flvcr1a	Heme exporter membrane protein: Feline leukemia virus subgroup C receptor 1	Erythropoiesis	Embryonic lethal, heme export control, severe macrocytic anemia, splenomegaly, iron accumulation in duodenum, liver and spleen, reduced BFU-E and CFU-E, defective erythroid differentiation	[101]
Rhau	RNA helicase: RNA helicase associated with AU-rich element	Hematopoiesis	Embryonic lethal, hemolytic anemia, differentiation defect at the proerythroblast stage	[102]

(continued)

Table 3 (continued)

Floxed allele	Gene product description	Process under investigation	Phenotype/process affected	Reference
Add3	Adducins: Skeletal protein involved in the assembly of spectrin-actin network	RBC membrane skeleton structure	Animals were viable and presented no obvious erythroid cell defects	[103]
cyclin $E^{T74A\ T393A}$ *knock-in*	Cell cycle regulator: Complexes with the CDK2 cyclin-dependent kinase subunit	Erythropoiesis	Links ubiquitin-proteasome pathway control of G1-to-S-phase progression to regulation of metabolism and gene expression in terminally differentiating bone marrow erythroid cells, abnormal erythropoiesis characterized by a large expansion of abnormally proliferating progenitors, impaired differentiation, dysplasia, and anemia, models early stage human refractory anemia/ myelodysplastic syndrome	[104]
Rps14	40S ribosomal subunit protein S14	Erythropoiesis	Erythroid differentiation defect dependent on p53 characterized by apoptosis during transition from polychromatic to orthochromatic erythroblasts, age-dependent progressive anemia, megakaryocyte dysplasia, block in erythroid differentiation, loss of HSC quiescence	[105]
Rps6	40s Ribosomal subunit protein S6	Erythropoiesis	Hypoproliferative macrocytic anemia, granulocytopenia, thrombocytosis, lymphopenia (model for Diamond-Blackfan Anemia and myelodysplastic syndrome)	[106]

(continued)

Table 3 (continued)

Floxed allele	Gene product description	Process under investigation	Phenotype/process affected	Reference
Ppp2ca	Enzyme: Serine/threonine phosphatase, protein phosphatase 2A	Erythropoiesis	Perturbed definitive erythropoiesis characterized by fetal liver atrophy, reduced Ter119^{+} cell number, atypical expression patterns of molecular markers, reduced colony formation, reduction in definitive globin gene expression	[107]
Dicer	Enzyme: Endoribonuclease	Hematopoietic Stem Cells	Down-regulation of erythroid genes, defective erythroid linage differentiation	[108]
Mll	Enzyme: Mixed-lineage leukemia	Hematopoietic Stem Cells	Embryonic lethal, reduced numbers of HSCs, abnormal HSC function	[109]
Ufm1	Ubiquitin-like modification system: Ubiquitin-fold modifier 1	Hematopoietic Stem Cells	Embryonic lethal, diminished hematopoietic development, increased HSC death, severe anemia, cytopenia, elevated endoplasmic reticulum stress, blocked autophagic degradation, increased ROS	[110]
Rac1, Rac2	Rac GTPases: Rac1 and Rac2 regulates many aspects of intracellular actin dynamics	Erythropoiesis	Abnormal BFU-E and CFU-E morphology, decreased megakaryocyte-erythrocyte progenitors in bone marrow, increased splenic erythropoiesis	[111]
Adar1	RNA-editing enzyme: Adenosine deaminase acting on RNA 1	Hematopoietic Stem Cells	Embryonic lethal, abnormal HSC function, increased HSC apoptosis	[112]

(continued)

Table 3 (continued)

Floxed allele	Gene product description	Process under investigation	Phenotype/process affected	Reference
Sod2	Enzyme: Superoxide dismutase 2	Hematopoietic Stem Cells	Increased ROS in erythroid progeny, diminished erythrocytes, decreased ferrochelatase activity, extramedullary hematopoiesis, systemic iron redistribution, abnormal gene expression (hematopoietic transcription factors, globins, and iron-response genes), changes in histone posttranslational modification	[113]

fused to the $^{G}\gamma$-globin promoter [40] and a human β-globin locus $^{A}\gamma$-GFP (or GPA/GFP) β-DsRed PAC transgenic mouse [41].

9 Future Directions

Transgenesis via classic microinjection of oocytes or knock-in replacement of murine genes with their human counterparts in mouse ES cells remains cutting-edge and are important techniques for producing humanized mouse models. However, with the advent of genome editing via CRISPR/Cas9, TALENS, and zinc-finger nucleases, existing humanized mouse lines can be modified without the arduous process of generating an entire new mouse line. iPSCs or ESCs can be derived from existing mice, the cells subjected to genome editing, and new mice created from the engineered cells. The human transgenes themselves could be modified or endogenous mouse genes that play a role in transgene expression or gene product function could be modified. All of this would dramatically reduce the time required for mouse production, in many instances eliminating transgene introduction, gene targeting in ES cells, and many generations of breeding to produce mice with a desired combination of human transgenes and endogenous gene knockouts or mutants. Part of the power of this approach would be the ability to introduce subtle alterations affecting phenotype via a variety of hypomorphic mutations or changes that affect protein function including those affecting posttranslational modification, allostery, protein-protein interaction, and protein folding.

References

1. Trimborn T, Gribnau J, Grosveld F, Fraser P (1999) Mechanisms of developmental control of transcription in the murine α- and β-globin loci. Genes Dev 13(1):112–124
2. Grosveld F, van Assendelft GB, Greaves DR, Kollias G (1987) Position-independent, high-level expression of the human β-globin gene in transgenic mice. Cell 51:975–985
3. Chada K, Magram J, Raphael K et al (1985) Specific expression of a foreign β-globin gene in erythroid cells of transgenic mice. Nature 314:377–380
4. Kollias G, Wrighton N, Hurst J, Grosveld F (1986) Regulated expression of human $^{A}\gamma$-, β-, and hybrid γβ-globin genes in transgenic mice: manipulation of the developmental expression patterns. Cell 46:89–94
5. Enver T, Raich N, Ebens AJ et al (1990) Developmental regulation of human fetal-to-adult globin gene switching in transgenic mice. Nature 344:309–313
6. Higgs DR, Wood WG, Jarman AP et al (1990) A major positive regulatory region located far upstream of the human α-globin gene locus. Genes Dev 4:1588–1601
7. Sharpe JA, Wells DJ, Whitelaw E et al (1993) Analysis of the human α-globin gene cluster in transgenic mice. Proc Natl Acad Sci U S A 90:11262–11266
8. Waggoner SA, Liebhaber SA (2003) Regulation of α-globin mRNA stability. Exp Biol Med 228:387–395
9. Kaufman RM, Pham CT, Ley TJ (1999) Transgenic analysis of a 100-kb human β-globin cluster-containing DNA fragment propagated as a bacterial artificial chromosome. Blood 94:3178–3184
10. Peterson KR (2003) Transgenic mice carrying yeast artificial chromosomes. Expert Rev Mol Med 2003:1–25
11. Katsantoni EZ, de Krom M, Kong-a-San J et al (2004) An embryonic-specific repressor element located 3′ to the $^{A}\gamma$-globin gene influences transcription of the human β-globin locus in transgenic mice. Exp Hematol 32:224–233
12. Gaensler KM, Kitamura M, Kan YW (1993) Germ-line transmission and developmental regulation of a 150-kb yeast artificial chromosome containing the human β-globin locus in transgenic mice. Proc Natl Acad Sci U S A 90:11381–11385
13. Peterson KR, Clegg CH, Huxley C et al (1993) Transgenic mice containing a 248-kb yeast artificial chromosome carrying the human β-globin locus display proper developmental control of human globin genes. Proc Natl Acad Sci U S A 90:7593–7597
14. Narayanan K, Williamson R, Zhang Y, Stewart AF, Ioannou PA (1999) Efficient and precise engineering of a 200 kb β-globin human/bacterial artificial chromosome in E. coli DH10B using an inducible homologous recombination system. Gene Ther 6:442–447
15. Harju S, McQueen KJ, Peterson KR (2002) Chromatin structure and control of β-like globin gene switching. Exp Biol Med 227:683–700
16. de Laat W, Grosveld F (2003) Spatial organization of gene expression: the active chromatin hub. Chromosome Res 11:447–459
17. Paw BH, Choe S-K, Costa FC, Sundar SV, Peterson KR (2006) Vertebrate models for sickle cell disease research. In: Pace BS (ed) Renaissance of sickle cell disease research in the genomic era. World Scientific and Imperial College Press, London, pp 237–257
18. Nagel RL, Fabry ME (2001) The panoply of animal models for sickle cell anaemia. Br J Haematol 112:19–25
19. Rubin EM, Witkowska HE, Spangler E et al (1991) Hypoxia-induced in vivo sickling of transgenic mouse red cells. J Clin Invest 87:639–647
20. Popp RA, Popp DM, Shinpock SG et al (1997) A transgenic mouse model of hemoglobin S Antilles disease. Blood 89:4204–4212
21. Fabry ME, Costantini F, Pachnis A et al (1992) High expression of human β^{S}- and α-globins in transgenic mice: erythrocyte abnormalities, organ damage, and the effect of hypoxia. Proc Natl Acad Sci U S A 89:12155–12159
22. Fabry ME, Nagel RL, Pachnis A, Suzuka SM, Costantini F (1992) High expression of human β^{S}- and α-globins in transgenic mice: hemoglobin composition and hematological consequences. Proc Natl Acad Sci U S A 89:12150–12154
23. Fabry ME, Sengupta A, Suzuka SM et al (1995) A second generation transgenic mouse model expressing both hemoglobin S (HbS) and HbS-Antilles results in increased phenotypic severity. Blood 86:2419–2428
24. Trudel M, Saadane N, Garel MC et al (1991) Towards a transgenic mouse model of sickle cell disease: hemoglobin SAD. EMBO J 10:3157–3165

25. Trudel M, De Paepe ME, Chretien N et al (1994) Sickle cell disease of transgenic SAD mice. Blood 84:3189–3197
26. Paszty C, Brion CM, Manci E et al (1997) Transgenic knockout mice with exclusively human sickle hemoglobin and sickle cell disease. Science 278:876–878
27. Ryan TM, Townes TM, Reilly MP et al (1990) Human sickle hemoglobin in transgenic mice. Science 247:566–568
28. Ryan TM, Ciavatta DJ, Townes TM (1997) Knockout-transgenic mouse model of sickle cell disease. Science 278:873–876
29. Chang JC, Lu R, Lin C et al (1998) Transgenic knockout mice exclusively expressing human hemoglobin S after transfer of a 240-kb β^S-globin yeast artificial chromosome: a mouse model of sickle cell anemia. Proc Natl Acad Sci U S A 95:14886–14890
30. Blouin MJ, Beauchemin H, Wright A et al (2000) Genetic correction of sickle cell disease: insights using transgenic mouse models. Nat Med 6:177–182
31. Fabry ME, Suzuka SM, Weinberg RS et al (2001) Second generation knockout sickle mice: the effect of HbF. Blood 97:410–418
32. Yang B, Kirby S, Lewis J et al (1995) A mouse model for β^0-thalassemia. Proc Natl Acad Sci U S A 92:11608–11612
33. Ciavatta DJ, Ryan TM, Farmer SC, Townes TM (1995) Mouse model of human beta zero thalassemia: targeted deletion of the mouse β^{maj}- and β^{min}-globin genes in embryonic stem cells. Proc Natl Acad Sci U S A 92:9259–9263
34. Paszty C, Mohandas N, Stevens ME et al (1995) Lethal α-thalassaemia created by gene targeting in mice and its genetic rescue. Nat Genet 11:33–39
35. Skow LC, Burkhart BA, Johnson FM et al (1983) A mouse model for β-thalassemia. Cell 34:1043–1052
36. Lewis J, Yang B, Kim R et al (1998) A common human β globin splicing mutation modeled in mice. Blood 91:2152–2156
37. Huo Y, McConnell SC, Liu S-R et al (2009) Humanized model of Cooley's anemia. J Biol Chem 284:4889–4896
38. Huo Y, McConnell SC, Ryan TM (2009) Preclinical transfusion-dependent humanized mouse model of β thalassemia major. Blood 113:4763–4770
39. Peterson KR, Costa FC, Fedosyuk F, Neades RY, Chazelle AM, Zelenchuk L, Fonteles AH, Dalal P, Roy A, Chaguturu R, Li B, Pace BS (2014) A cell-based high-throughput screen for novel inducers of fetal hemoglobin for treatment of hemoglobinopathies. PLoS One 9:e107006
40. McColl B, Kao BR, Lourthai P et al (2014) An *in vivo* model for analysis of developmental erythropoiesis and globin gene regulation. FASEB J 28:2306–2317
41. Papadopoulos P, Gutierrez L, van der Linden R et al (2012) A dual reporter mouse model of the human b-globin locus: applications and limitations. PLoS One 7(12):e51272
42. Heinrich AC, Pelanda R, Klingmuller U (2004) A mouse model for visualization and conditional mutations in the erythroid lineage. Blood 104(3):659–666. doi:10.1182/blood-2003-05-1442
43. Georgiades P, Ogilvy S, Duval H, Licence DR, Charnock-Jones DS, Smith SK, Print CG (2002) VavCre transgenic mice: a tool for mutagenesis in hematopoietic and endothelial lineages. Genesis 34(4):251–256. doi:10.1002/gene.10161
44. Yao YC, Zi XY, Li JX, Xiong J, Xie QD, Wang XM, Hu YP (2001) Generation of Mx-cre transgenic mice. Yi Chuan Xue Bao 28 (4):313–316
45. Jasinski M, Keller P, Fujiwara Y, Orkin SH, Bessler M (2001) GATA1-Cre mediates Piga gene inactivation in the erythroid/megakaryocytic lineage and leads to circulating red cells with a partial deficiency in glycosyl phosphatidylinositol-linked proteins (paroxysmal nocturnal hemoglobinuria type II cells). Blood 98(7):2248–2255. doi:10.1182/blood.V98.7.2248
46. Soriano P (1999) Generalized lacZ expression with the ROSA26 Cre reporter strain. Nat Genet 21(1):70–71. doi:10.1038/5007
47. Kisanuki YY (2001) Tie2-Cre transgenic mice: a new model for endothelial cell-lineage analysis in vivo. Dev Biol 230(2):230–242. doi:10.1006/dbio.2000.0106
48. Wagner KU, Wall RJ, St-Onge L, Gruss P, Wynshaw-Boris A, Garrett L, Li M, Furth PA, Hennighausen L (1997) Cre-mediated gene deletion in the mammary gland. Nucleic Acids Res 25(21):4323–4330
49. Peterson KR, Fedosyuk H, Zelenchuk L, Nakamoto B, Yannaki E, Stamatoyannopoulos G, Ciciotte S, Peters LL, Scott LM, Papayannopoulou T (2004) Transgenic Cre expression mice for generation of erythroid-specific gene alterations. Genesis 39(1):1–9. doi:10.1002/gene.20020
50. Holzenberger M, Lenzner C, Leneuve P, Zaoui R, Hamard G, Vaulont S, Le Bouc Y (2000) Cre-mediated germline mosaicism: a method allowing rapid generation of several

alleles of a target gene. Nucleic Acids Res 28 (21):e92

51. Shimshek DR, Kim J, Hubner MR, Spergel DJ, Buchholz F, Casanova E, Stewart AF, Seeburg PH, Sprengel R (2002) Codon-improved Cre recombinase (iCre) expression in the mouse. Genesis 32(1):19–26
52. Maeda T, Ito K, Merghoub T, Poliseno L, Hobbs RM, Wang G, Dong L, Maeda M, Dore LC, Zelent A, Luzzatto L, Teruya-Feldstein J, Weiss MJ, Pandolfi PP (2009) LRF is an essential downstream target of GATA1 in erythroid development and regulates BIM-dependent apoptosis. Dev Cell 17 (4):527–540. doi:10.1016/j.devcel.2009.09.005
53. Trumpp A, Refaeli Y, Oskarsson T, Gasser S, Murphy M, Martin GR, Bishop JM (2001) c-Myc regulates mammalian body size by controlling cell number but not cell size. Nature 414(6865):768–773. doi:10.1038/414768a
54. Mikkola HKA, Klintman J, Yang H, Hock H, Schlaeger TM, Fujiwara Y, Orkin SH (2003) Haematopoietic stem cells retain long-term repopulating activity and multipotency in the absence of stem-cell leukaemia SCL/tal-1 gene. Nature 421(6922):547–551. doi:10.1038/nature01345
55. Abdel-Wahab O, Gao J, Adli M, Dey A, Trimarchi T, Chung YR, Kuscu C, Hricik T, Ndiaye-Lobry D, LaFave LM, Koche R, Shih AH, Guryanova OA, Kim E, Li S, Pandey S, Shin JY, Telis L, Liu J, Bhatt PK, Monette S, Zhao X, Mason CE, Park CY, Bernstein BE, Aifantis I, Levine RL (2013) Deletion of Asxl1 results in myelodysplasia and severe developmental defects in vivo. J Exp Med 210 (12):2641–2659. doi:10.1084/jem.20131141
56. Unnisa Z, Clark JP, Roychoudhury J, Thomas E, Tessarollo L, Copeland NG, Jenkins NA, Grimes HL, Kumar AR (2012) Meis1 preserves hematopoietic stem cells in mice by limiting oxidative stress. Blood 120 (25):4973–4981. doi:10.1182/blood-2012-06-435800
57. Alhashem YN, Vinjamur DS, Basu M, Klingmüller U, Gaensler KML, Lloyd JA (2011) Transcription factors KLF1 and KLF2 positively regulate embryonic and fetal β-globin genes through direct promoter binding*. J Biol Chem 286(28):24819–24827. doi:10.1074/jbc.M111.247536
58. Sankaran VG, Xu J, Ragoczy T, Ippolito GC, Walkley CR, Maika SD, Fujiwara Y, Ito M, Groudine M, Bender MA, Tucker PW, Orkin SH (2009) Developmental and species-divergent globin switching are driven by BCL11A. Nature 460(7259):1093–1097. doi:10.1038/nature08243
59. Costa FC, Fedosyuk H, Chazelle AM, Neades RY, Peterson KR (2012) Mi2β is required for γ-globin gene silencing: temporal assembly of a GATA-1-FOG-1-Mi2 repressor complex in β-YAC transgenic mice. PLoS Genet 8(12): e1003155. doi:10.1371/journal.pgen.1003155
60. Mortensen M, Ferguson DJP, Edelmann M, Kessler B, Morten KJ, Komatsu M, Simon AK (2010) Loss of autophagy in erythroid cells leads to defective removal of mitochondria and severe anemia in vivo. Proc Natl Acad Sci U S A 107(2):832–837. doi:10.1073/pnas.0913170107
61. Stumpf M, Yue X, Schmitz S, Luche H, Reddy JK, Borggrefe T (2010) Specific erythroid-lineage defect in mice conditionally deficient for Mediator subunit Med1. Proc Natl Acad Sci U S A 107(50):21541–21546. doi:10.1073/pnas.1005794107
62. Mohan IW, Scheer E, Wendling O, Metzger D, Tora L (2003) TAF10 (TAFII30) Is Necessary for TFIID Stability and Early Embryogenesis in Mice. Mol Cell Biol 23 (12):4307–4318. doi:10.1128/mcb.23.12.4307-4318.2003
63. Khandanpour C, Sharif-Askari E, Vassen L, Gaudreau M-C, Zhu J, Paul WE, Okayama T, Kosan C, Möröy T (2010) Evidence that Growth factor independence 1b regulates dormancy and peripheral blood mobilization of hematopoietic stem cells. Blood 116 (24):5149–5161. doi:10.1182/blood-2010-04-280305
64. Knight ZA, Schmidt SF, Birsoy K, Tan K, Friedman JM (2014) A critical role for mTORC1 in erythropoiesis and anemia. eLife 3:e01913. doi:10.7554/eLife.01913
65. Li L, Lee JY, Gross J, Song SH, Dean A, Love PE (2010) A requirement for Lim domain binding protein 1 in erythropoiesis. J Exp Med 207(12):2543–2550. doi:10.1084/jem.20100504
66. Liu L, Sánchez-Bonilla M, Crouthamel M, Giachelli C, Keel S (2013) Mice lacking the sodium-dependent phosphate import protein, PiT1 (SLC20A1), have a severe defect in terminal erythroid differentiation and early B cell development. Exp Hematol 41 (5):432–443. e437. doi:10.1016/j.exphem.2013.01.004
67. Obeng EA et al (2016) Physiologic expression of Sf3b1K700E causes impaired erythropoiesis, aberrant splicing, and sensitivity to therapeutic spliceosome modulation. Cancer Cell 30:404. doi:10.1016/j.ccell.2016.08.006

68. Kosan C, Rashkovan M, Ross J, Schaffer AM, Saba I, Lemsaddek W, Trudel M, Möröy T (2014) The transcription factor Miz-1 is required for embryonic and stress-induced erythropoiesis but dispensable for adult erythropoiesis. Am J Blood Res 4(1):7–19
69. Griffin CT, Brennan J, Magnuson T (2008) The chromatin-remodeling enzyme BRG1 plays an essential role in primitive erythropoiesis and vascular development. Development 135(3):493–500. doi:10.1242/dev.010090
70. Konstantinidis DG, Giger KM, Risinger M, Pushkaran S, Zhou P, Dexheimer P, Yerneni S, Andreassen P, Klingmüller U, Palis J, Zheng Y, Kalfa TA (2015) Cytokinesis failure in RhoA-deficient mouse erythroblasts involves actomyosin and midbody dysregulation and triggers p53 activation. Blood 126 (12):1473–1482. doi:10.1182/blood-2014-12-616169
71. Sathyanarayana P, Dev A, Pradeep A, Ufkin M, Licht JD, Wojchowski DM (2012) Spry1 as a novel regulator of erythropoiesis, EPO/EPOR target, and suppressor of JAK2. Blood 119(23):5522–5531. doi:10.1182/blood-2011-11-392571
72. Drogat B, Kalucka J, Gutiérrez L, Hammad H, Goossens S, Ghahremani MF, Bartunkova S, Haigh K, Deswarte K, Nyabi O, Naessens M, Ferrara N, Klingmüller U, Lambrecht BN, Nagy A, Philipsen S, Haigh JJ (2010) Vegf regulates embryonic erythroid development through Gata1 modulation. Blood 116 (12):2141–2151. doi:10.1182/blood-2010-01-264143
73. Zhu B-M, McLaughlin SK, Na R, Liu J, Cui Y, Martin C, Kimura A, Robinson GW, Andrews NC, Hennighausen L (2008) Hematopoietic-specific Stat5-null mice display microcytic hypochromic anemia associated with reduced transferrin receptor gene expression. Blood 112(5):2071–2080. doi:10.1182/blood-2007-12-127480
74. Yang L, Wang L, Kalfa TA, Cancelas JA, Shang X, Pushkaran S, Mo J, Williams DA, Zheng Y (2007) Cdc42 critically regulates the balance between myelopoiesis and erythropoiesis. Blood 110(12):3853–3861. doi:10.1182/blood-2007-03-079582
75. Roetto A, Cunto FD, Pellegrino RM, Hirsch E, Azzolino O, Bondi A, Defilippi I, Carturan S, Miniscalco B, Riondato F, Cilloni D, Silengo L, Altruda F, Camaschella C, Saglio G (2010) Comparison of 3 Tfr2-deficient murine models suggests distinct functions for Tfr2-α and Tfr2-β isoforms in different tissues. Blood 115(16):3382–3389. doi:10.1182/blood-2009-09-240960
76. Troadec MB, Warner D, Wallace J, Thomas K, Spangrude GJ, Phillips J, Khalimonchuk O, Paw BH, Ward DM, Kaplan J (2011) Targeted deletion of the mouse Mitoferrin1 gene: from anemia to protoporphyria. Blood 117(20):5494–5502. doi:10.1182/blood-2010-11-319483
77. Akada H, Yan D, Zou H, Fiering S, Hutchison RE, Mohi MG (2010) Conditional expression of heterozygous or homozygous Jak2V617F from its endogenous promoter induces a polycythemia vera-like disease. Blood 115(17):3589–3597. doi:10.1182/blood-2009-04-215848
78. Takeda K, Aguila HL, Parikh NS, Li X, Lamothe K, Duan LJ, Takeda H, Lee FS, Fong GH (2008) Regulation of adult erythropoiesis by prolyl hydroxylase domain proteins. Blood 111(6):3229–3235. doi:10.1182/blood-2007-09-114561
79. Zeigler BM, Vajdos J, Qin W, Loverro L, Niss K (2010) A mouse model for an erythropoietin-deficiency anemia. Dis Model Mech 3(11-12):763–772. doi:10.1242/dmm.004788
80. Sankaran VG, Orkin SH, Walkley CR (2008) Rb intrinsically promotes erythropoiesis by coupling cell cycle exit with mitochondrial biogenesis. Genes Dev 22(4):463–475. doi:10.1101/gad.1627208
81. Kim J, Kaartinen V (2008) Generation of mice with a conditional allele for Trim33. Genesis 46(6):329–333. doi:10.1002/dvg.20401
82. Wagner KU, Claudio E, Rucker EB 3rd, Riedlinger G, Broussard C, Schwartzberg PL, Siebenlist U, Hennighausen L (2000) Conditional deletion of the Bcl-x gene from erythroid cells results in hemolytic anemia and profound splenomegaly. Development 127 (22):4949–4958
83. Diwan A, Koesters AG, Odley AM, Pushkaran S, Baines CP, Spike BT, Daria D, Jegga AG, Geiger H, Aronow BJ, Molkentin JD, Macleod KF, Kalfa TA, Dorn GW (2007) Unrestrained erythroblast development in Nix−/− mice reveals a mechanism for apoptotic modulation of erythropoiesis. Proc Natl Acad Sci U S A 104(16):6794–6799. doi:10.1073/pnas.0610666104
84. Cui S, Tanabe O, Sierant M, Shi L, Campbell A, Lim KC, Engel JD (2015) Compound loss of function of nuclear receptors Tr2 and Tr4 leads to induction of murine embryonic β-type globin genes. Blood 125 (9):1477–1487. doi:10.1182/blood-2014-10-605022

85. Li X, Romain RD, Park D, Scadden DT, Merchant JL, Arnaout MA (2014) Stress hematopoiesis is regulated by the Krüppel-like transcription factor ZBP-89. Stem Cells 32 (3):791–801. doi:10.1002/stem.1598
86. Hosoya T, Clifford M, Losson R, Tanabe O, Engel JD (2013) TRIM28 is essential for erythroblast differentiation in the mouse. Blood 122(23):3798–3807. doi:10.1182/blood-2013-04-496166
87. Zou Z, Enis DR, Bui H, Khandros E, Kumar V, Jakus Z, Thom C, Yang Y, Dhillon V, Chen M, Lu M, Weiss MJ, Kahn ML (2013) The secreted lymphangiogenic factor CCBE1 is essential for fetal liver erythropoiesis. Blood 121(16):3228–3236. doi:10.1182/blood-2012-10-462689
88. Leung CG, Xu Y, Mularski B, Liu H, Gurbuxani S, Crispino JD (2007) Requirements for survivin in terminal differentiation of erythroid cells and maintenance of hematopoietic stem and progenitor cells. J Exp Med 204 (7):1603–1611. doi:10.1084/jem.20062395
89. Growney JD, Shigematsu H, Li Z, Lee BH, Adelsperger J, Rowan R, Curley DP, Kutok JL, Akashi K, Williams IR, Speck NA, Gilliland DG (2005) Loss of Runx1 perturbs adult hematopoiesis and is associated with a myeloproliferative phenotype. Blood 106 (2):494–504. doi:10.1182/blood-2004-08-3280
90. Scott LM, Priestley GV, Papayannopoulou T (2003) Deletion of α4 Integrins from Adult Hematopoietic Cells Reveals Roles in Homeostasis, Regeneration, and Homing. Mol Cell Biol 23(24):9349–9360. doi:10.1128/mcb.23.24.9349-9360.2003
91. Kolbus A, Pilat S, Husak Z, Deiner EM, Stengl G, Beug H, Baccarini M (2002) Raf-1 antagonizes erythroid differentiation by restraining caspase activation. J Exp Med 196(10):1347–1353. doi:10.1084/jem.20020562
92. Park SO, Wamsley HL, Bae K, Hu Z, Li X, Choe S, Slayton WB, Oh SP, Wagner KU, Sayeski PP (2013) Conditional deletion of Jak2 reveals an essential role in hematopoiesis throughout mouse ontogeny: implications for Jak2 inhibition in humans. PLoS One 8(3):e59675. doi:10.1371/journal.pone.0059675
93. Pondarre C, Campagna DR, Antiochos B, Sikorski L, Mulhern H, Fleming MD (2007) Abcb7, the gene responsible for X-linked sideroblastic anemia with ataxia, is essential for hematopoiesis. Blood 109(8):3567–3569. doi:10.1182/blood-2006-04-015768
94. Cai Y, Pi W, Sivaprakasam S, Zhu X, Zhang M, Chen J, Makala L, Lu C, Wu J, Teng Y, Pace B, Tuan D, Singh N, Li H (2015) UFBP1, a key component of the Ufm1 conjugation system, is essential for ufmylation-mediated regulation of erythroid development. PLoS Genet 11(11):e1005643. doi:10.1371/journal.pgen.1005643
95. Mochizuki-Kashio M, Mishima Y, Miyagi S, Negishi M, Saraya A, Konuma T, Shinga J, Koseki H, Iwama A (2011) Dependency on the polycomb gene Ezh2 distinguishes fetal from adult hematopoietic stem cells. Blood 118(25):6553–6561. doi:10.1182/blood-2011-03-340554
96. Yao H, Goldman DC, Nechiporuk T, Kawane S, McWeeney SK, Tyner JW, Fan G, Kerenyi MA, Orkin SH, Fleming WH, Mandel G (2014) Corepressor Rcor1 is essential for murine erythropoiesis. Blood 123 (20):3175–3184. doi:10.1182/blood-2013-11-538678
97. Malik J, Getman M, Lillis J, Gallagher PG, Steiner LA (2015) The histone methyltransferase Setd8 is essential for mammalian erythropoiesis. Blood 126:3577
98. Pan D, Schomber T, Kalberer CP, Terracciano LM, Hafen K, Krenger W, Hao-Shen H, Deng C, Skoda RC (2007) Normal erythropoiesis but severe polyposis and bleeding anemia in Smad4-deficient mice. Blood 110 (8):3049–3055. doi:10.1182/blood-2007-02-074393
99. Kundu M, Lindsten T, Yang CY, Wu J, Zhao F, Zhang J, Selak MA, Ney PA, Thompson CB (2008) Ulk1 plays a critical role in the autophagic clearance of mitochondria and ribosomes during reticulocyte maturation. Blood 112(4):1493–1502. doi:10.1182/blood-2008-02-137398
100. Tubi LQ, Nunes SC, Casellato A, Mandato E, Zaffino F, Brancalion A, Filhol-Cochet O, Boldyreff B, Manni S, Semenzato G, Piazza F (2014) Csnk2β knockout during hematopoiesis results in lethality at mid/late gestation mostly due to impaired fetal erythropoiesis. Blood 124:4329
101. Mercurio S, Petrillo S, Chiabrando D, Bassi ZI, Gays D, Camporeale A, Vacaru A, Miniscalco B, Valperga G, Silengo L, Altruda F, Baron MH, Santoro MM, Tolosano E (2015) The heme exporter Flvcr1 regulates expansion and differentiation of committed erythroid progenitors by controlling intracellular heme accumulation. Haematologica 100

(6):720–729. doi:10.3324/haematol.2014.114488

102. Lai JC, Ponti S, Pan D, Kohler H, Skoda RC, Matthias P, Nagamine Y (2012) The DEAH-box helicase RHAU is an essential gene and critical for mouse hematopoiesis. Blood 119 (18):4291–4300. doi:10.1182/blood-2011-08-362954
103. Sahr KE, Lambert AJ, Ciciotte SL, Mohandas N, Peters LL (2009) Targeted deletion of the γ-adducin gene (Add3) in mice reveals differences in α-adducin interactions in erythroid and nonerythroid cells. Am J Hematol 84 (6):354–361. doi:10.1002/ajh.21427
104. Minella AC, Loeb KR, Knecht A, Welcker M, Varnum-Finney BJ, Bernstein ID, Roberts JM, Clurman BE (2008) Cyclin E phosphorylation regulates cell proliferation in hematopoietic and epithelial lineages in vivo. Genes Dev 22(12):1677–1689. doi:10.1101/gad.1650208
105. Schneider RK, Schenone M, Ferreira MV, Kramann R, Joyce CE, Hartigan C, Beier F, Brümmendorf TH, Germing U, Platzbecker U, Büsche G, Knüchel R, Chen MC, Waters CS, Chen E, Chu LP, Novina CD, Lindsley RC, Carr SA, Ebert BL (2016) Rps14 haploinsufficiency causes a block in erythroid differentiation mediated by S100A8 and S100A9. Nat Med 22:288–297. doi:10.1038/nm.4047
106. Keel S, Phelps S, Sabo K, O'Leary MN, Kirn-Safran CB, Abkowitz JL (2012) Establishing Rps6 hemizygous mice as a model for studying how ribosomal protein haploinsufficiency impairs erythropoiesis. Exp Hematol 40 (4):290–294. doi:10.1016/j.exphem.2011.12.003
107. Chen W, Gu P, Jiang X, Ruan HB, Li C, Gao X (2011) Protein Phosphatase 2A Catalytic Subunit α (PP2Acα) Maintains Survival of Committed Erythroid Cells in Fetal Liver Erythropoiesis through the STAT5 Pathway. Am J Pathol 178(5):2333–2343. doi:10.1016/j.ajpath.2011.01.041
108. Buza-Vidas N, Cismasiu VB, Moore S, Mead AJ, Woll PS, Lutteropp M, Melchiori L, Luc S, Bouriez-Jones T, Atkinson D, O'Carroll D, Jacobsen SEW, Nerlov C (2012) Dicer is selectively important for the earliest stages of erythroid development. Blood 120 (12):2412–2416. doi:10.1182/blood-2011-10-383653
109. McMahon KA, Hiew SY, Hadjur S, Veiga-Fernandes H, Menzel U, Price AJ, Kioussis D, Williams O, Brady HJ (2007) Mll has a critical role in fetal and adult hematopoietic stem cell self-renewal. Cell Stem Cell 1 (3):338–345. doi:10.1016/j.stem.2007.07.002
110. Zhang M, Zhu X, Zhang Y, Cai Y, Chen J, Sivaprakasam S, Gurav A, Pi W, Makala L, Wu J, Pace B, Tuan-Lo D, Ganapathy V, Singh N, Li H (2015) RCAD|[sol]|Ufl1, a Ufm1 E3 ligase, is essential for hematopoietic stem cell function and murine hematopoiesis. Cell Death Differ 22(12):1922–1934. doi:10.1038/cdd.2015.51
111. Kalfa TA, Pushkaran S, Zhang X, Johnson JF, Pan D, Daria D, Geiger H, Cancelas JA, Williams DA, Zheng Y (2010) Rac1 and Rac2 GTPases are necessary for early erythropoietic expansion in the bone marrow but not in the spleen. Haematologica 95(1):27–35. doi:10.3324/haematol.2009.006239
112. XuFeng R, Boyer MJ, Shen H, Li Y, Yu H, Gao Y, Yang Q, Wang Q, Cheng T (2009) ADAR1 is required for hematopoietic progenitor cell survival via RNA editing. Proc Natl Acad Sci U S A 106(42):17763–17768. doi:10.1073/pnas.0903324106
113. Case AJ et al (2013) Manganese superoxide dismutase depletion in murine hematopoietic stem cells perturbs iron homeostasis, globin switching, and epigenetic control in erythrocyte precursorcells. Free Radic Biol Med 56:17–27. doi:10.1016/j.freeradbiomed.2012.11.018

Chapter 4

Dissecting Regulatory Mechanisms Using Mouse Fetal Liver-Derived Erythroid Cells

Skye C. McIver, Kyle J. Hewitt, Xin Gao, Charu Mehta, Jing Zhang, and Emery H. Bresnick

Abstract

Multipotent hematopoietic stem cells differentiate into an ensemble of committed progenitor cells that produce the diverse blood cells essential for life. Physiological mechanisms governing hematopoiesis, and mechanistic aberrations underlying non-malignant and malignant hematologic disorders, are often very similar in mouse and man. Thus, mouse models provide powerful systems for unraveling mechanisms that control hematopoietic stem/progenitor cell (HSPC) function in their resident microenvironments in vivo. Ex vivo systems, involving the culture of HSPCs generated in vivo, allow one to dissociate microenvironment-based and cell intrinsic mechanisms, and therefore have considerable utility. Dissecting mechanisms controlling cellular proliferation and differentiation is facilitated by the use of primary cells, since mutations and chromosome aberrations in immortalized and cancer cell lines corrupt normal mechanisms. Primary erythroid precursor cells can be expanded or differentiated in culture to yield large numbers of progeny at discrete maturation stages. We described a robust method for isolation, culture, and analysis of primary mouse erythroid precursor cells and their progeny.

Key words Hematopoiesis, Erythropoiesis, Ex vivo culture, Flow cytometry, Signaling

1 Introduction

Hematopoietic stem cells (HSCs), the essential stem cells responsible for generating and sustaining the blood system, are initially formed in the aorta-gonad-mesonephros region of the embryo (mice: embryonic day 10.5 or E10.5). Subsequently, the cells undergo a complex journey involving migration to and colonization of the fetal liver (mice: E11.5 onward). While the fetal liver is the primary site of hematopoiesis in late fetal development, after birth the HSCs migrate to and colonize the bone marrow [1–3]. Fetal liver-resident HSCs have a high proliferative capacity and generate large numbers of multi potent and committed progenitor cells [4, 5]. Therefore, the fetal liver provides a unique source of HSPCs for ex vivo analyses to elucidate mechanisms.

Joyce A. Lloyd (ed.), *Erythropoiesis: Methods and Protocols*, Methods in Molecular Biology, vol. 1698,
DOI 10.1007/978-1-4939-7428-3_4,

During hematopoiesis, HSCs generate progeny with unique signatures of cell surface proteins that are routinely used to distinguish distinct cell types. The signatures can be exploited using antibodies specific for the cell surface proteins and flow cytometry to analyze the composition of complex mixtures of cells and to isolate or enrich for unique cell types or populations. Alternatively, separations are also accomplished by incubating cell mixtures with antibody-bound beads, which can be isolated with a magnet, or based on other physical properties. Examples of instructive markers include: Ter119, erythroid [6]; CD45R (B220), B-cells [7]; and GR-1, myeloid [8]. The protocol presented in this chapter uses antibodies against cell-specific markers to bind to and remove differentiated myeloid, lymphoid, erythroid, and hepatic cells from a heterogeneous mixture of fetal liver cells, thus yielding an HSPC-enriched cell population deficient in terminally differentiated cells (Lineage$^-$ or Lin$^-$ cells). Lin$^-$ cells can be cultured in medium that supports rapid erythroid cell proliferation and differentiation, while suppressing the survival of other lineages [9].

Immortalized cell lines, which are often extremely easy to culture, have catalyzed many scientific advances and therefore can have high utility for certain experimental analysis. However, they often differ from primary cells in their growth, differentiation, and/or survival requirements, which is not surprising given mutations that may disrupt mechanisms governing fundamental cellular processes. The ability to culture, expand, and differentiate primary cells ex vivo is therefore a powerful approach for mechanistic analyses—either as an essential system to validate cell culture studies or as a front-line system for asking mechanistic questions. We describe herein a well-optimized method to isolate and culture primary murine fetal liver hematopoietic precursor cells, conduct loss-of-function analysis using retrovirus expressing an shRNA to target a specific mRNA, and quantitate molecular and cellular phenotypes (Fig. 1). The protocol can also be used for gain-of-function analyses, including rescue experiments, comparisons of mechanisms with cells from wild type and mutant mice, and pharmacological studies. The culture system can generate predominantly erythroid precursor cells (expansion culture) or mature erythroid cells (differentiation culture). This method can easily yield millions of the respective cells, consistent with requirements for diverse analytical approaches including genome-wide transcriptomics (RNA-seq), protein occupancy at chromatin sites (chromatin immunoprecipitation (ChIP), qChIP or ChIP-seq), protein analysis by mass spectrometry or Western blotting, and assessment of cellular attributes and/or signaling pathway status via flow cytometry. The application of these protocols has been described in multiple recent publications [10–16].

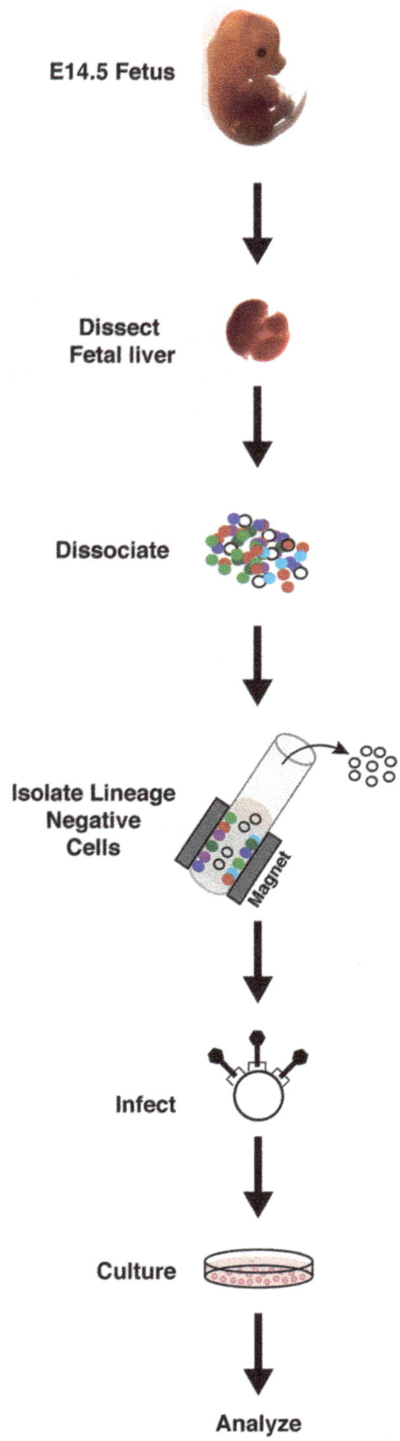

Fig. 1 Isolation of lineage negative hematopoietic precursor cells

2 Materials

2.1 shRNA Cloning

1. Phusion high-fidelity DNA polymerase (ThermoFisher).
2. Restriction enzymes: BglII 10,000 U/ml, XhoI 20,000 U/ml.
3. T4 DNA ligase 400,000 U/ml.
4. Alkaline Phosphatase, Calf Intestinal (CIP) 10,000 U/ml.
5. Big Dye 3.1 Sequencing kit (ThermoFisher).
6. Primers: BglII-miR30sh-F: AGATCTAGATCTTGCTGTTGACAGTGAGCG, XhoI-miR30sh-R: CTCGAGCTCGAGTCCGAGGCAGTAGGC, PIG 5′Seq: CCCTTGAACCTCCTCGTTCGACC, PIG 3′Seq: GAGACGTGCTACTTCCATTTGTC.
7. PCR purification kit.
8. Gel purification kit.
9. Plasmid Miniprep kit.
10. Big Dye clean up kit.
11. Plasmid Endotoxin Free Maxiprep kit.
12. Expression vector: MSCV-PIG.
13. Apparatus and reagents to run DNA agarose gels.
14. Chemically competent bacteria, for example Jm109.
15. SOC medium.
16. LB broth and plates with 100 μg/ml Ampicillin.
17. 15 ml round bottom snap top tubes.

2.2 Retrovirus Preparation

1. shRNA expression vector MSCV-PIG, purified via endotoxin-free maxiprep.
2. Retroviral packaging vector: pCL-Eco, purified via endotoxin-free maxiprep.
3. 293T cells.
4. 293T media: DMEM with glutamine and sodium pyruvate, 10% fetal bovine serum (FBS) and 1% penicillin/streptomycin.
5. 2× transfection buffer, 5 ml: 500 μl 0.5 M HEPES, 450 μl 2 M NaCl, 100 μl 100 mM Na_2HPO_4, 4.05 ml water.
6. 10 cm culture plates.
7. 10× NTE buffer: 1.5 M NaCl, 100 mM Tris-HCl (pH 7.4), 10 mM EDTA.
8. 0.5 M Hepes (pH 7.1).
9. 100 mM Na_2HPO_4 (pH 7.0 with phosphoric acid).
10. 2 M $CaCl_2$.
11. 2 M NaCl.

2.3 Fetal Liver Isolation and Lineage Depletion

1. Dissection tools: forceps and scissors.
2. Dissection microscope.
3. 12-well culture plates.
4. 5 or 10 cm culture dishes.
5. 5 mL round bottom tubes, solid and 40 μm filter cap.
6. PBS without calcium or magnesium.
7. Fetal Liver Juice (FLJ): 2% FBS, 2.5 mM EDTA and 10 mM Glucose in PBS.
8. Antibodies, all 0.5 mg/ml, anti-mouse and biotinylated: CD3e (clone 145-2C11), CD11b (clone M1/70), CD19 (clone 6D5), CD45R (B220) (clone RA3-6B2), GR-1 (clone RB6-8C5), CD71 (clone RI7217) and Ter119 (clone Ter119).
9. Streptavidin magnetic beads: MojoSort™ Streptavidin Nanobeads (BioLegend).
10. Magnet suitable for 5 ml tubes: MojoSort™ magnet (BioLegend) or EasySep™ magnet (StemCell Technologies).

2.4 Retroviral Infection

1. 12-well culture plates.
2. Viral supernatant.
3. Polybrene: 2 mg/ml stock in water.
4. 1 M HEPES.
5. Temperature-controlled centrifuge with swinging bucket rotor capable of spinning plates.

2.5 Culture of Lineage Negative Fetal Liver Cells

1. Expansion media: StemPro-34 (ThermoFisher) with 1× nutrient supplement, 2 mM L-Glutamine, 1% penicillin/streptomycin, 100 μM monothioglycerol, 1 μM dexamethasone, 0.5 U/ml of erythropoietin (Epo), and 1% conditioned medium from a stem cell factor (SCF) producing CHO cell line (or 100 ng/ml recombinant SCF).
2. Differentiation media: IMDM (glutamine-free) with 10% FBS, 10% PDS (Plasma Derived Serum; Animal Technologies), 5% PFHM II, 2 mM L-glutamine, 1% penicillin/streptomycin, 100 μM monothioglycerol, and 6 U/ml erythropoietin.

2.6 Quantitating Erythroid Differentiation by Flow Cytometry

1. Antibodies: Fluorescent-conjugated anti-mouse CD71, Ter119, c-Kit.
2. Fetal Liver juice (FLJ): 2% FBS, 2.5 mM EDTA and 10 mM Glucose in PBS.
3. PBS without calcium or magnesium.
4. Cell-impermeable DNA dye, such as DAPI or PI, for live/dead cell discrimination.

5. 5 mL tubes with 40 μm filter cap.
6. Multichannel analysis cytometer, e.g., BD LSR II or BD LSR Fortessa, or cell sorter, e.g., BD FACSAria III.

2.7 Quantitating Erythroid Differentiation by Morphology

1. 50% FBS in PBS.
2. Glass slides.
3. Cytospin funnel.
4. Cytospin.
5. Giemsa stain.

2.8 Quantitating Colony-Forming Unit Activity

1. 3 ml aliquots of MethoCult™ GF M3434 (StemCell Technologies).
2. 16-gauge needles.
3. 3 ml syringes.
4. Expansion media.
5. Sterile water.
6. 35 mm × 10 mm culture dishes.
7. 10 cm culture dishes.
8. Benzidine stain stock solution: 0.3 g benzidine, 9 ml acetic acid, 1 ml H_2O. The stock is stable for 2 weeks.
9. Benzidine stain: 5 parts H_2O, 0.1 part H_2O_2 (30%), 1 part benzidine stock (*see* **item 8**). The stain should be made just before use.

2.9 Quantitating Intracellular Signaling with Antibodies

1. Biotinylated anti-mouse Ter119.
2. Phospho-specific antibody against signing pathway components. For example, rabbit anti-mouse phospho-Akt, or rabbit anti-mouse phospho-ERK.
3. Fluorescent-conjugated secondary antibody. For example goat anti-rabbit-APC.
4. Fetal Liver juice (FLJ): 2% FBS, 2.5 mM EDTA and 10 mM Glucose in PBS.
5. Streptavidin magnetic beads: MojoSort™ Streptavidin Nanobeads (BioLegend).
6. Magnet suitable for 5 ml tubes: MojoSort™ magnet (BioLegend) or EasySep™ magnet (StemCell Technologies).
7. 5 mL tubes, solid and 40 μm filter cap.
8. Cytokines of interest. For example SCF or Epo.
9. Fixable live/dead stain such as Zombie Dye (BioLegend).
10. 1% BSA solution in IMDM.
11. PBS without calcium or magnesium.

12. 4% FBS in HBSS.
13. 16% paraformaldehyde.
14. Methanol.

3 Methods

3.1 Constructing shRNA-Expressing Retrovirus

1. Design shRNA sequences specific for your gene using a publically accessible database. For example: http://portals.broadinstitute.org/gpp/public/ is searchable using NCBI gene accession numbers. Select at least three sequences that target distinct regions of the gene of interest (*see* **Note 1**).
2. To plan how to integrate the shRNA into a miR-30 context, use the Ravi Lab resource: http://katahdin.cshl.org/html/scripts/resources.pl
3. Pick the RNAi project and siRNA/shRNA design and select shRNA psm2 design, then select mouse.
4. State the information you have available, for example, a 21mer. Input the sequence obtained in **step 1** and the target gene NCBI accession number. If shRNA sequences are not available, the Ravi Lab resource will design shRNA from accession numbers. Order the 97mer oligonucleotide.
5. Add BglII and XhoI cut sites to the 97mer. To do this, set up 50 μl PCR reactions for each construct, according to the instructions provided with the high-fidelity DNA polymerase. Use a final concentration of 200 μM dNTPs, 150 μM of each primer (BglII-miR30sh-F and XhoI-miR30sh-R) and 300 μM of the shRNA 97mer oligonucleotide.
6. Run PCR: 98 °C 5 min, (98 °C 15 s, 72 °C 50 s) × 10, 72 °C 7 min.
7. Check for the presence of a ~100 bp band on a 1.5% agarose gel.
8. Purify the product with the PCR purification kit, and elute the shRNA construct with 30 μl water.
9. Cut the shRNA construct with BglII and XhoI, 28 μl of DNA from **step 5**, 1.5 μl each enzyme, 5 μl 10× restriction enzyme buffer, and water in 50 μl.
10. Cut 4 μg of MSCV-PIG vector with BglII and XhoI, 1.5 μl each enzyme, 5 μl 10× restriction enzyme buffer, and water in 50 μl.
11. Incubate the restriction digest at 37 °C for 4 h (at 3.5 h, add 0.5 μl CIP to the MSCV-PIG vector).
12. Run the digested MSCV-PIG vector on a 1% agarose gel, visualize DNA, and excise the vector band (~7.5 kb).

13. Use the gel purification kit to purify the MSCV-PIG vector, and elute in 30 μl water.
14. Use the PCR purification kit to purify the shRNA products, and elute in 30 μl water.
15. Set up a ligation reaction with 16 μl of purified shRNA construct (from **step 14**), 1 μl purified cut MSCV-PIG vector (from **step 13**), 1 μl T4 DNA ligase, 2 μl 10× ligation buffer, in a 20 μl reaction. Incubate overnight at 16 °C.
16. Thaw competent bacteria on ice for 5 min (25–40 μl per reaction).
17. Add 7 μl of the ligation mix and incubate on ice for 20 min.
18. Heat-shock bacteria for 40 s at 42 °C, and then incubate on ice for 2 min.
19. Add 200 μl SOC medium and incubate for 1 h at 37 °C.
20. Plate 200 μl of bacteria on LB-Agar/Ampicillin plates and incubate overnight at 37 °C.
21. To screen colonies, make PCR mastermix sufficient for the number of colonies to test (25 μl per colony) according to the instructions provided with the DNA polymerase: Use a final concentration of 400 μM dNTPs, 400 μM each primer (PIG 5′Seq, PIG 3′Seq). Aliquot into PCR tubes.
22. Using a pipette tip, pick colonies of interest. First, dip tip into 2.5 ml LB broth/ampicillin and then into PCR master mix.
23. Perform PCR analysis: 95°C 5 min, (95°C 15 s, 54°C 30 s, 72°C 30 s) × 30, 72°C 7 min and analyze on a 1% agarose gel. Plasmids containing a single insert should have a 500 bp amplicon, plasmids with no insert will have a 400 bp amplicon.
24. Incubate LB broth samples from positive colonies at 37°C (shaking) for 10–24 h and perform minipreps to purify plasmids of interest.
25. To prepare the minipreps for sequencing, make a Big Dye reaction with 200 ng DNA template, 1.5 μl PIG 3′seq (10 mM stock), 3 μl Big Dye Buffer, 2 μl Big Dye Enzyme, water up to 20 μl. Amplify at 96 °C 3 min, (96 °C 10 s, 55 °C 15 s, 60 °C 3 min) × 34, 72 °C 7 min (*see* **Note 2**).
26. Purify product using a Big Dye clean-up kit and analyze by Sanger sequencing.
27. Once correct sequence is confirmed, inoculate 100 ml LB/ Ampicillin broth with bacteria, and incubate at 37 °C (shaking) for 16–18 h.
28. Purify plasmid using an endotoxin-free plasmid maxiprep kit.

3.2 Virus Preparation

1. Seed 3.5–4 million 293T cells in 10 ml medium in a 10 cm tissue culture dish, and incubate at 37 °C 5% CO_2 overnight (*see* **Note 3**).
2. Make 2× transfection buffer (*see* Subheading 2.2, **item 5**).
3. Make DNA cocktail in eppendorf tubes: In this order, add water up to 500 μl, 62.5 μl 2 M $CaCl_2$, 50 μl 10× NTE, 15 μg of shRNA expression plasmid (MSCV-PIG vector from Subheading 3.1, **step 28**), 15 μg retroviral packaging vector (pCL-Eco).
4. Aspirate the medium from the 10 cm 293T plates, and replace it with a 5 ml pre-warmed fresh medium (*see* **Note 4**).
5. Add 500 μl 2× transfection buffer to the DNA cocktail drop-wise. Bubble air through the liquid for 30 s to mix.
6. Slowly add transfection reaction to 293T cells on plate, and tilt to mix.
7. Incubate at 37°C, 5% CO_2 for 6 h, and then replace the medium with a 6 ml pre-warmed fresh medium.
8. Collect viral supernatant by collecting medium from 293T at 24 h, replace the medium with 6 ml pre-warmed fresh medium, and repeat collection at 48 h.
9. Spin medium containing virus at 2000 rpm (670 × *g*) for 5 min at room temperature to pellet cell debris.
10. Aliquot the supernatant (0.5–1 ml). Snap freeze in liquid nitrogen and store at −80 °C.

3.3 Isolate Fetal Livers

1. Set up timed mating. Sacrifice a pregnant mouse 14 days after the observation of a vaginal plug. Embryos are considered E14.5 (*see* **Note 5**).
2. Place mouse on its back and grip the stomach, near the final set of nipples with forceps with the handle pointing toward the head, and pull upward gently.
3. Cut the skin and underlying muscular layer with a scissors in a V shape, starting underneath the region you are gripping with the forceps, until the uterus is easily accessible.
4. Grip the uterus between two fetuses and pull gently. Each of the two arms of the uterus will be attached at the ovary, near the ribs, and to the vagina near the legs with some blood vessels and fat in-between. Cut these connections and the uterus is extracted easily.
5. Place uteri containing the fetuses (usually around 9) into cold sterile PBS in a petri dish, wash once, and place in fresh PBS.
6. Isolate fetuses one at a time, using fine tipped forceps to tear uterus in the gap between the fetuses and squeeze out the fetus. If the yolk sac remains around the fetus, tease it off with two

pairs of forceps. Immediately transfer fetus into cold FLJ in a new petri dish.

7. If working with mutant animals, obtain a portion of the tail and leg of each fetus and store separately to conduct genotyping.
8. Isolate livers by removing connective tissue surrounding the liver, under sterile conditions if possible. Place into a 12-well dish containing 1.5 ml FLJ and wash once. Intact fetal livers can be maintained on ice for several hours if required for genotyping, although minimizing this time is recommended. Wild-type fetal livers can be pooled for analysis.
9. Dissociate the fetal livers by pipetting in FLJ to generate a single-cell suspension (10–20 million cells per liver).
10. Pass the cell suspension through a 40 μm strainer top into a 5 ml tube.
11. Wash the strainer cap with cold FLJ until the tubes are full and replace the strainer cap with a solid cap.
12. Centrifuge for 5 min at 1200 rpm (250 × *g*) at 4°C, and proceed immediately to progenitor purification.

3.4 Lineage Depletion

This protocol is designed for the isolation of myelo-erythroid progenitors from three biological replicate cell suspensions, each containing three pooled wild-type fetal livers. Volumes can be scaled up or down depending on the desired cell number or number of replicates. The minimum volume per isolation is 100 μl, while the maximum volume is 1.5 ml.

1. Assemble master mix of all antibodies: 3 μl/ml each of CD3e, CD11b, CD19, CD45R, GR-1 and CD71, and 5 μl/ml of Ter119. For three replicates, each containing three fetal livers, use 2.25 ml FLJ, 6.75 μl each of CD3e, CD11b, CD19, CD45R, GR-1 and CD71, and 11.25 μl of Ter119 (*see* **Note 6**).
2. Resuspend fetal liver cells at 0.5–1 × 10^8 cells/ml (generally 250 μl/E14.5 liver) in antibody master mix (**step 1**).
3. Incubate on ice for 15 min, and then wash off unbound antibody by adding 3–4 ml FLJ, and centrifuge for 5 min at 1200 rpm (250 × *g*) at 4 °C. Discard the supernatant.
4. Assemble master mix of streptavidin beads: 75 μl/ml in FLJ. For three replicates each containing three fetal livers: 168.75 μl streptavidin beads in 2.25 ml FLJ.
5. Resuspend fetal liver cells in the bead master mix. Use a volume equal to that used in **step 2**.
6. Incubate on ice for 15 min, wash by adding 3–4 ml FLJ, and centrifuge for 5 min at 1200 rpm (250 × *g*) at 4 °C. Discard the supernatant.

7. Resuspend cells in 3 ml of FLJ using a P1000 or other pipetteman.
8. Remove the cap and incubate each tube in the magnet for 5 min.
9. Pour out cells of interest into a new tube in one continuous motion without removing the tube from the magnet. Do not shake/touch tube to the new tube etc. to obtain the last drop. This will cause the beads to contaminate your sample.
10. It is recommended that **steps 7–9** are repeated once to increase yield.
11. Pass the cell suspension through a 40 μm strainer.
12. Count cells, isolate by centrifugation for 5 min at 1200 rpm (250 × *g*) at 4 °C, and resuspend in an appropriate volume of expansion media. A typical yield from three wild-type E14.5 fetal livers is 1.5–2 × 10^6 cells.

3.5 Viral Infection

Infections should be conducted immediately after lineage depletion. Generally, this includes retroviruses harboring a control shRNA (*see* **Note** 7) or an experimental shRNA. For flow cytometry applications, culture uninfected cells for single-stain controls.

1. Add 1–3 × 10^5 cells in 400 μl expansion medium to each well of a 12-well plate.
2. It is recommended that more than one well be infected per condition. Cells can be used as technical replicates or pooled to increase yield.
3. Add 100 μl viral supernatant, 2 μl polybrene solution, and 5 μl 1 M HEPES to each well and mix (*see* **Note 8**).
4. Centrifuge in a swinging bucket rotor at 2800 rpm (1300 × *g*) at 30 °C for 90 min.
5. Add 500 μl expansion medium to each well and incubate at 37 °C and 5% CO_2.

3.6 Expansion Culture

The expansion medium used in this protocol is formulated to promote rapid proliferation of early erythroid cells, while preventing their terminal differentiation. Cells cultured in the expansion medium rarely differentiate beyond the stage of basophilic erythroblasts. SCF (Kit receptor agonist) and dexamethasone (glucocorticoid receptor agonist) stimulate expansion of immature erythroid precursor cells (burst forming unit - erythroid (BFU-E), colony forming unit - erythroid (CFU-E) and proerythroblasts) (*see* **Note 9**) [17]. A small amount of Epo in the medium supports the growth of proerythroblasts and the slightly more mature erythroblasts termed polychromatophilic erythroblasts. The expansion phase of the culture system permits analysis of early erythroid precursor function, e.g., to identify mechanisms that promote or suppress erythroblast proliferation, maturation, or survival.

1. Plate the cells from Subheading 3.5, **step 5** at a concentration of 0.25–1 × 10^6 cells/ml in expansion media (*see* **Note 10**).
2. Cells are cultured 37 °C and 5% CO_2 and double every 6–12 h (*see* **Note 11**).
3. It is recommended that the volume of medium is doubled each morning, and each afternoon count and dilute the cells to 0.25 × 10^6 cells/ml to ensure the cell concentration does not exceed one million cells/ml (*see* **Note 12**).

3.7 Differentiation Culture

The differentiation medium used in this protocol supports the terminal differentiation of erythroid cells. Cells cultured in this medium mature into basophilic and orthochromatic erythroblasts and about 20–30% enucleate after 48 h. This medium does not contain SCF or dexamethasone, and accordingly, early erythroid cells stop proliferating and differentiate. The high Epo concentration supports the erythroid cells upon terminal differentiation. The differentiation culture system enables the analysis of factors and signals that promote or inhibit terminal erythroid differentiation.

1. Collect greater than 0.5 × 10^6 cells in a 15 ml tube.
2. Centrifuge the cells for 5 min at 1200 rpm (250 × *g*) at 4 °C, and wash twice with 5 ml of PBS.
3. Resuspend cells in an appropriate volume of differentiation media. Cells must be maintained at a concentration of 1–2 × 10^6 cells/ml. Cells cultured at 37 °C and 5% CO_2 double every 12–24 h, and proliferation slows during the culture period.
4. Phenotypic analyses of the differentiated erythroid cells can include flow cytometry, morphology, colony forming unit (CFU) assay, cell signaling, and cell cycle analyses, as described in the sections below. The optimal number of days in the differentiation medium will vary depending on the assay.

3.8 Quantitation of Erythroid Maturation by Flow Cytometry

During erythroid maturation, major changes occur in the expression of cell surface markers, which can be exploited to establish the maturation stage. For example, the earliest erythroid cells express c-Kit and then acquire CD71 expression. As maturation progresses these cells gain Ter119 expression. Finally, these cells lose CD71 expression [1]. The dynamic expression of cell surface proteins, along with the availability of fluorescently-conjugated antibodies recognizing these proteins, allows the quantitation and sorting of populations enriched for specific erythroid cell types. A common nomenclature for the erythroid maturation series is as follows: R1

(BFU-E, CFU-E, proerythroblast), R2 (proerythroblasts and early basophilic erythroblasts), R3 (early and late basophilic erythroblasts), R4 (chromatophilic and orthochromatic erythroblasts), and R5 (late orthochromatic erythroblasts and reticulocytes) [6]. During erythroid maturation, the expression of hundreds of genes increases or decreases. Thus, phenotypic changes (molecular or cellular) occurring between cells infected with control shRNA vs. a specific shRNA must be carefully interpreted (including the quantitation of knockdown efficiency), as such changes might be an indirect consequence of alterations in maturation. Typically, we isolate R1–R4/5 cells, using the cell surface markers CD71 and Ter119 by fluorescence activated cell sorting (FACS), to purify RNA from the specific cell populations. This minimizes the possibility that changes in mRNA expression reflect differences in cell maturation between control and treatment samples.

shRNA-mediated downregulation of specific mRNAs occurs by 16 h post-infection. Perturbations in erythroid maturation are particularly prominent after 48–72 h culture in expansion media. After culturing for 24 h, cells reside predominantly in R1, R2, and R3 populations. Further culture (48 and 72 h) decreases the percentage of cells in R1 and increases the R3 population (Fig. 2). After culturing for 48 h in differentiation media, the erythroid cells are predominantly in the R3 population, ~20–30% are in the R4 and ~1–5% are in the R5 population. It is important to evaluate the impact of downregulating factors on erythroid maturation at various times, while carefully considering the relationship between kinetics for shRNA-mediated downregulation of the target mRNA and the respective protein, and potential phenotypic alterations (Fig. 2).

1. Harvest 0.5–1 × 10^6 cells for flow cytometry analysis (FCM) or 10–20 × 10^6 for flow sorting (FACS). Rinse in PBS.
2. Assemble antibody master mix with 100 μl FLJ per sample. For FCM, add 1:100 (or 0.2 μg (1 μl)) each antibody: CD71, Ter119, c-Kit. For FACS, add 1:25 (or 0.8 μg (4 μl)) each antibody: CD71, Ter119, c-Kit (*see* **Notes 13** and **14**).
3. For either FCM or FACS, do single-stain controls using uninfected cells and resuspend each control 100 μl FLJ. Add each antibody individually to separate tubes and leave some cells unstained for use as a no-stain compensation control (*see* **Note 15**).
4. Resuspend fetal liver cells in 100 μl of antibody master mix, and incubate in the dark at 4 °C for 30 min with occasional mixing.
5. Add 2 ml PBS and isolate cells by centrifugation at 1200 rpm (250 × *g*) for 5 min at 4 °C.

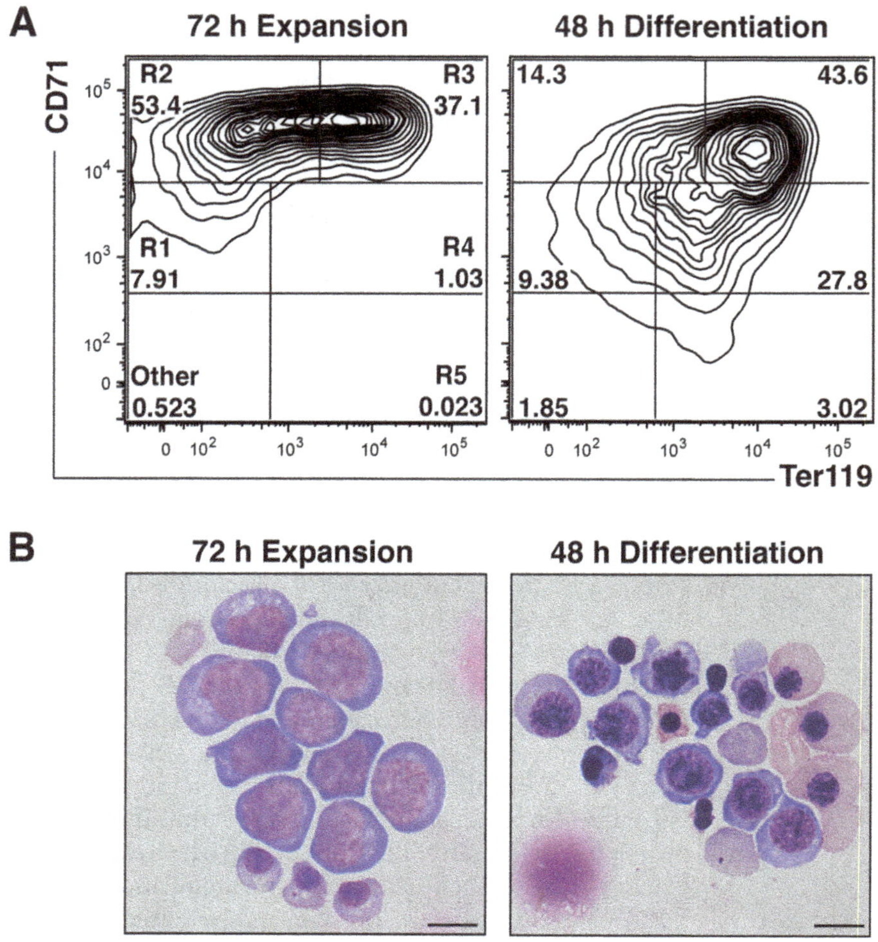

Fig. 2 Cell analysis using flow cytometry and Wight-Giemsa staining. (**a**) Representative flow cytometry plots of lineage negative cells stained with CD71-PE and Ter119-APC after 72 h expansion culture and then 48 h differentiation culture. R1 through R5 gates are denoted. (**b**) Representative Wright-Giemsa stain after 72 h expansion culture and then 48 h differentiation culture

6. Resuspend cells in 500 μl of PBS containing cell-impermeable DNA dye for live/dead discrimination. For example 1 μg/ml DAPI.
7. Pass the cell suspension through a 40 μm strainer.
8. Analyze or sort with a flow cytometer.

3.9 Morphological Analyses

The identification and quantitation of cells at distinct stages of erythroid maturation is readily accomplished by applying cells to a slide with a Cytospin and staining with Wright-Giemsa. Immature erythroid cells are large with large purple nuclei and a dark blue cytoplasm. As the cells mature, the erythroblasts get progressively smaller and their nuclei condense and become darker. The cytoplasm becomes hemoglobinized, attaining a light blue and then pink color. Terminal erythroid maturation involves enucleation. Wright-Giemsa stain also has utility for distinguishing between erythroid and non-erythroid lineages. For example, neutrophils have a lobed or donut-shaped nucleus, while macrophages are very large multi-nucleated cells [18, 19].

Large immature erythroblasts are abundant during the expansion culture. In the differentiation culture, which supports erythroblast terminal differentiation, cell size decreases, cytoplasm becomes hemoglobinized, and a significant percentage of the cells enucleate (Fig. 2).

1. Harvest 1×10^5 cells per slide, isolate by centrifugation, and resuspend in 500 μl of 50% FBS/PBS.
2. Load slide into the cytospin funnel apparatus, and add the cells.
3. Spin for 10 min at 700 rpm ($55 \times g$) in the cytospin.
4. Remove the slides as soon as possible after the spin, and air dry for 10 min.
5. Store in the dark (up to 1 week) until staining (*see* **Note 16**).

3.10 Hematopoietic Progenitor Quantitation Using a Colony-Forming Unit Assay

The competence of hematopoietic precursors to differentiate into myeloid and erythroid cell types can be quantitated using CFU assays. The number, size, and morphology of colonies formed in semisolid medium provides a highly instructive metric of their differentiation and proliferation capabilities [20, 21]. This assay is performed within 24 h of lineage negative cell isolation, and varies depending on application. The GFP selection marker on the MSCV-PIG vector is easily detectable by 16 h post-infection. FACS can be used to isolate the GFP^+ infected population, or a more specific cell population for analysis. CD71 and Ter119 markers are used to identify Rl–R5 populations. The Rl population contains a high proportion of BFU-E and CFU-E. Total cells or GFP^+ cells after 24 h expansion culture produce CFU-GM, BFU-E, CFU-E, with only a few CFU-GEMM.

1. Make a 10× cell suspension in expansion medium with total or GFP^+ cells at 2×10^5 cells/ml. If using sorted Rl cells, suspend at $2–5 \times 10^4$ cells/ml in expansion medium.
2. Mix 300 μl cells with 3 ml MethoCult™ and vortex 5 s (*see* **Note 17**).
3. Incubate for 5 min to allow any bubbles to settle.
4. Use a 16-gauge needle and 3 ml syringe to dispense 1.1 ml Methocult™ into 35 mm × 10 mm dishes each (make two technical replicates). Distribute evenly by tilting the dishes.
5. Place dishes in a 10 cm plate, and add a third uncovered 35 mm × 10 mm dish containing 3 ml sterile H_2O (The use of a water dish maintains humidity).

6. Score for CFU-E colonies (groups of 8–30 cells that appear to be fused together) after 2 days of culture at 37 °C and 5% CO_2.
7. Score for BFU-E colonies (multiple cell clusters consisting of bright red, tiny, irregularly shaped cells that may appear to be fused together), CFU-GM colonies (colorless colonies consisting of large, oval cells (macrophage) and smaller, round cells (granulocytes)) and CFU-GEMM (typically large colonies consisting of red erythroid cells mixed with colorless granulocytes, macrophages, and megakaryocytes) after 7–10 days of culture at 37 °C and 5% CO_2. Before scoring for BFU-Es, plates can be stained with 500–700 μl of benzidine stain, which stains hemoglobin. Hemoglobinized BFU-E colonies will instantly stain dark blue. Benzidine stain will also stain CFU-E colonies, but is not compatible with continued culture to subsequently generate BFU-E, or for optional morphological analyses in **step 8**.
8. If colonies are to be harvested for morphological analysis, incubate plates at 4 °C for 2–3 h and add 1 ml of cold PBS per dish, pipetting up and down gently to mix the methylcellulose-based medium with the liquid medium. Move to a 15 ml conical tube. Wash the culture plate, and add an additional 5 ml PBS to the 15 ml conical tube. Centrifuge cells at 1200 rpm (250 × g) for 5 min at 4 °C, and wash twice with 5 ml PBS. Incubate for 10 min on a shaker at room temperature before spinning down after each wash. Count cells, and proceed with the morphological analysis procedure (Subheading 3.9).

3.11 Quantitation of Signaling

Cytokines control the survival, growth, and differentiation of erythroid precursor cells. Proliferation of immature erythroid precursors requires SCF [22, 23], while intermediate erythroblasts respond to SCF and Epo [24, 25] and terminally differentiated erythroid cells require Epo for survival [26]. A variety of mechanisms are in place to control the erythroid cell response to cytokines, and it would not be surprising if major additional mechanisms remain to be discovered. As cellular responses to cytokines can depend on the maturation status of the cells [27–29], this protocol provides a sensitive assay to quantitate cellular responses to cytokine signals in specific cell types delineated by their ensemble of cell surface proteins detected by flow cytometry.

Cytokines rapidly activate kinase cascades, which at least in certain contexts have a very restricted duration. To measure cell responses to a particular cytokine, it is commonplace to serum-starve cells, which decreases their basal signaling. Quantitation of intracellular signaling molecules by flow cytometry requires cellular fixation and permeabilization to allow appropriate antibodies to access their target, while retaining the target epitope. We find that common fixation procedures result in loss of the erythroid cell surface antigen recognized by the anti-Ter119 antibody. Thus in a typical experiment, we separate the erythroid cell population into early (Ter119$^-$) and late (Ter119$^+$) erythroblasts using magnetic beads prior to cytokine stimulation [14]. Typically, we compare the capacity of a cytokine to stimulate a control cell versus a cell in which a particular factor has been downregulated using shRNA (or cells from wild-type vs. mutant animals).

1. Centrifuge 5–20 × 10^6 cells at 1200 rpm (250 × *g*) for 5 min at 4 °C, aspirate medium, wash in ice-cold PBS, and transfer to a FACS tube.
2. Separate sample into Ter119$^+$ and Ter119$^-$ populations using the lineage depletion protocol described in Subheading 3.4. Use 500 μl FLJ/sample, 2 μl/ml anti-Ter119, and 50 μl/ml streptavidin beads.
3. Count cells, wash and resuspend in 1 ml IMDM/1% BSA with a cell concentration that does not exceed 5 × 10^6 cells/ml.
4. Serum Starve the cells by incubating in IMDM/1% BSA for 60 min at 37 °C.
5. Isolate cells by centrifugation, and wash cells in PBS to remove all serum proteins.
6. Stain with fixable live/dead marker. For example, resuspend cells in 100 μl of a 1:100 dilution of Zombie Dye (BioLegend) in PBS, incubate at room temperature for 15 min, and then inactivate live/dead stain by adding pre-warmed IMDM/1% BSA.
7. Centrifuge cells at 1200 rpm (250 × *g*) for 5 min at room temperature and resuspend in IMDM/1% BSA at 37 °C, 1 ml/condition (vehicle and cytokine treated).
8. Incubate for 10 min at 37 °C.
9. Stimulate with vehicle or cytokine of interest for 10 min at 37 °C (for example, 10 ng/ml SCF or 2 U/ml Epo). For precise timing, add SCF or vehicle every 20 s to individual cell suspensions.
10. Fix cells by adding 16% fresh paraformaldehyde (PFA) to a final concentration of 2% (143 μl/ml), and incubate for 10 min at 37 °C (*see* **Note 18**).
11. Add an equal volume of PBS, isolate cells by centrifugation at 2200 rpm (800 × *g*) for 5 min at room temperature and wash twice with 4 ml PBS.
12. While vortexing, slowly add 2 ml ice-cold 95% methanol dropwise to each cell suspension. Place each sample on ice immediately thereafter (no longer than 30 min). Tighten the lid, and store samples at −20 °C for at least 4–6 h and preferably overnight.

13. Add an equal volume of PBS. Do not mix, and isolate cells by centrifugation at 2200 rpm (800 × *g*) for 5 min at 4 °C.
14. Aspirate the supernatant carefully, leaving ~100–200 μl, and wash twice with 3 ml PBS.
15. Resuspend samples in 1 ml of cold HBSS/4% FBS, vortex, and incubate at 4 °C for 60 min.
16. Isolate cells by centrifugation at 2200 rpm (800 × *g*) for 5 min at 4 °C.
17. Resuspend in 100 μl primary antibody master mix and incubate for 30 min at room temperature. For example, rabbit anti-mouse phospho-AKT or phospho-ERK at 1:200 or biotinylated antibodies 1:200 in FLJ.
18. Wash cells with 4 ml PBS, and isolate cells by centrifugation at 2200 rpm (800 × *g*) for 5 min.
19. Resuspend in 100 μl fluorescently conjugated antibody master mix, and incubate for 30 min at room temperature in the dark. This master mix will contain both secondary antibodies: anti-rabbit 1:200 (e.g., APC), Streptavidin 1:400 (or 0.05 μg (0.25 μl), e.g., PE-Cy7) and any pre-conjugated primary antibodies: 1:50 (or 0.4 μg (2 μl), e.g., CD71-PE) in FLJ.
20. For single-stain controls, use fixed uninfected cells, and add antibodies separately to different tubes leaving an aliquot of cells unstained for a no-stain compensation control.
21. Wash once with 4 ml PBS, and resuspend in 400 μl PBS.
22. Pass the cell suspension through a 40 μm strainer.
23. Analyze data on flow cytometer as soon as possible. Use low/middle pressure to collect the data.
24. The intensity of the cellular response to the cytokine is measured via the change in Median Fluorescence Intensity (MFI) of the phosphorylated kinase, e.g., phospho-Akt (vehicle vs. cytokine).
25. Compared to the vehicle control, cytokine stimulation will shift the histogram of fluorescence intensity of phospho-Akt to the right (Fig. 3).

3.12 Cell Cycle Analyses

Erythroid maturation is intimately linked to the cell cycle [30], and therefore cell cycle perturbation can have a large impact on maturation.

1. Isolate 5–20 × 10^6 cells by centrifugation at 1200 rpm (250 × *g*) for 5 min at 4 °C, aspirate the medium, and wash in cold PBS. Transfer to a FACS tube.
2. Separate the sample into Ter119$^+$ and Ter119$^-$ populations using the lineage depletion protocol in Subheading 3.4. Use

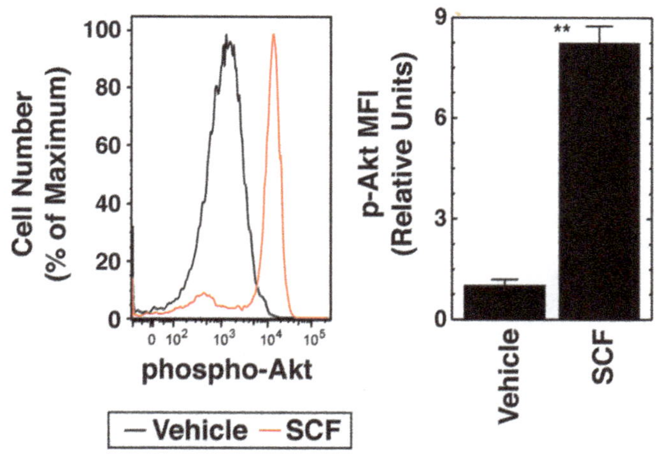

Fig. 3 Phospho-flow cytometric analysis of cytokine signaling. *Left*: Flow cytometry plots of phospho-Akt after 10 min treatment with vehicle (*black*) or 10 ng/ml SCF (*red*) in control Ter119^{-}/CD71^{high} cells cultured in expansion medium for 24 h. *Right*: Phospho-Akt quantitation. MFI is expressed relative to unstimulated cells (mean ± SE, three biological replicates)

500 μl FLJ/sample, 2 μl/ml anti Ter119, and 50 μl/ml streptavidin beads.

3. Count cells, and wash in PBS.
4. Alternatively, use FACS to isolate R1–R4 populations.
5. Resuspend cells in 100 μl of PBS and add ice-cold 70% ethanol dropwise.
6. Add DAPI at 5 μg/ml to the cells.
7. Incubate cells overnight at −20 °C.
8. Wash twice with PBS.
9. Collect DAPI fluorescence intensity on a linear scale on a flow cytometer.
10. Use Modfit software to analyze data. This software will attempt to fit the signal intensity of DAPI into a cell cycle model based on the quantity of DNA in the cell nucleus. Typically, this data is seen as two peaks, with the low-intensity peak representing G_1/G_0, the high-intensity peak representing G_2/M, and S-phase representing the region between the peaks (Fig. 4).

4 Notes

1. The cloning procedure described in Subheading 3.1 can be adapted to overexpress a gene of interest, for example to conduct a rescue experiment. A MSCV-PIG vector map is available at https://www.addgene.org/21654/.

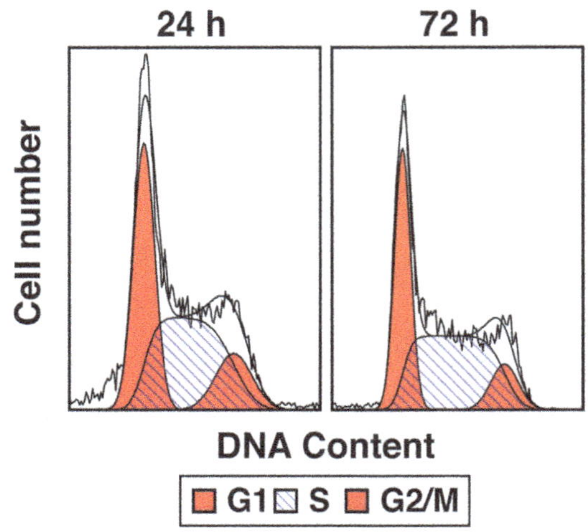

Fig. 4 Cell cycle analysis of control Ter119$^+$ cells. Flow cytometry plots of DAPI stained Ter119$^+$ cells after 24 or 72 h culture in expansion media. G_1, S and G_2/M populations denoted

2. The 3′PIG sequencing primer is best for sequencing short products. For longer products, run two sequencing reactions using the 3′PIG sequencing primer and the 5′PIG sequencing primer to cover the entire insert.
3. Ensure that the 293T cells do not become fully confluent during maintenance culture, which may negatively impact virus generation.
4. 293T cells do not adhere strongly to the bottom of flasks and plates. Thus, to avoid disrupting the cells, add the medium and the buffers to the side of the plate slowly and gently.
5. E13.5–15.5 fetuses can also be used for the isolation of lineage negative cells. E14.5 and E15.5 livers give a greater yield than E13.5.
6. There are several commercial kits for the negative selection of lineage negative hematopoietic precursor cells. These include EasySep™ Mouse Hematopoietic Progenitor Cell Enrichment Kit (StemCell Technologies) and MojoSort™ Mouse Hematopoietic Progenitor Cell Isolation Kit (BioLegend). However, the hematopoietic selection cocktail must be supplemented with anti-CD71 biotin.
7. We use shRNA against luciferase as a negative control in knockdown experiments.
8. Fetal liver lineage negative cells are very easy to infect with an infection efficiency of over 90%. Inhibitors/activators can be used in the culture as an alternative to gene knockdown or overexpression.

9. Expansion culture: Cells grow better in SCF-containing conditioned medium than in medium containing recombinant SCF.
10. Expansion culture: Cells do not grow well if the concentration is less than 1×10^5 cells/ml.
11. shRNA-mediated knockdown of factors is observed by 16 h of culture.
12. In general, we grow the lineage negative cells for 72 h in the expansion medium and then move them to the differentiation medium for 48 h. By the end of the 72 h expansion culture, 0.2×10^6 cells will expand to at least 2×10^6 cells.
13. In general, our flow cytometric analysis panel includes any or all of following fluorochromes, PE, APC, PE-cy7, APC-cy7. A live dead marker, DAPI (or Zombie UV) and an infection marker, GFP.
14. An alternative measure of erythroid maturation replaces CD71 with CD44 in the flow cytometry panel [31].
15. Single-stain controls during flow cytometry are essential for compensation.
    (a) It is very important to save uninfected cells for compensation controls.
16. We subject our slides to Wright-Giemsa staining. A detailed protocol can be found in ref. 32.
17. Colony assays can also be performed with Methocult™ GF M3334 for exclusive analysis of CFU-Es and BFU-Es.
18. Although Ter119 staining is lost during cell fixation, CD71 staining and GFP signals are generally maintained when cells are fixed in paraformaldehyde/methanol. Antigen-recognition sites for antibodies targeting Ter119 and CD71, as well as GFP signal, are lost when cells are fixed/permeabilized in one step using 70% ethanol.

Acknowledgments

The lineage negative cell isolation and culture protocols were adapted from those originally provided by M.J. Weiss. This work was supported by NIH grants DK50107 and DK68634.

References

1. Dzierzak E, Philipsen S (2013) Erythropoiesis: development and differentiation. Cold Spring Harb Perspect Med 3(4):a011601. doi:10.1101/cshperspect.a011601
2. Dzierzak E, de Pater E (2016) Regulation of blood stem cell development. Curr Top Dev Biol 118:1–20. doi:10.1016/bs.ctdb.2016.01.001

3. Tober J, Maijenburg MW, Speck NA (2016) Taking the leap: Runx1 in the formation of blood from endothelium. Curr Top Dev Biol 118:113–162. doi:10.1016/bs.ctdb.2016.01.008
4. Morrison SJ, Hemmati HD, Wandycz AM, Weissman IL (1995) The purification and characterization of fetal liver hematopoietic stem cells. Proc Natl Acad Sci U S A 92 (22):10302–10306
5. Rebel VI, Miller CL, Eaves CJ, Lansdorp PM (1996) The repopulation potential of fetal liver hematopoietic stem cells in mice exceeds that of their liver adult bone marrow counterparts. Blood 87(8):3500–3507
6. Zhang J, Socolovsky M, Gross AW, Lodish HF (2003) Role of Ras signaling in erythroid differentiation of mouse fetal liver cells: functional analysis by a flow cytometry-based novel culture system. Blood 102(12):3938–3946. doi:10.1182/blood-2003-05-1479
7. Coffman RL, Weissman IL (1981) B220: a B cell-specific member of the T200 glycoprotein family. Nature 289(5799):681–683
8. Fleming TJ, Fleming ML, Malek TR (1993) Selective expression of Ly-6G on myeloid lineage cells in mouse bone marrow. RB6-8C5 mAb to granulocyte-differentiation antigen (Gr-1) detects members of the Ly-6 family. J Immunol 151(5):2399–2408
9. Khandros E, Thom CS, D'Souza J, Weiss MJ (2012) Integrated protein quality-control pathways regulate free alpha-globin in murine beta-thalassemia. Blood 119(22):5265–5275. doi:10.1182/blood-2011-12-397729
10. DeVilbiss AW, Sanalkumar R, Hall BD, Katsumura KR, de Andrade IF, Bresnick EH (2015) Epigenetic determinants of erythropoiesis: role of the histone methyltransferase SetD8 in promoting erythroid cell maturation and survival. Mol Cell Biol 35(12):2073–2087. doi:10.1128/MCB.01422-14
11. McIver SC, Kang YA, DeVilbiss AW, O'Driscoll CA, Ouellette JN, Pope NJ, Camprecios G, Chang CJ, Yang D, Bouhassira EE, Ghaffari S, Bresnick EH (2014) The exosome complex establishes a barricade to erythroid maturation. Blood 124(14):2285–2297. doi:10.1182/blood-2014-04-571083
12. Hewitt KJ, Kim DH, Devadas P, Prathibha R, Zuo C, Sanalkumar R, Johnson KD, Kang YA, Kim JS, Dewey CN, Keles S, Bresnick EH (2015) Hematopoietic signaling mechanism revealed from a stem/progenitor cell cistrome. Mol Cell 59(1):62–74. doi:10.1016/j.molcel.2015.05.020
13. Gao X, Wu T, Johnson KD, Lahvic JL, Ranheim EA, Zon LI, Bresnick EH (2016) GATA factor-G-protein-coupled receptor circuit suppresses hematopoiesis. Stem Cell Rep 6 (3):368–382. doi:10.1016/j.stemcr.2016.01.008
14. McIver SC, Katsumura KR, Davids E, Liu P, Kang Y-A, Yang D, Bresnick EH (2016) Exosome complex orchestrates developmental signaling to balance proliferation and differentiation during erythropoiesis. eLife 5: e17877. doi:10.7554/eLife.17877
15. Hewitt KJ, Katsumura KR, Matson DR, Devadas P, Tanimura N, Hebert AS, Coon JJ, Kim JS, Dewey CN, Keles S, Hao S, Paulson RF, Bresnick EH (2017) GATA factor-regulated Samd14 enhancer confers red blood cell regeneration and survival in severe anemia. Dev Cell 42(3):213–225.e4. https://doi.org/10.1016/j.devcel.2017.07.009.7
16. Mehta C, Johnson KD, Gao X, Ong IM, Katsumura KR, McIver SC, Ranheim EA, Bresnick EH (2017) Integrating enhancer mechanisms to establish a hierarchical blood development program. Cell Rep (In Press)
17. Lee HY, Gao X, Barrasa MI, Li H, Elmes RR, Peters LL, Lodish HF (2015) PPAR-alpha and glucocorticoid receptor synergize to promote erythroid progenitor self-renewal. Nature 522 (7557):474–477. doi:10.1038/nature14326
18. Williams DA, Bunn HF, Sieff C, Zon LI (2003) Hematopoiesis. In: Handin RI, Lux SE, Stossel TP (eds) Blood: principles and practice of hematology, 2nd edn. Lippincott Williams and Wilkins, Philadelphia, PA, pp 147–208
19. McGarry MP, Protheroe CA, Lee JJ (2010) Mouse hematology: a laboratory manual. Cold Spring Harbor Laboratory Press, Woodbury, NY
20. Broxmeyer HE (1984) Colony assays of hematopoietic progenitor cells and correlations to clinical situations. Crit Rev Oncol Hematol 1 (3):227–257
21. Gregory CJ, Eaves AC (1978) Three stages of erythropoietic progenitor cell differentiation distinguished by a number of physical and biologic properties. Blood 51(3):527–537
22. Munugalavadla V, Kapur R (2005) Role of c-Kit and erythropoietin receptor in erythropoiesis. Crit Rev Oncol Hematol 54(1):63–75. doi:10.1016/j.critrevonc.2004.11.005
23. Muta K, Krantz SB, Bondurant MC, Dai CH (1995) Stem cell factor retards differentiation of normal human erythroid progenitor cells

while stimulating proliferation. Blood 86 (2):572–580

24. Sui X, Krantz SB, You M, Zhao Z (1998) Synergistic activation of MAP kinase (ERK1/2) by erythropoietin and stem cell factor is essential for expanded erythropoiesis. Blood 92(4):1142–1149
25. Wu H, Klingmuller U, Acurio A, Hsiao JG, Lodish HF (1997) Functional interaction of erythropoietin and stem cell factor receptors is essential for erythroid colony formation. Proc Natl Acad Sci U S A 94(5):1806–1810. doi:10.1160/TH08-08-0556
26. Wu H, Liu X, Jaenisch R, Lodish HF (1995) Generation of committed erythroid BFU-E and CFU-E progenitors does not require erythropoietin or the erythropoietin receptor. Cell 83(1):59–67
27. Wu H, Klingmuller U, Besmer P, Lodish HF (1995) Interaction of the erythropoietin and stem-cell-factor receptors. Nature 377 (6546):242–246. doi:10.1038/377242a0
28. Hattangadi SM, Wong P, Zhang L, Flygare J, Lodish HF (2011) From stem cell to red cell: regulation of erythropoiesis at multiple levels by multiple proteins, RNAs, and chromatin modifications. Blood 118(24):6258–6268. doi:10.1182/blood-2011-07-356006
29. Haas N, Riedt T, Labbaf Z, Bassler K, Gergis D, Frohlich H, Gutgemann I, Janzen V, Schorle H (2015) Kit transduced signals counteract erythroid maturation by MAPK-dependent modulation of erythropoietin signaling and apoptosis induction in mouse fetal liver. Cell Death Differ 22(5):790–800. doi:10.1038/cdd.2014.172
30. Pop R, Shearstone JR, Shen Q, Liu Y, Hallstrom K, Koulnis M, Gribnau J, Socolovsky M (2010) A key commitment step in erythropoiesis is synchronized with the cell cycle clock through mutual inhibition between PU.1 and S-phase progression. PLoS Biol 8(9): e1000484. doi:10.1371/journal.pbio.1000484
31. Chen K, Liu J, Heck S, Chasis JA, An X, Mohandas N (2009) Resolving the distinct stages in erythroid differentiation based on dynamic changes in membrane protein expression during erythropoiesis. Proc Natl Acad Sci U S A 106(41):17413–17418. doi:10.1073/pnas.0909296106
32. Dunning K, Safo AO (2011) The ultimate Wright-Giemsa stain: 60 years in the making. Biotech Histochem 86(2):69–75. doi:10.3109/10520295.2010.515496

Chapter 5

Stress Erythropoiesis Model Systems

Laura F. Bennett, Chang Liao, and Robert F. Paulson

Abstract

Bone marrow steady-state erythropoiesis maintains erythroid homeostasis throughout life. This process constantly generates new erythrocytes to replace the senescent erythrocytes that are removed by macrophages in the spleen. In contrast, anemic or hypoxic stress induces a physiological response designed to increase oxygen delivery to the tissues. Stress erythropoiesis is a key component of this response. It is best understood in mice where it is extramedullary occurring in the adult spleen and liver and in the fetal liver during development. Stress erythropoiesis utilizes progenitor cells and signals that are distinct from bone marrow steady-state erythropoiesis. Because of that observation many genes may play a role in stress erythropoiesis despite having no effect on steady-state erythropoiesis. In this chapter, we will discuss in vivo and in vitro techniques to study stress erythropoiesis in mice and how the in vitro culture system can be extended to study human stress erythropoiesis.

Key words Stress erythropoiesis, Hypoxia, Anemia, In vitro culture, Bone marrow transplant, Phenylhydrazine induced acute hemolytic anemia, Methylcellulose colony assays

1 Introduction

Under most conditions bone marrow steady-state erythropoiesis is able to maintain erythroid homeostasis; however, there are times when bone marrow erythropoiesis is inhibited or is unable to maintain erythrocyte numbers. At these times an alternative pathway for erythropoiesis, stress erythropoiesis, is activated [1]. It functions at times of anemic stress and produces a wave of new erythrocytes to maintain tissue oxygenation until steady-state erythropoiesis can recover [2–4]. It is best understood in the mouse where it is extra-medullary. Stress erythropoiesis is observed during development in the fetal liver and in adults in the spleen and liver [5, 6]. The features that define stress erythropoiesis are the use of signals and progenitor cells that are distinct from bone marrow steady-state erythropoiesis. Some of the signals utilized are not associated with erythropoiesis, but are commonly observed in stem cell-mediated tissue regeneration, for example in the intestinal epithelium [7]. Among these signals are bone morphogenetic

Joyce A. Lloyd (ed.), *Erythropoiesis: Methods and Protocols*, Methods in Molecular Biology, vol. 1698,
DOI 10.1007/978-1-4939-7428-3_5, © Springer Science+Business Media LLC 2018

protein 4 (BMP4), Hedgehog (HH), and growth and differentiation factor 15 (GDF15), which function in concert with stem cell factor (SCF) and erythropoietin (Epo) [3, 4, 8, 9]. The progenitor cells in stress erythropoiesis are also distinct from steady-state erythropoiesis. Steady-state erythroid progenitors have been shown to develop from a defined hierarchy of myeloid differentiation which results in the development of erythroid committed progenitors from megakaryocyte/erythroid progenitors (MEPs) [10, 11]. In contrast, our analysis has demonstrated that stress erythroid progenitors in the spleen are derived directly from short-term reconstituting hematopoietic stem cells (ST-HSCs, CD34+Kit+Sca1+Lin−), which migrate to the spleen [2]. Once in the spleen, these cells become committed to the stress erythroid fate and undergo an initial expansion stage where cells maintain their stem cell characteristics, but are unable to differentiate. The increase in serum Epo concentration causes the progenitor cells to transition to committed erythroid progenitors that will form BFU-E colonies when plated in methylcellulose media. During the differentiation stage progenitor cells continue to proliferate and undergo terminal erythroid maturation [9]. This developmental pathway is associated with a change in cell surface markers as well as changes in the growth factors required (Fig. 1). The experimental readout of stress erythropoiesis in vitro is the stress BFU-E. Unlike steady-state BFU-E formation that requires two signals Epo and a second signal, a burst promoting activity or BPA, stress BFU-E require only Epo. Colony assays done in media containing Epo only quantitate mature stress BFU-E. A second readout measures maximal stress BFU-E potential, which measures immature stress BFU-E that do not respond to Epo only media but are induced to form stress BFU-E when additional growth factors are present in the media. Maximum stress BFU-E potential can be measured by plating cells in media containing BMP4 + SCF + Epo at 1% oxygen [3, 8]. By assaying cells using both experimental conditions, one can delineate defects in stress BFU-E formation at different stages of development. An important distinction between stress erythropoiesis and steady-state erythropoiesis is that steady-state BFU-E do not respond to BMP4. Another feature of stress erythropoiesis is that it is not continuous like steady-state erythropoiesis. It occurs in a coupled wave of expansion, transition, and differentiation [2, 9]. Afterward, new cells need to migrate into the spleen to start the cycle again. These properties are characteristic of a stress response pathway, where stress induces a short-term biochemical or cellular response designed to alleviate that stress until the homeostatic machinery can resume and maintain homeostasis.

Because of these unique properties, many of the genes involved in stress erythropoiesis have little or no effect on steady-state erythropoiesis [4, 12, 13]. Therefore, the roles of different genes in

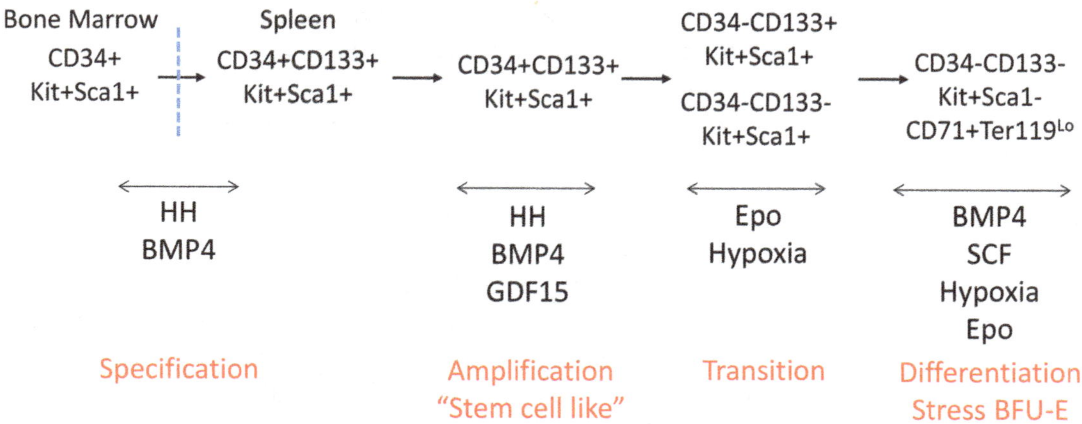

Fig. 1 Schematic of stress erythropoiesis. The cell surface markers of the progenitors at each stage of development are shown, and growth factors required at each stage are indicated below

stress erythropoiesis must be directly assayed under conditions that model stress erythropoiesis. Over the years, we have utilized two in vivo assays for stress erythropoiesis that complement one another and allow us to test all stages of stress erythropoiesis. Below, we describe the analysis of donor-derived erythroid short-term radioprotection immediately following bone marrow transplant as a method to study all the stages of stress erythropoiesis [2]. This technique allows for the analysis of the migration of progenitor cells to the spleen, the expansion of early stress progenitors, their transition to differentiation, and the terminal differentiation of new erythrocytes, as all of these events occur prior to stem cell engraftment and the establishment of donor-derived steady-state erythropoiesis. In addition, we have adapted the well-known protocol for phenylhydrazine-induced acute hemolytic anemia as a means to study the expansion and differentiation of stress progenitors already resident in the spleen [3]. Coupled with these in vivo assays, we have developed an in vitro culture system that recapitulates the development of stress erythroid progenitors. It uses growth factors that are known to be present in the spleen and whole unfractionated bone marrow as its starting material. By manipulating the culture conditions, we can maintain and expand cells in the early expansion stage and then by adding Epo and growing cells at low oxygen, we can induce the transition and differentiation stage. We can quantitate the ability of the cells to generate stress progenitors by plating for stress BFU-E colony formation in methylcellulose assays. In addition, this culture protocol generates functional early stress erythroid progenitors that can be isolated by flow cytometry sorting and transplanted into irradiated recipient mice to test their ability to provide erythroid short-term radioprotection. The ability of the in vitro culture system to recapitulate stress erythropoietic development can also be extended to the culture of human bone marrow to analyze human stress erythropoiesis [9].

2 Materials

2.1 Phenylhydrazine Treatment

1. 8–10 week old mice. This protocol is designed for C57BL/6 mice (*see* **Note 1**). All mice should be approximately the same weight (for example 24–26 g).
2. Phenylhydrazine: Calculate the concentration of phenylhydrazine needed for a dose of 100 mg/kg in 100 μL of sterile PBS. The solution must be made fresh immediately prior to use and protected from light.

2.2 Stress BFU-E Assays

All the cytokine stock solutions are stored frozen until diluted in media for use in assays.

1. Methylcellulose media with 3 U/mL Epo (Stem Cell Technologies), stored frozen in 1.8 mL aliquots. BMP4 and SCF are also added for assays to assess maximum stress BFU-E potential (*see* Subheading 3.2, **step 11**).
2. BMP4: 10 ng/μL (Used at a final concentration of 15 ng/mL).
3. SCF: 50 ng/μL (Used at a final concentration of 50 ng/mL).
4. Insulin from bovine pancreas (1000× stock): 10 mg/mL (dissolved in H_2O, pH = 3).
5. Bovine holo-transferrin (100× stock): 20 mg/mL.
6. Ciprofloxacin HCL (1000× stock): 10 mg/mL.
7. Iscove's Modified Dulbecco's Media (IMDM) base medium: 500 mL IMDM containing 2 mM L-Glutamine and 25 mM Hepes buffer, 5 g Bovine Serum Albumin (BSA), 3.5 μL β-mercaptoethanol. Filter sterilize.
8. Final IMDM (per 100 mL): 85 mL IMDM base medium (from **item 8**), 15 mL Fetal bovine serum (FBS), 1 mL 100× transferrin, 0.1 mL 1000× Insulin, 0.1 mL 1000× Ciproflaxin.
9. Trypan blue.
10. Red Cell Lysis Buffer: 8.26 g Ammonium chloride (NH_4Cl), 1.0 g Potassium bicarbonate ($KHCO_3$), 0.037 g EDTA. Add water to total of 1 L of solution. Filter sterilize before use. It can be stored for up to 6 months.
11. Benzidine stock solution: 0.3 g benzidine, 9 mL acetic acid, 1 mL H_2O. Avoid light, store in a foil-covered tube.
12. Acid Benzidine stain: 5 parts H_2O, 1 part H_2O_2 (Fresh), 1 part benzidine stock solution (from **item 10**). Avoid light, store in a foil-covered tube.

2.3 Bone Marrow Transplant and In Vitro Culture of Stress Erythroid Progenitors from Bone Marrow

1. Acidified water: Add glacial acetic acid to double distilled water until pH 2.5–3.0.
2. Epo stock solution: 10 U/μL.
3. BMP4 stock solution: 10 ng/μL.
4. SCF stock solution: 50 ng/μL.
5. GDF15 stock solution: 10 ng/μL.
6. Shh stock solution: 25 ng/μL.
7. Stress erythropoiesis expansion media (SEEM): Final IMDM (*see* Subheading 2.2, **item 8**), 15 ng/mL BMP4, 30 ng/mL GDF15, 25 ng/mL Shh, 50 ng/mL SCF.
8. Stress erythropoiesis differentiation media (SEDM): SEEM containing 3 U/mL Epo.

3 Methods

3.1 Phenylhydrazine-Induced Acute Hemolytic Anemia

Phenylhydrazine (PHZ) induces erythrocyte lysis by membrane oxidation leading to acute hemolytic anemia [14]. This technique is used to examine the expansion and differentiation of stress erythroid progenitors resident in the spleen. We have developed a protocol that utilizes a single dose of PHZ (100 mg/kg mouse) which induces an acute anemia with reproducible kinetics of recovery. In approximately 24 h, 50% of the erythrocytes are lysed, and the mice will develop anemia. The stress erythropoiesis response occurs over a 7 day period. Control C57BL/6 mice fully recover their hematocrits in 5–7 days.

1. Weigh mice. Concentration of PHZ depends on weight of mice. All mice should be approximately of the same weight (*see* **Note 1**).
2. Dissolve PHZ in sterile PBS. Dose is 100 mg/kg and total volume given per mouse is 100 μL.
3. Inject 100 μL of PHZ subcutaneously.
4. Typical analysis of anemic mice would include complete blood counts during the recovery period (*see* **Note 2**). Euthanize mice on different days posttreatment and analyze the expansion of stress erythroid progenitors in the spleen by stress BFU-E colony assays (*see* Subheading 3.2 below). Analysis of gene expression in the spleen or bone marrow may also be performed using established techniques.

3.2 Stress BFU-E Assay

Stress erythroid progenitors can be quantitated by using methylcellulose colony assays. These assays are similar to those used to analyze bone marrow steady state but differ in the growth factors required and in the culture conditions [8]. Unlike steady-state

BFU-E, stress BFU-E can form when cultured with Epo alone. These stress BFU-E are considered mature stress BFU-E. In addition, we also plate cells in media containing BMP4 + SCF + Epo at 1–2% O_2. Under these conditions the number of stress BFU-E increases because these signals allow for the maximum stress BFU-E potential to be measured. We typically do both conditions when analyzing mutant mice.

1. To isolate splenocytes for colony assay, sacrifice mouse and harvest the spleen, removing as much connective tissue as possible.
2. Place spleen in 1 mL of sterile PBS, and mash the spleen with the flat end of a syringe until you have a homogeneous suspension.
3. Pass cell suspension through a 70 μM filter into a 50 mL collection tube. Centrifuge the cells and collect the pellet.
4. Add 10 mL of cold red cell lysis buffer and incubate on ice for 5 min.
5. Add 10 mL PBS after incubation to stop cell lysis.
6. Spin cells at 1700 × *g* for 5 min to pellet.
7. Remove the supernatant and resuspend cells in Final IMDM (*see* Subheading 2.2, **item 8**). For a normal spleen, 1 mL is sufficient but this volume should be increased in the case of larger spleens (i.e., 5 mL is more appropriate for mice treated with phenylhydrazine. Cells from phenylhydrazine-treated spleens tend to stick to each other. If necessary filter cells through a cell strainer to remove any small clumps.).
8. Count live cells using a hemocytometer after staining with trypan blue.
9. Thaw methylcelluose aliquots (for colony assays done in a 12-well plate, aliquots should be 1.8 mL).
10. Add 4×10^5 cells (enough for four wells).
11. There are two possible cytokine conditions that can be used. Condition 1 is for mature stress BFU-E, and contains 3 U/mL Epo only. Condition 2 is for maximum stress BFU-E potential and for this condition the media contains 15 ng/mL BMP4, 50 ng/mL SCF, and 3 U/mL Epo.
12. For plating Colony Assays, cells, cytokines, and IMDM should be added to methylcellulose for a total volume of 2.0 mL to allow for 500 μL/well. The volume of cells, cytokines, and IMDM should equal 200 μL.
13. Once cells and cytokines are added to methylcellulose media, vortex briefly to mix. The viscosity of the media does not allow for pipetting.

14. Plate 0.5 mL/well of methylcellulose mixture in a 12-well plate using a 16 g needle and a thin 1 mL syringe (*see* **Note 3**). Each condition should be plated in triplicate. This will give 1×10^5 cells/well. *For best results, begin plating in the center and spiral outward to evenly disperse the methylcellulose. If there are uncovered portions of the well, tilt the plate until the entire well is covered.
15. Let methylcellulose sit until the bubbles have risen to the top (approximately 15 min).
16. Any large bubbles should be popped before plates are put in the incubator since they will affect colony formation and will not pop on their own.
17. Incubate at 37 °C for 5–7 days at 20% O_2 for Epo only Stress BFU-E, or at 1% O_2 for cells plated in media containing BMP4 + SCF + Epo for maximum stress BFU-E potential (*see* **Note 4**).
18. Stress BFU-E colonies can be easily visualized and counted by staining with acid benzidine stain, which reacts with hemoglobin to form a dark blue color (*see* **Note 5**). Carefully add enough stain to each well to cover methylcellulose medium without disturbing the colonies (approximately 500 μL/well is sufficient). Stain acts instantly and colonies should be counted immediately. Acid benzidine stain should be made fresh and used immediately. BFU-Es will be a dark and distinct blue color from the benzidine stain.

3.3 Bone Marrow Transplant

A second in vivo method to study stress erythropoiesis is to analyze the erythroid recovery in the immediate post-transplant period. Donor-derived stress erythropoiesis generates a wave of new erythrocytes that maintains the recipient mouse until stem cells engraft and begin donor-derived steady-state erythropoiesis [3]. Bone marrow transplant (BMT) has a few advantages over PHZ-induced acute anemia. It allows one to analyze the initial migration to the spleen and expansion stress erythroid progenitors. BMT is compatible with flow cytometry. Mutant donor cells can be transplanted into wild-type control recipients and vice versa so that cell intrinsic defects and microenvironment defects can be analyzed. Below is the standard transplant protocol that is used at The Penn State University.

1. Acidified water—Animals to be irradiated should be placed on acidified water (pH 2.5–3.0) from 1 week prior to 2 weeks post irradiation.
2. Antibiotic treatment: Prior to irradiation animals must be treated with antibiotic with the intent to eliminate *Pseudomonas aeruginosa* from the gastrointestinal tract. The antibiotics are selected by the investigator in consultation with the Attending

Veterinarian and are specified in the IACUC Protocol. Directions for preparation and administration of the antibiotic are provided by the Attending Veterinarian. The following list contains alternatives for antibiotic treatment (concentrations are in drinking water): 1–1.2 mg/mL Gentamycin, 1.1 mg/mL Neomycin, 1.1 mg/mL Neomycin and 1000 U/mL Polymyxin B sulfate, 0.17 mg/mL Baytril, or between 0.65 and 1.6 mg/mL Sulfamethoxazole (Sulfadiazine) with Trimethoprim. This last drug comes as a mixture of the two drugs and the dose is for the mixture.

3. Three different approaches can be used for irradiating mice, depending upon the mouse strain and whether full reconstitution is needed. For the single exposure protocol, mice are irradiated with 800–950 rads (exact dose is strain dependent, we commonly use 950 rads for C57BL/6 mice) at a rate of approximately 60–80 rads/min. For split exposure, the total dose is delivered in two 5–6 min periods separated by 3 h. The split dose is often used with mice that might have defect in erythroid recovery after transplant. Strains highly sensitive to irradiation such as SCID mice receive a reduced total exposure dose of 350–400 rads. The reduced exposure can also be used when full reconstitution is not needed, such as adoptive transfer of mature cells or transfer of leukemia stem cells.
4. Transplant procedure: Bone marrow donors are euthanized with CO_2 for marrow collection. Bone marrow cells are collected in sterile media, and injected into the tail vein or retro-orbital sinus of the irradiated recipients. The number of cells will depend on the experiment and will range from 2×10^2 to 5×10^6 in 150–200 μL. Although we routinely transplant $1–5 \times 10^5$ cells into C57BL/6 recipients up to 5×10^6 cells can be transplanted (*see* **Note 6**).
5. Whole body irradiation damages the bone marrow and potentially the GI tract leading to increased susceptibility to opportunistic infections. Although successful bone marrow transplant should be protective, the transfer will occasionally be unsuccessful. Potential signs of complication are nonspecific and may include the following: Poor body condition/weight loss; rough coat, inactivity, hunched posture; and death without prior signs of illness. Animals need to be monitored daily by the investigator following irradiation and bone marrow transplant. Mildly affected animals (rough coat, inactive, hunched posture) but still in good body condition are monitored for further deterioration. More severely affected animals displaying obvious weight loss and poor body condition (body condition score of 2 minus) are removed from the study. A veterinarian may need to be consulted for possible diagnostic testing.

6. Transplanted mice can be analyzed in a number of ways (*see* **Notes 7** and **8**). Complete blood counts and reticulocyte assays can be done. In addition mice that constitutively express GFP (for example C57BL/6-Tg(UBC-GFP)30Scha/J) (15) could be used so that the donor or the recipient is GFP+. The usual dose of 1–5 × 10^5 unfractionated bone marrow cells leads to erythroid recovery in approximately 16–20 days.

3.4 In Vitro Analysis of Stress Erythropoiesis

1. We have developed an in vitro culture system that recapitulates the signals and microenvironment present in the spleen. This system uses unfractionated bone marrow cells as starter material. During culture, the cells form a microenvironment that supports the expansion and development of stress erythroid progenitors that can be assayed in stress BFU-E assays or transplant assays. The culture can be done in two ways. The first way is to include all factors necessary for the expansion and differentiation of cells, which includes BMP4, Shh, SCF, GDF15, Epo and the culture is performed at 1–2% O_2. The production of stress BFU-E can be assayed on different days of culture. Alternatively, the culture can be separated into an expansion phase and a differentiation phase, which mimics the development in vivo. For this type culture, cells are first grown in stress erythropoiesis expansion media (SEEM), which contains BMP4, Shh, GDF15, and SCF. The cells are cultured in this media for 5–7 days at 20% O_2. The production of stress BFU-E is minimal in these culture conditions. The second stage of culture is a switch to stress erythropoiesis differentiation media (SEDM). In addition to the growth factors in SEEM, SEDM contains Epo and the cells are grown at 1–2% O_2. The production of stress BFU-E can be assayed from days 1–5 of SEDM culture. In addition, the two-stage culture system can allow investigators to examine the expansion of early stress progenitors characterized by the expression of CD34+CD133+Kit+Sca1+ cell surface markers. This analysis can be extended to examine the transition of these early cells to more mature cell populations which lose the expression of CD34, CD133, and Sca1, as described previously [9]. This culture system can be readily adapted to human cells. We purchase whole unfractionated bone marrow cells for these experiments, but unfractionated cord blood will also work. Purified subsets like CD34+ cells do not work in these cultures. When culturing human cells, obviously human growth factors need to be used and we usually purchase fetal calf serum that has been tested for human cultures.

1. For a single-stage culture: Isolate cells from bone marrow and culture at a concentration of 1 × 10^6 cells/mL in SEDM for 5–7 days at 1–2% O_2. Cells can be harvested any time after 24 h of culture depending on the experiment and plated for stress

BFU-E as described above or analyzed by flow cytometry as previously described.

2. For a two-stage culture:

First stage—Expansion of early progenitors: Isolate cells from bone marrow and culture at a concentration of 1×10^6 cells/mL in SEEM for 5–7 days at 20% O_2. At this stage, early stress progenitors (CD34+CD133+Kit+Sca1+ and CD34−CD133+Kit+Sca1+ cells) can be isolated by flow cytometry and analyzed (*see* **Note 9**).

Second stage transition to differentiation: After first stage culture, count cells and shift to SEDM at a concentration of 1×10^5 cells/mL for 1–5 days at 1% O_2. When changing the media, return the cells in SEDM to the same dish that contains the adherent microenvironment.

4 Notes

1. This dose of PHZ was designed using C57BL/6 mice. The kinetics of the response is well characterized (for example *see* ref. 3). The kinetics in other mouse strains is different, so it is always important to have a strain specific wild-type control. For some mutant mice, 100 mg/kg mouse is too strong a dose and results in lethality. If we encounter lethality we test mice at a 75% and a 50% dose, so mice survive the treatment and the phenotype in stress erythroid progenitors can be analyzed.
2. PHZ treatment can complicate flow cytometry. In our hands, stress erythroid progenitors from mice treated with PHZ exhibit high levels of autofluorescence, which make analysis of cell populations difficult.
3. Our use of 12-well plates is designed to conserve methylcellulose and other reagents. The assays can be done in 6-well plates, but in that case the number of cells should be increased to $0.5–1 \times 10^6$ cells per well in 1–1.5 mL of media. Methylcellulose is viscous. We always set up enough media and cells for four wells so that each well of the triplicate set receives the full 0.5 mL of media and cells. When plating in a 12-well dish we always fill at least three wells with sterile distilled water, which prevents the methylcellulose media from drying out during the culture.
4. Hypoxic culture can be done in several ways. The most economical is a hypoxia chamber (an example is the modular incubator chamber from Billups-Rothenberg). Plates of cells are placed in the chamber, the chamber is flushed with a custom gas mixture (we use 94% N_2, 5% CO_2, and 1% O_2) and then the chamber is placed in the incubator. The hypoxia chamber is flushed with gas daily during the culture period. The second option is the use of a chamber that fits in an incubator, but independently controls

CO_2 and O_2 levels (an example of this system is the Biospherix Proox C21 CO_2/O_2 controller and C-Chamber for hypoxia culture). The third option is an incubator that controls both CO_2 and O_2 levels.

5. Benzidine staining is very efficient. If the stain does not work, the most likely problem is old H_2O_2. When counting colonies, stain wells as you count them. The stain will dilute and fade with time. Also once you start staining wells on a plate you have to count the whole plate. If you place a plate with stained wells in the incubator the cells will die in the other wells.
6. The radiation dose and number of cells needed for transplant are optimized for C57BL/6 mice. Other strains may be more or less tolerant to radiation doses and number of donor cells. We routinely transplant with 1×10^5 cells, which is on the lower end of doses. At that dose we feel that we have a sensitized screen for defects in erythroid short-term radioprotection. If you encounter lethality at that dose, increase the dose of donor cells to 5×10^5 or 1×10^6. At the higher doses of cells, more mutant mice will survive so that analysis will be possible. Similarly, using a split dose of irradiation can improve the survival of recipient mice.
7. We use a number of genetic methods to mark donor versus recipient cells. The classic method is to use C57BL/6 (CD45.2+) and congenic B6.SJL mice that carry the CD45.1 allele as donors and recipients. In addition, mice that constitutively express GFP (for example C57BL/6-Tg(UBC-GFP) 30Scha/J) [15] could be used so that the donor or the recipient is GFP+.
8. For the analysis of mice following transplant, we use complete blood counts. Investigators should be aware that if you continuously sample blood from the recovering animals that is an anemic stress. Control transplanted mice are not affected by the blood sampling, but for some mutant donor transplants the added stress is enough to cause transplant failure. To circumvent that problem you can euthanize mice at each time point or you can set up several cohorts where the blood sampling occurs less frequently; for example, four cohorts with sampling every other day will only require two blood samples per mouse over the 16 day recovery period.
9. It is normal for cell numbers to decrease in the cultures. The progenitor cells not supported by the growth factors are lost by apoptosis. By day 3 of the initial culture in SEEM, the numbers of cells will begin to increase. Hypoxic cultures can be done as described in **Note 4**, but the investigator should realize that under hypoxic conditions cells grow faster. Care should be given so that the cultures do not over-grow the media.

References

1. Paulson RF, Shi L, Wu DC (2011) Stress erythropoiesis: new signals and new stress progenitor cells. Curr Opin Hematol 18(3):139–145
2. Harandi OF, Hedge S, Wu DC, McKeone D, Paulson RF (2010) Murine erythroid short-term radioprotection requires a BMP4-dependent, self-renewing population of stress erythroid progenitors. J Clin Invest 120 (12):4507–4519
3. Lenox L, Perry J, Paulson R (2005) BMP4 and Madh5 regulate the erythroid response to acute anemia. Blood 105:2741–2748
4. Perry JM, Harandi OF, Porayette P, Hegde S, Kannan AK, Paulson RF (2009) Maintenance of the BMP4-dependent stress erythropoiesis pathway in the murine spleen requires hedgehog signaling. Blood 113(4):911–918. PMCID: 2630276
5. Porayette P, Paulson RF (2008) BMP4/Smad5 dependent stress erythropoiesis is required for the expansion of erythroid progenitors during fetal development. Dev Biol 317(1):24–35
6. Lenox LE, Shi L, Hegde S, Paulson RF (2009) Extramedullary erythropoiesis in the adult liver requires BMP-4/Smad5-dependent signaling. Exp Hematol 37(5):549–558
7. Ishizuya-Oka A, Hasebe T (2008) Sonic hedgehog and bone morphogenetic protein-4 signaling pathway involved in epithelial cell renewal along the radial axis of the intestine. Digestion 77(Suppl 1):42–47
8. Perry JM, Harandi OF, Paulson RF (2007) BMP4, SCF, and hypoxia cooperatively regulate the expansion of murine stress erythroid progenitors. Blood 109(10):4494–4502
9. Xiang J, Wu DC, Chen Y, Paulson RF (2015) In vitro culture of stress erythroid progenitors identifies distinct progenitor populations and analogous human progenitors. Blood 125 (11):1803–1812
10. Pronk CJ, Bryder D (2011) Flow cytometry-based identification of immature myeloerythroid development. Methods Mol Biol 699:275–293
11. Pronk CJ, Rossi DJ, Mansson R, Attema JL, Norddahl GL, Chan CK et al (2007) Elucidation of the phenotypic, functional, and molecular topography of a myeloerythroid progenitor cell hierarchy. Cell Stem Cell 1 (4):428–442
12. Gao J, Graves S, Koch U, Liu S, Jankovic V, Buonamici S et al (2009) Hedgehog signaling is dispensable for adult hematopoietic stem cell function. Cell Stem Cell 4(6):548–558
13. Hofmann I, Stover EH, Cullen DE, Mao J, Morgan KJ, Lee BH et al (2009) Hedgehog signaling is dispensable for adult murine hematopoietic stem cell function and hematopoiesis. Cell Stem Cell 4(6):559–567
14. Goldberg B, Stern A (1977) The mechanism of oxidative hemolysis produced by phenylhydrazine. Mol Pharmacol 13(5):832–839
15. Schaefer BC, Schaefer ML, Kappler JW, Marrack P, Kedl RM (2001) Observation of antigen-dependent CD8+ T-cell/ dendritic cell interactions in vivo. Cell Immunol 214 (2):110–122

Chapter 6

Approaches for Analysis of Erythroid Cell Parameters and Hemoglobinopathies in Mouse Models

Marie Trudel and Josepha-Clara Sedzro

Abstract

Over the last decades several mouse models of human hemoglobin disorders have been and continue to be generated. This chapter aims at describing various approaches to evaluate the global red blood cell properties in mouse models of human hemoglobin disorders, in particular thalassemia and sickle cell disease. Analysis of erythroid parameters provides insights into the RBC physiologic or pathophysiologic status. Mice expressing both murine and human globin genes can be investigated using adapted protocols provided herein.

Key words Hemoglobin, Globin chain, Urea-triton polyacrylamide gel electrophoresis, Cellulose acetate electrophoresis, Smear

1 Introduction

In transgenic mice, human globin RNA is often not as efficiently translated relative to the endogenous globin genes, and is a poor predictor of the protein level in circulating red blood cells. Differential expression likely results from epistatic effects and possibly from competition of human-mouse chimeric α-β globin dimer or tetramer assembly. Nevertheless, it is crucial to directly assess protein levels in the red blood cells (RBC) of murine models of hemoglobinopathies. This chapter describes the most frequently used methods to monitor RBC, individual globin chains, and hemoglobin.

Clinical parameter monitoring by typical complete blood cell count performed in the species-specific Advia hematology provides for comprehensive blood profiles. These include RBC (count, mean corpuscular volume or MCV, hematocrit, total hemoglobin, mean corpuscular hemoglobin or MCH, mean corpuscular hemoglobin concentration or MCHC, hemoglobin distribution width, red cell volume distribution width, macrocytic, microcytic, hypochromic, hyperchromic), reticulocyte analysis (percentage, count, volume,

Joyce A. Lloyd (ed.), *Erythropoiesis: Methods and Protocols*, Methods in Molecular Biology, vol. 1698,
DOI 10.1007/978-1-4939-7428-3_6,

hemoglobin content, and concentration), in addition to indices of platelet and total and individual white blood cell counts [1–3]. It is important to note that in comparison to human, red cells are much smaller in size in mice. In the context of hemoglobinopathies, in particular, this is likely to influence RBC physiologic responses. Interpretation of complete blood cell count combined with complementary technical approaches of RBC analysis defines a global picture of erythroid cells in mice.

Cytology analyses of peripheral blood smears are a fast and simple method to visualize blood cells [4]. This is a manual approach, so the number of cell samples that can be evaluated is low. However, assessment of the appearance and morphology of all blood cells can determine the cell status and the nature of anomalies that can be best detected by direct visualization by eye. While both β-thalassemia and sickle cell disease cause anemia, these different hemoglobinopathies each result in a distinct typical red cell morphology and appearance. In cases of β-thalassemia, RBC frequently display inclusions of denatured globin chains or Heinz bodies that precipitate on the RBC membrane. In cases of sickle cell disease, there are a significant proportion of irreversibly sickled cells due to repeated cycles of deoxygenation. Both criteria are hallmarks of the hemoglobinopathies and among key determinants of clinical severity for the illness.

Since hemoglobin and globin chains are the most abundant protein in the RBC, they can be readily separated from whole blood samples by gel electrophoresis and visualized by various staining techniques. Electrophoresis allows detection of several hemoglobin variants. Although high-resolution techniques by isoelectric focusing (IEF), high-performance liquid chromatography (HPLC), mass spectroscopy (MS) can be achieved [5], the use of standard methods of cellulose acetate to separate hemoglobin and of urea triton polyacrylamide gels to separate globin chains are easy, inexpensive, and rapid.

Cellulose acetate electrophoresis in alkaline conditions provides a superb resolution of hemoglobin that identifies globin locus genotypes in mice and human-mouse chimeric forms in transgenic mice [6, 7]. This method is rapid and separates hemoglobins based on their isoelectric point. Cellulose acetate fractionates hemoglobins qualitatively, but can be quantitative, provided loading is within an appropriate linear range for densitometric analysis. To identify the multiple mouse hemoglobins from inbred lines, modification with a disulfide reagent (cystamine) allows for discrimination between the mouse single and diffuse alleles/haplotype. This modification relies on the presence of two cysteine residues in the β-major and β-minor chains of hemoglobin of the diffuse haplotype in comparison to one present in the β-single chain. Hemoglobin molecules are separated by electrophoresis according to the charge on modified residues, as a function of their milieu. Hemoglobin at pH 8.5 is negatively charged and migrates under an electric field toward the positive pole anode. Transgenic mice generated with human globin genes can produce human-mouse chimeric α-β tetrameric forms in addition to mouse and human hemoglobin that can be identified with this approach (Fig. 1).

Urea-triton polyacrylamide gel electrophoresis (UT-PAGE) [5, 8–10] is a procedure developed to investigate fetal Gγ- and Aγ-globin chain ratios, the presence of abnormal globin chains and for prenatal diagnosis of thalassemia. This approach is extremely useful for the identification and assessment of murine and human globin chain levels in genetically modified mouse models [2, 5]. The relative proportion of globin chains can be determined following denaturation of the hemoglobin tetramer (Fig. 2). The presence of urea and mercaptoethanol solubilizes the heme group and fractionates the globin chains into monomers. Addition of Triton, a nonionic detergent, to globin chains with hydrophobic regions causes micelle formation that reduces the mobility of the individual globin

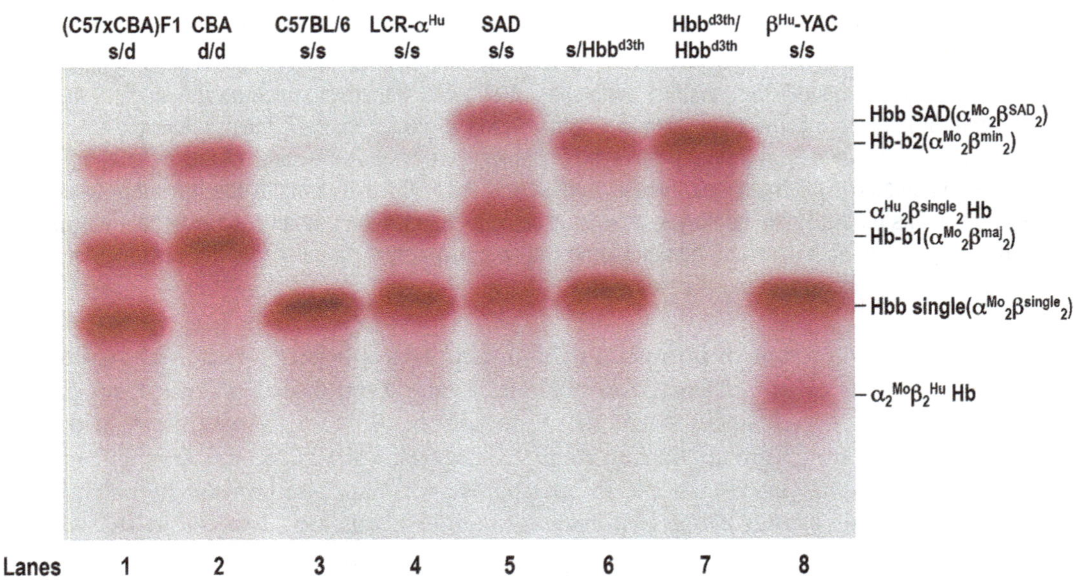

Fig. 1 Mouse and human hemoglobin analysis. Total blood lysates were separated in native condition on cellulose acetate membrane. Controls for Hb separation includes mouse adult peripheral blood samples of two inbred mouse strains individually C57BL/6J (*lane 3*) with the Hbb single haplotype (s/s), CBA/J (*lane 2*) with the Hbb diffuse haplotype (d/d) consisting of Hb major and Hb minor or (C57BL/6J × CBA/J)F1 (*lane 1*) (s/d) comprise all three Hb single, major and minor. Blood samples from chimeric human-mouse hemoglobin were analyzed. A human α-globin transgenic mouse on the single haplotype (*lane 4*) produces Hb single ($\alpha^{Mo}{}_2\beta^{single}{}_2$) and the chimeric Hb ($\alpha^{Hu}{}_2\beta^{single}{}_2$). The SAD transgenic mouse model of SCD that contains the human α-globin and β^{SAD}-globin chains on the single haplotype (*lane 5*) produces Hb single ($\alpha^{Mo}{}_2\beta^{single}{}_2$), chimeric Hb ($\alpha^{Hu}{}_2\beta^{single}{}_2$), and Hb SAD ($\alpha^{Hu}{}_2\beta^{SAD}{}_2$). A thalassemic mouse mutant Hbbd^{3th} heterozygote on a single haplotype (*lane 6*) produces the Hb single ($\alpha^{Mo}{}_2\beta^{single}{}_2$) and Hb minor ($\alpha^{Mo}{}_2\beta^{minor}{}_2$), or homozygote (*lane 7*) yields only the Hb minor ($\alpha^{Mo}{}_2\beta^{minor}{}_2$). Blood from a β^{Hu}-YAC transgenic mouse on the single haplotype (*lane 8*) generates the Hb single ($\alpha^{Mo}{}_2\beta^{single}{}_2$) and the chimeric Hb ($\alpha^{Mo}{}_2\beta^{Hu}{}_2$) that can easily be detected

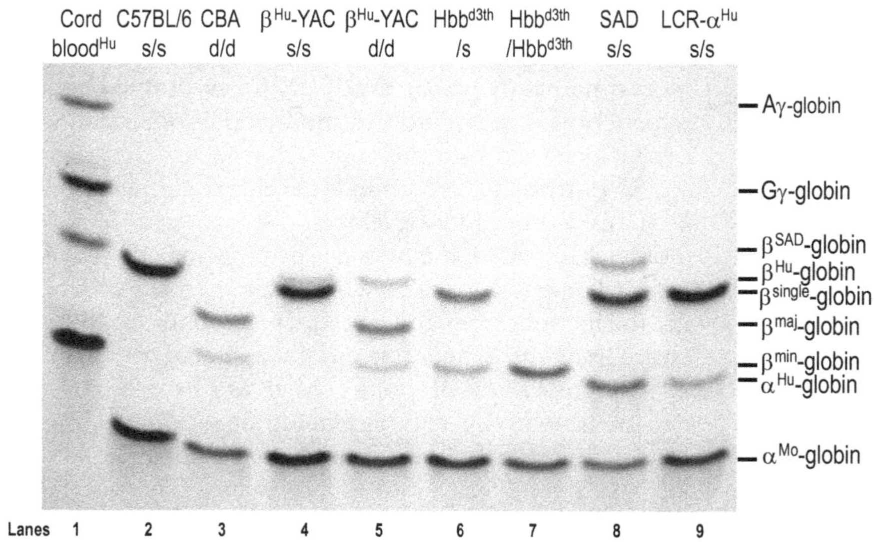

Fig. 2 Mouse and human globin chain analysis. Total blood lysates were separated in denaturing conditions on urea-triton polyacrylamide gel electrophoresis. Adult peripheral blood samples from two inbred mouse strains C57BL/6J (*lane 2*) with the Hbb single haplotype (s/s) and CBA/J (*lane 3*) with the Hbb diffuse haplotype (d/d) were included as standard controls in addition to human fetal control, cord blood (*lane 1*). Blood samples were migrated from various genetically modified genotypes. A β^{Hu}-YAC transgenic mice on the single (*lane 4*) or diffuse (*lane 5*) haplotype allow visualizing the human adult β-globin chain. A thalassemic mouse mutant Hbbd^{3th} heterozygote (*lane 6*) on a single haplotype or homozygote (*lane 7*) provides evidence that the β major globin chain is deleted. SAD transgenic mouse model (*lane 8*) of SCD that contains the human α-globin and β^{SAD}-globin chains on the single haplotype; four transgenic mouse for the human α-globin chain only on the single haplotype (*lane 9*). Mouse and human globin chains are clearly distinguishable from each other. *Right margin*, positions of human and murine globin chains

chains proportionally to their hydrophobicity. After gel densitometry analysis, an imbalance in the α- to β-globin chain ratio due to deficient synthesis or stability results in a thalassemic disorder. In human and murine β-thalassemia, unbalanced globin chain expression is toxic as excess chains precipitate, and release heme and iron that cause oxidative damage. Hematology analysis in thalassemia exhibits target, hypochromic, and microcytic RBC with reduced mean cell hemoglobin (MCH) that contrasts with normal MCH in human sickle cell disease. In sickle cell disease transgenic mice, most models have imbalance of chain synthesis and very low MCH and MCHC, typical of a co-existing thalassemia defect [11–14]. The sickle cell SAD mouse model however displays balanced α- to β-globin chain ratio, normal MCH, irreversibly sickled cells, and vasoocclusions resulting from the primary cause of hemoglobin polymerization, and develops very severe organ pathologies [5, 15, 16]. Thorough analysis of mouse models from complementary approaches is important in diagnosis toward evaluating gene-cell therapy and drug discovery. Guidance in performing different approaches is provided with the following detailed protocols.

2 Materials

2.1 Peripheral Blood Cell Smear

1. Glass microscope slides.
2. Heparinized capillary tubes.
3. Methanol.
4. Wright Stain solution, Giemsa Stain solution, or Wright-Giemsa Stain solution.
5. Light Microscope.

2.2 Hemoglobin Separation by Cellulose Acetate Electrophoresis

1. Heparinized Micro-Hematocrit Capillary Tubes.
2. Crito-seal Capillary Tube Sealant (Fisher).
3. 5× Cystamine solution: 225 mg Cystamine diHCl, 1.0 ml NH_4OH (concentrated stock diluted 1/10), 0.2 ml of 0.1 M DTT, 1.8 ml distilled H_2O. Aliquots can be stored at −20 °C.
4. Micro-Capillary centrifuge, Model MB.
5. Diamond Marking Pencil.
6. 1.5 ml eppendorf Tubes.
7. 5× TBE: 842 mM Tris–HCl, 16 mM EDTA and 517 mM Boric Acid. Adjust pH to 8.5.
8. Migration chamber with applicator, sample well plate, aligning base (Helena Laboratories or others).
9. Disposable wicks for migration chamber.
10. Titan III-H Cellulose Acetate gel, 94 × 76 mm.
11. Power Supply.
12. Staining solution for membranes: 0.5 g Ponceau S, 5 g TCA, and 100 ml H_2O. Store away from light. This solution can be recycled a few times.
13. Plastic containers (*see* **Note 1**).
14. Forceps.
15. 5% Acetic Acid.
16. 100% ethanol.
17. Solution of methanol and acid acetic: 70% methanol and 30% acetic acid.
18. 55 °C incubator.
19. Plastic wrap.

2.3 Globin Chain Separation by Urea-Triton Polyacrylamide Gel Electrophoresis

1. Heparinized Micro-Hematocrit Capillary Tubes.
2. Sample buffer: 3 g Urea, 2.5 ml H_2O, 625 μl Glacial Acetic Acid, 1.25 μl β-mercaptoethanol and 2.5 mg Pyronin Y (optional). This solution must be made fresh each time.
3. 0.9% NaCl.

4. Drabkin solution, purchased or homemade with 50 mg KCN, 200 mg $K_3Fe(CN)_6$, 140 mg KH_2PO_4, 1 ml Triton X-100, water to 1 l. This solution must be kept away from light.
5. Spectrophotometer.
6. Vertical slab gel unit with accessories: a set of glass plates 18 × 16 or 20 × 20 cm, two clamp assemblies and two cams, ten well comb and two spacers 1.5 mm thick.
7. 16G 1½ and 21G ½ needles.
8. 10 and 50 ml plastic syringe needle.
9. 100% ethanol.
10. Parafilm.
11. Urea.
12. Acrylamide:Bis acrylamide (30:0.2): 60 g Acrylamide, 0.4 g Bis acrylamide, 200 ml distilled water. This solution can be stored at 4 °C.
13. 10% Triton X-100.
14. 5% Glacial Acetic Acid.
15. *N*,*N*,*N*′,*N*′-tetramethylethylenediamine (Temed).
16. 10% ammonium persulfate (APS).
17. Hot Plate.
18. 10 and 15 ml Pipettes.
19. 30 °C incubator.
20. 3 M filter paper, 0.35 mm thickness.
21. Cysteamine solution: 0.227 g cysteamine hydrochloride and 1.9 ml of water. This solution should be made fresh each time (*see* **Note 2**).
22. Power supply.
23. Migration buffer: 5% glacial acetic acid.
24. Staining solution: 0.25% coomassie blue R-250, 7% glacial acetic acid and 30% methanol. This solution must be kept away from light and can be recycled a few times.
25. Large Tupperware container.
26. Destaining solution: 5% glacial acetic acid and 25% methanol.
27. Sponges.
28. Drying solution: 2% glycerol and 40% methanol.
29. Cellophane.
30. One glass plate 19 × 20 cm.
31. Clamps.
32. Instrument for densitometric analysis (for example, AlphaImager 2200).

3 Methods

3.1 Peripheral Blood Smear

1. Retrieve peripheral blood from mouse tail, submandibular or orbital sinus veins in heparinized capillary tube.
2. Place a drop of blood at one extremity of a slide.
3. Bring a second slide or a spreader slide, held at an angle (~40° to 50°) on the short edge of the sample slide, toward the drop of blood. This action will spread the drop of blood in a thin line.
4. In a single motion, drag the second slide rapidly and smoothly along the entire length of the sample slide. This should produce a feathered smear.
5. Air-dry the sample slide.
6. Add one drop of 100% Methanol over the cells for fixation and air-dry slides for 5 min.
7. Stain blood cells, using Wright, Giemsa or Wright-Giemsa stain (*see* **Note 1**). Apply a few drops of chosen stain to the slide for 5–15 min.
8. Rinse the slide in distilled water to remove excess stain and air-dry.
9. Add a few drops of phosphate buffer solution on slides for 5 min, rinse with distilled water and air-dry to enhance stain color (optional step).
10. Microscopy visualization for blood and red cell analysis, morphology, e.g., counts of irreversibly sickle cells (ISC) or cells with Heinz bodies. For the identification of the blood cells, refer to **Note 3**.

3.2 Hemoglobin Separation by Cellulose Acetate Electrophoresis

1. Retrieve peripheral blood from mouse tail, submandibular or orbital sinus veins in a capillary tube (>1 cm) (*see* **Note 4**).
2. Seal the capillary by inserting into a Critoseal Capillary Tube Sealant System.
3. Capillaries containing fresh blood cells are centrifuged for 3 min in a Micro-Capillary centrifuge (*see* **Note 5**).
4. Label one 1.5 ml tube containing 50 μl of 1× cystamine solution for each capillary.
5. Cut the capillary (packed RBC) with a Diamond Marking Pencil from the end of the sealant to 0.5 cm length further.
6. Add the 0.5 cm capillary to the 1.5 ml tube containing the cystamine.
7. Mix or vortex each tubes to obtain a homogeneous blood cell suspension out of the capillary before loading the blood on the sample well-plate.

8. Slide the TITAN III-H cellulose acetate gradually into 1× TBE from one end to wet evenly without introducing air bubbles and soak for 5 min.
9. Meanwhile pipet 8 μl of each sample into the Sample Well Plate.
10. Once all samples have been pipetted on the well plate, cellulose acetate is retrieved with flat forceps and blotted between two paper towels to remove excess liquid.
11. Lay down the cellulose on the Aligning Base.
12. The Applicator is aligned in the guides of the well plate and depressed into the wells to pick up aliquots. The applicator is transferred on the aligning base and depressed on the cellulose acetate. This procedure can be repeated three times to optimize sample application to the cellulose acetate.
13. The loaded cellulose acetate is placed sample side down on the paper wicks of the chamber rails. Buffer is added to each reservoir chamber and migration is carried out from negative to positive at 300 V for 30–35 min (*see* **Note 6**).
14. Cellulose acetate membrane is immersed in the Red Ponceau solution for 10 min (*see* **Note 7**).
15. Destain the cellulose acetate in 5% acetic acid for 5 min using a plastic container with occasional agitation. This step is repeated two additional times.
16. Cellulose acetate is then soaked in 100% EtOH in another plastic container for 10 min without agitation.
17. Cellulose acetate is transferred into 70% MeOH/30% Acetic acid solution for 5 min without agitation to become transparent.
18. Cellulose acetate membrane is incubated at 55 °C vertically on an absorbent paper towel for 6–8 min (*see* **Note 8**).
19. When dry, the membrane becomes brittle and should be tightly wrapped in a plastic wrap.
20. Densitometry scanning is performed to determine the proportion of each hemoglobin fraction.

3.3 Globin Chain Separation on Urea-Triton Polyacrylamide Gel Electrophoresis

1. Apply a stretch of Parafilm M on the gasket of the dual casting stand of the gel unit in order to prevent leakage; two gels can be prepared simultaneously.
2. Clean two 18 × 16 cm glass plates with soap and water and then with 100% ethanol.
3. Insert spacers on right and left sides between the two glass plates and ensure that the plates are well aligned.
4. Attach a set of clamps on each side of the glass plates to hold plates together tightly.

5. Assembled glass plates are inserted on one of the gaskets of the dual casting stand.
6. Seal the bottom of the assembled glass plates into the casting stand by turning the black cams. It is important to verify that unit assembly is properly sealed off (*see* **Note 9**).
7. Repeat the same procedure to produce a second gel and insert the assembled glass plate on the basis of the dual casting stand. If only one gel is necessary, use the acrylic buffer dam on the base of the dual casting stand.
8. Separating (resolving) gel solution is prepared by adding 10.8 g Urea and 12 ml Acrylamide:Bis (30:0.2) to 2.4 ml double distilled water. Stir and heat gently until urea dissolves.
9. Add 6 ml 10% Triton X-100, 1.5 ml 5% glacial acetic acid, 0.18 ml Temed, and 0.18 ml 10% APS (*see* **Note 10**).
10. Cast the separating gel solution immediately between the glass plates with a 50 ml syringe. Overlay a small film of water to prevent oxidization and allow to polymerize for 3 h to overnight at 30 °C (*see* **Note 11**).
11. The stacking gel is prepared with: 3.6 g Urea, 4 ml Acrylamide: Bis (30:0.2) and double distilled water to 3 ml. Stir and heat gently until urea dissolves.
12. Add to the mixture 1 ml 10% Triton X-100, 0.55 ml Glacial Acetic Acid, 0.05 ml Temed, and 0.06 ml 10% APS.
13. Remove water over the resolving gel with 3M paper and cast the stacking gel.
14. Insert the comb carefully to avoid bubbles in and around the wells.
15. Cover the comb with wet paper towels and plastic wrap and allow it to polymerize for at least 1.5 h at 30 °C.
16. Remove the comb and transfer the assembled glass plates into the migration apparatus.
17. Pre-run the gel with the migration buffer alone (5% acetic acid) at 200 V (constant voltage), 4 °C for 1 h minimum with inverted anode-cathode setup (*see* **Note 12**).
18. Empty the upper chamber completely and fill the wells with cysteamine solution (approximately 200 μl/well).
19. Carefully add the migration buffer (5% acetic acid) to the upper chamber without displacing the cysteamine solution from the wells.
20. Pre-run the gel at 150 V (constant voltage) for 1 h minimum at 4 °C (or run cold water within the apparatus circulating system) with inverted anode-cathode setup (*see* **Note 12**).

21. Samples should be prepared during the pre-run. Collected peripheral blood cells from mouse tail, submandibular or orbital sinus veins in a capillary tube are flushed out into a 1.5 ml eppendorf containing 200 μl of cold saline. Centrifuge for 2 min at 1500 × *g* (3000–4000 rpm) and wash two additional times with cold saline (*see* **Note 13**).
22. Resuspend the pellet (frozen or fresh) in approximately four pellet volumes of water (usually 60–80 μl). The final volume (Vf) of this lysate should be measured precisely.
23. 20 μl of the lysate is added to 980 μl of Drabkin to determine the optical density at 540 nm. Calculate the final concentration C (μg/μl) = OD540 × 11.9 (*see* **Note 14**).
24. Sample buffer is prepared under the hood by dissolving 6 g of urea in 5 ml of H_2O. Stir and heat gently until urea dissolves (*see* **Note 15**). Add 1.25 ml of glacial acetic acid, 2.5 ml of β-mercaptoethanol and 2.5 mg Pyronin Y (optional). Sample buffer should be made fresh or kept for few days (*see* **Note 16**).
25. To the volume of the lysate (Vf − 20 μl) determined, add 5 volumes of sample buffer. The loading range of protein to visualize globin chains is 5–50 μg and is calculated with the following formula: Volume to load = μg of protein/ Concentration.
26. Empty the upper chamber of the gel apparatus and fill half the upper chamber with fresh migration buffer.
27. Wash the wells at least five times with a 21G ½ needle containing 10 ml of 5% acetic acid to remove all remaining cysteamine.
28. Immediately before loading each well with lysate-sample, wash wells with 5% acetic acid. Once all wells are loaded, fill the upper chamber of gel apparatus with 5% acetic acid.
29. Run at 10 mA (constant current) per gel (20 mA for two gels) for 16–17 h at 4 °C with inverted anode-cathode setup (*see* **Note 12**).
30. Remove the gel from between the glass plates and transfer into a large tupperware.
31. Add sufficient staining solution to cover the gel completely.
32. Stain 1 h with gentle agitation at room temperature.
33. Remove the staining solution. Wash the gel two times with migration buffer to remove excess stain.
34. Add destaining solution and small sponge to the container with the gel and agitate gently for 1 h at 30 °C. Keep the tupperware covered.
35. Replace the destaining solution with fresh solution and leave overnight at room temperature to completely remove background stain. The gel will become transparent.

36. Densitometric analysis is performed to determine the proportion of each globin chain per total chains, ratio of β-like globin per total β-globin chains or ratio of total α-globin per total β-globin chains.
37. Soak the gel in a drying solution and wrap in cellophane that has previously been wet with drying solution.
38. Maintain on a glass plate with clips and dry overnight under the hood.

4 Notes

1. The plastic containers we use for this purpose are the covers of yellow tips boxes.
2. Fresh cysteamine solution is required to prevent disulfide binding of molecule to adult β-chains.
3. Giemsa will color erythrocyte pink, platelet light pink, lymphocyte blue, monocyte cytoplasm pale blue, eosinophil bright pink, leukocyte nucleus magenta.
4. Red blood cells are stored for short term on ice or 4 °C but cannot be kept at room temperature or −20 °C due to high instability.
5. Centrifuge the capillary with the sealed side of the capillary on the outer edge of the capillary rotor.
6. Hemoglobin is negatively charged at this pH and will migrate toward the anode or positive pole. Purified hemoglobin controls for example Hb A, Hb S, HbD Punjab, and other variants can be purchased from commercial sources (e.g., Perkin Elmer, Canterbury) to provide standards for migration distance.
7. Cellulose acetate can be stained more than 10 min but must be kept in the dark. The staining solution (Red Ponceau) can be recycled a few times.
8. Do not over-dry the cellulose acetate, it should be removed when transparent.
9. Assure there is no leakage by adding water between the two glass plates using a syringe with a 16G 1½ needle. Water is then removed with 3M paper.
10. An optional step is to degas the gel mixtures by vacuum for 10 min that will remove small air bubbles and enhance the gel resolution. If this step is carried out, Temed and APS are added after degassing, when ready to polymerize.
11. The separating gel should be <2 cm from the bottom of the wells.

12. Globin protein chains have an overall negative charge in contrast to most proteins. Therefore, the electric field or current should be applied from negative to positive pole for gel electrophoresis. For this purpose, it is important to reverse the cables when plugging into the power supply.
13. Blood cell pellets can be stored at −80 °C or in liquid nitrogen for several months after washes. If the blood samples are frozen, the protocol will continue at Subheading 3.3, **step 22** of the sample preparation.
14. The final concentration is then determined as follows: $C = OD_{540} \div 0.7 \times 50$ (dilution factor 20 μl/1000 μl total volume) ÷ 6 (dilution factor of the lysate with fivefold with sample buffer). The final concentration and the quantity of protein will serve to calculate the volume to load.
15. It is critical to ensure that the urea is totally dissolved before adding other reagents to avoid further precipitation.
16. The sample buffer should be kept in a tube with minimal air to volume ratio since β-mercaptoethanol is reactive with oxygen.

Acknowledgment

This work was supported by Canadian Blood Services and Canadian Institute for Health Research grants to M.T.

References

1. Beauchemin H, Blouin MJ, Trudel M (2004) Differential regulatory and compensatory responses in hematopoiesis/erythropoiesis in a- and b-globin hemizygous mice. J Biol Chem 279(19):19471–19480
2. Blouin M-J, Beauchemin H, Wright A, DePaepe M, Sorette M, Bleau A-M, Nakamoto B, Ou C-N, Stamatoyannopoulos G, Trudel M (2000) Genetic correction of sickle cell disease: insights using transgenic mouse models. Nat Med 6:177–182
3. Felfly H, Trudel M (2010) Successful correction of murine sickle cell disease with reduced stem cell requirements reinforced by fractionated marrow infusions. Br J Haematol 148 (4):646–658
4. Pang CJ, Lemsaddek W, Alhashem YN, Bondzi C, Redmond LC, Ah-Son N, Dumur CI, Archer KJ, Haar JL, Lloyd JA, Trudel M (2012) Kruppel-like factor 1 (KLF1), KLF2, and Myc control a regulatory network essential for embryonic erythropoiesis. Mol Cell Biol 32 (13):2628–2644. doi:10.1128/mcb.00104-12
5. Trudel M, Saadane N, Garel M-C, Bardakdjian-Michau J, Blouquit Y, Guerquin-Kern J-L, Rouyer-Fessard P, Vidaud D, Pachnis A, Romeo P-H, Beuzard Y, Costantini F (1991) Towards a transgenic mouse model of sickle cell disease: hemoglobin SAD. EMBO J 10(11):3157–3165
6. Whitney JB III (1978) Simplified typing of mouse hemoglobin (Hbb) phenotypes using cystamine. Biochem Genet 16(7–8):667–672
7. Stoyanova E, Trudel M, Felfly H, Garcia D, Cloutier G (2007) Characterization of circulatory disorders in b-thalassemic mice by noninvasive ultrasound biomicroscopy. Physiol Genomics 29:84–90
8. Alter BP, Goff SC, Efremov GD, Gravely ME, Huisman THJ (1980) Globin chain electrophoresis: a new approach to the determination of the $^{G}g^{G}g$ ratio in fetal haemoglobin and to studies of globin synthesis. Br J Haematol 44:527–534
9. Beauchemin H, Trudel M (2009) Evidence for bigenic chromatin subdomain in fetal to adult

hemoglobin switch regulation. Mol Cell Biol 29(6):1635–1648

10. Rovera G, Magarian C, Borun TW (1978) Resolution of hemoglobin subunits by electrophoresis in acid urea polyacrylamide gels containing Triton X-100. Anal Biochem 85 (2):506–518
11. Ryan TM, Ciavatta DJ, Townes TM (1997) Knockout-transgenic mouse model of sickle cell disease. Science 278:873–876
12. Paszty C, Brion CM, Manci E, Witkowska HE, Stevens ME, Mohandas N, Rubin EM (1997) Transgenic knockout mice with exclusively human sickle hemoglobin and sickle cell disease. Science 278:876–878
13. Fabry ME, Bouhassira EE, Suzuka SM, Nagel RL (2003) Transgenic mice and hemoglobinopathies. Methods Mol Med 82:213–241. doi:10.1385/1-59259-373-9:213
14. Manci EA, Hillery CA, Bodian CA, Zhang ZG, Lutty GA, Coller BS (2006) Pathology of Berkeley sickle cell mice: similarities and differences with human sickle cell disease. Blood 107(4):1651–1658. doi:10.1182/blood-2005-07-2839
15. Trudel M, De Paepe ME, Chretien N, Saadane N, Jacmain J, Sorette M, Hoang T, Beuzard Y (1994) Sickle cell disease of transgenic SAD mice. Blood 84(9):3189–3197
16. De Paepe ME, Trudel M (1994) The transgenic SAD mouse: a model of human sickle cell glomerulopathy. Kidney Int 46 (5):1337–1345

Chapter 7

Functional Analysis of Erythroid Progenitors by Colony-Forming Assays

James Palis and Anne Koniski

Abstract

The capacity of erythroid-lineage progenitors to form colonies of maturing red blood cells in semisolid media has provided a functional assay for these progenitors and has greatly contributed to our understanding of erythropoiesis. Studies since the 1970s have led to the development of a model of the erythron, whereby the earliest erythroid-committed progenitor, the immature burst-forming unit erythroid (BFU-E), gives rise sequentially to late-stage BFU-E and to colony-forming units erythroid (CFU-E). CFU-E give rise, in turn, to maturing erythroblast precursors that hemoglobinize. It is these terminal cells that comprise the mature colonies of erythroid cells derived from the progenitors cultured in semisolid media. The in vitro generation of erythroid colonies requires cytokine support, most notably erythropoietin (EPO), which is critical for CFU-E survival and for promoting erythroblast maturation.

During mouse embryogenesis, a transient population of primitive erythroid colony-forming progenitors (EryP-CFC) emerges in the yolk sac and gives rise to a wave of maturing primitive erythroblasts in the fetal bloodstream. This wave of EryP-CFC is followed closely by a wave of BFU-E in the yolk sac that enter the bloodstream and seed the fetal liver to generate the first definitive red cells in the fetus. BFU-E in the fetal liver, unlike those in the adult bone marrow, can give rise to colonies in vitro when cultured with EPO alone and also are more sensitive to EPO levels. Here, we describe methods for the in vitro culture of murine embryonic (primitive) and fetal/adult (definitive) erythroid progenitors in semisolid media.

Key words Erythropoiesis, Progenitor, Yolk sac, Colony-forming cell, Erythropoietin

1 Introduction

Hematopoietic stem cells differentiate into multipotential progenitor cells that generate multiple lineage-restricted progenitors that, in turn, mature through progressive precursor stages into mature blood cells. We produce several million new cells every second to maintain steady-state levels of circulating red blood cells, platelets, neutrophils, monocytes, and lymphoid cells. Recent studies indicate that this massive output of mature blood cells is functionally maintained, not by hematopoietic stem cells, but by multipotent hematopoietic progenitor cells [1, 2]. The capacity of hematopoietic progenitors to form colonies of maturing blood cells in

Joyce A. Lloyd (ed.), *Erythropoiesis: Methods and Protocols*, Methods in Molecular Biology, vol. 1698,
DOI 10.1007/978-1-4939-7428-3_7, © Springer Science+Business Media LLC 2018

semisolid media provides a functional assay of these progenitors and has greatly contributed to our understanding of lineage relationships within the hematopoietic system. This colony-forming assay requires the placement of a single-cell suspension of cells in semisolid media to allow for the clonal growth of maturing progeny. Initially, soft agar was used to grow "colony-forming units in culture" (CFU-C), which gave rise to colonies of myeloid cells [3, 4]. This was followed by the use of plasma clots for the growth of colonies of erythroid cells [5, 6]. Currently, methylcellulose is the most commonly used semisolid medium because it supports the clonal differentiation of erythroid [7], as well as multiple myeloid progenitors.

The in vitro growth of colonies in semisolid media also requires the action of cytokines to support the survival and proliferative maturation of plated progenitors. While conditioned media from various cell types was the intial source of growth factors to support hematopoietic colony-forming assays, specific recombinant cytokines are now widely used to support colony formation. Erythropoietin (EPO), cloned in 1985, is the major cytokine that supports erythropoiesis by signaling through its specific receptor expressed on maturing erythroid cells to mediate their survival and multiple other aspects of erythroid maturation [8–10].

The first erythroid colonies recognized in semisolid media were small, compact colonies containing 16–64 late-stage hemoglobinizing erythroblasts (Fig. 1) [5, 11]. These small colonies are derived from colony-forming units erythroid (CFU-E) that peak in number at 2 and 7 days of culture, respectively, for mouse and human systems [5, 12, 13]. CFU-E require optimal concentrations of EPO, as well as the presence of reducing agents, to generate colonies in vitro [11, 14].

Soon after the discovery of CFU-E, immature "burst-forming units capable of erythroid colony formation" (BFU-E) were recognized that generated much larger colonies consisting of 500–10,000 erythroblasts that peak in maturation at 7–10 and 14–20 days of culture, respectively, for mouse and human systems [7, 12, 15]. BFU-E-derived colonies have variable morphology, but typically consist of a collection of one or more large colonies surrounded by smaller colonies that comprise a "burst" (Fig. 1). An intermediate late-stage BFU-E has also been identified that produces intermediate sized erythroid colonies at 3–4 and 10–12 days of culture, respectively, in mouse and human systems (Fig. 1) [12, 14]. The correlation between colony size (immature BFU-E > late stage BFU-E > CFU-E) and the time it takes for the colony to peak in culture, i.e., when it is comprised of terminal, hemoglobinized erythroid cells, supports the concept that in vivo immature BFU-E generate late-stage BFU-E that lie 1–2 cell divisions upstream of CFU-E that, in turn, generate maturing erythroblasts that undergo terminal maturation (Fig. 2).

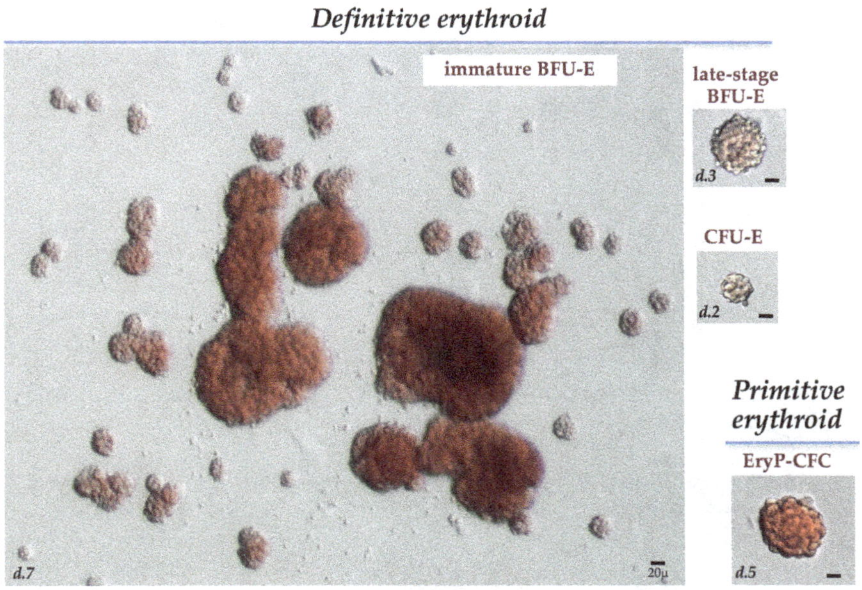

Fig. 1 Examples of colonies derived from murine definitive and primitive erythroid progenitors. Immature BFU-E-derived colonies, counted at day 7 of culture, typically consist of one or more large colonies surrounded by smaller colonies, which constitute a "burst." Late-stage BFU-E-derived colonies, counted at day 3 of culture, consist of smaller single colonies. CFU-E-derived colonies, counted at day 2 of culture, consist of a single small colony containing 16–64 hemoglobinizing cells. Primitive erythroid progenitors (EryP-CFC) derived from the mouse embryo or from differentiating mouse embryonic stem cells generate single compact grape-like colonies of large primitive erythroblasts at 5 days of in vitro culture. *d* day

Unlike CFU-E that only require EPO for cytokine support, the optimal in vitro growth of BFU-E-derived colonies requires a combination of two cytokines—EPO as well as some form of "burst-promoting activity" (BPA). It was subsequently recognized that interleukin 3 (IL-3), GM-CSF, IL-6, or stem cell factor can function as BPA, with IL-3 being the most potent of these [16–18]. CFU-E do not respond to these BPA-containing factors. Interestingly, immature BFU-E numbers in the bone marrow, unlike the numbers of late-stage BFU-E and CFU-E, do not change in response to changes in serum EPO levels [19] or to exogenously administered EPO [20]. Taken together, these findings suggest that immature BFU-E sequentially require burst-promoting activity followed by EPO to generate erythroid colonies in vitro (Fig. 2). These findings also indicate that the initial cellular commitment to the erythroid lineage (i.e., immature BFU-E) occurs prior to the development of EPO-responsive erythroid progenitors (late-stage BFU-E and CFU-E). Small numbers of immature BFU-E, but not CFU-E, normally circulate in the bloodstream [21, 22].

During embryogenesis, the yolk sac is the first site of blood cell development, which is characterized by the appearance of pools of erythroblasts in blood islands. Studies of the spatial and temporal emergence of hematopoietic progenitors in the mouse embryo

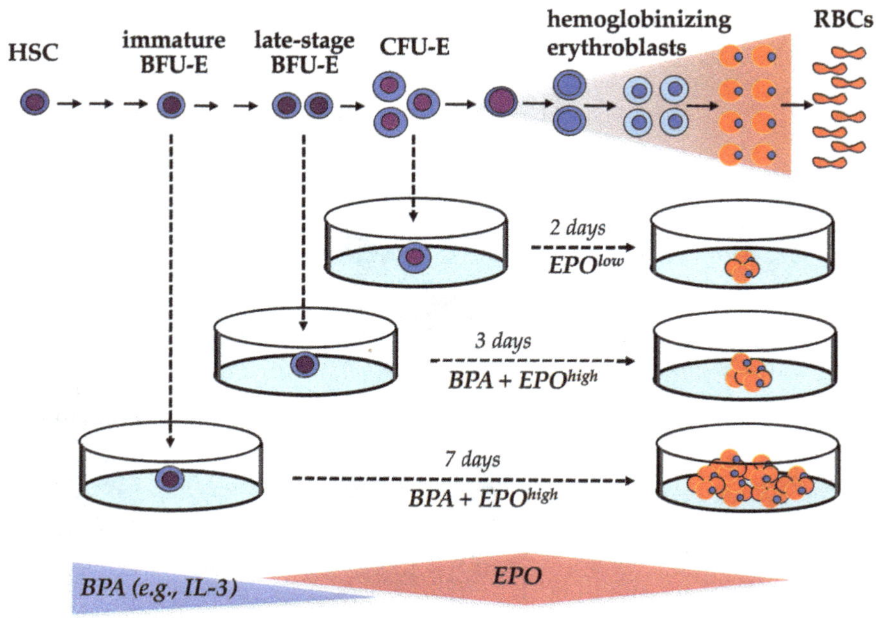

Fig. 2 Schematic overview of definitive erythropoiesis, which consists of the sequential formation of erythroid progenitors, as defined by their potential to form distinctive colonies of erythroid cells in vitro, a wave of hemoglobinizing erythroblasts, defined by morphological criteria, and mature red blood cells. While CFU-E require EPO alone for cytokine support, immature and late-stage BFU-E require both EPO, as well as some form of BPA (e.g., IL-3, IL-6, SCF) to form colonies in vitro. *BFU-E* burst-forming unit erythroid, *CFU-E* colony-forming unit erythroid, *EPO* erythropoietin, *BPA* burst-promoting activity, *RBC* red blood cells, *HSC* hematopoietic stem cells

indicate that the initial commitment to hematopoietic fates occurs soon after the start of gastrulation with two overlapping waves of erythroid progenitors in the yolk sac [23, 24]. The first wave consists of primitive erythroid progenitors (EryP-CFC), which give rise to the embryo's first circulating primitive erythroblasts. When cultured in vitro, murine EryP-CFC generate an intermediate sized colony of large erythroblasts that peaks at day 5 of culture (Fig. 1). The primitive erythroid cells comprising these colonies are large, round cells that express predominantly embryonic hemoglobins [24, 25].

Subsequently, a wave of definitive erythroid progenitors (BFU-E) also emerges in the yolk sac of the mouse embryo prior to the emergence of hematopoietic stem cells from large arterial vessels [23, 24]. Yolk sac-derived BFU-E rapidly enter the newly formed bloodstream and seed the liver where they give rise to large numbers of CFU-E and the first circulting definitive red blood cells in the fetus [24, 26, 27]. CFU-E in the murine fetal liver are more sensitive to EPO than CFU-E in the adult bone marrow [28]. Unlike adult marrow-derived BFU-E that require burst-promoting activity and EPO, fetal liver-derived BFU-E can generate colonies when cultured in EPO alone [29, 30]. Interestingly, in response to acute anemia, adult BFU-E can also give rise to colonies in vitro in the presence of EPO alone [31], consistent with other findings that stress erythropoiesis share features associated with fetal erythropoiesis [32]. Closer examination of erythroid progenitors in the murine fetal liver has led to the identification of two populations of BFU-E, one population that require burst-promoting activity and another population, like adult stress BFU-E, that require EPO alone [33].

This chapter contains detailed materials and methods to assay embryonic, fetal, and adult erythroid lineage-specific hematopoietic progenitors by their in vitro colony-forming potential. While the focus is on the in vitro culture of murine erythroid progenitors, information for the culture of human erythroid progenitors is also provided.

2 Materials

2.1 Mice

For studies of erythroid progenitors during embryogenesis, we primarily use outbred strains of mice (ICR or CD-1), since they have higher fecundity and increased litter size compared to inbred strains (*see* **Note 1**). Timed pregnant mice can be obtained from all major mouse suppliers. Alternatively, in-house colonies of mice can be maintained to generate timed pregnancies (*see* Subheading 3.2 below).

2.2 Laboratory Equipment

1. Dissection forceps (#5 straight forceps with fine tips). Forceps can be sharpened using an Arkansas oil stone and mineral oil.
2. Dissection stereomicroscopes with external fiber-optic light source are available from many manufacturers, including Leica, Nikon, and Olympus.
3. Microscopes.
4. Laminar flow hood.
5. Microcentrifuge and centrifuge for 15 ml conical tubes.
6. Vortexor.
7. 37 °C incubator with 5% CO_2.

2.3 Plastic Ware

2.3.1 Dishes

The 35 mm dishes must be non-tissue culture treated for methylcellulose cultures, but the other dishes can be either treated or non-treated.

1. 35 mm non-tissue culture treated (*see* **Note 2**).
2. 60 mm.
3. 60 mm gridded.
4. 245 × 245 square.

2.3.2 Tubes

1. 1.5 ml microcentrifuge.
2. 15 ml conical centrifuge.
3. 50 ml conical centrifuge.
4. 5 ml round-bottom snap cap.

2.3.3 Syringes

1. 3 ml.
2. 5 ml.
3. 10 ml.
4. 16 gauge needles.
5. 22 gauge needles.

2.4 Buffers, Media, and Supplements

1. Protein Free Hybridoma Medium II (PFHM II; GibcoBRL) (*see* **Note 3**).
2. Glutamax (GibcoBRL).
3. MTG (1-monothioglycerol).
4. 2ME (2-mercaptoethanol, 55 mM stock).
5. PB-2 solution: Dulbecco's phosphate-buffered saline (DPBS) with calcium chloride, 1.0 g Glucose, 3.0 g Bovine serum albumin (BSA, fraction V-Cohn), tissue culture grade water to 1 l. Sterile filter through a 0.2 μm filter and store at 4 °C.
6. 2× Iscove's Modified Dulbecco's Medium (2× IMDM): IMDM powder for 1 l, 10 ml Penicillin/Streptomycin, 3.025 g $NaHCO_3$, 500 ml of tissue culture grade water. Sterile filter through a 0.2 μm filter.
7. 1× IMDM: Dilute 2× IMDM with tissue culture grade water.
8. 1× IMDM/10% PDS (Platelet-poor Plasma Derived Serum).
9. PBS, calcium and magnesium free/1 mM EDTA.
10. 0.25% trypsin and 1 mM EDTA solution in calcium and magnesium-free PBS (*see* **Note 4**). Sterile filter and store at 4 °C.
11. 0.25% Collagenase Type 1 diluted 1:1 with PB2 or 1× IMDM/10% PDS.
12. The final concentrations of specific cytokines used for the culture of mouse and human erythroid progenitors are included in Tables 1 and 2, respectively.

2.5 Sera and Serum Substitutes

Different lots of serum must be tested to ensure appropriate biological activity, especially for primitive erythroid progenitors [23]. Heat inactivation of sera is not necessary.

1. Stem Cell Accelerator-PDS (Platelet-poor Plasma Derived Serum) (Animal Technologies). PDS supports the growth of murine primitive erythroid progenitors [24] (*see* **Note 5**).

Table 1
Media, supplements, and cytokines for the growth of primitive and definitive murine erythroid progenitor-derived colonies in methylcellulose (*see* Note 6)

	Definitive		Primitive
Components	BFU-E	CFU-E	EryP-CFC
IMDM (1×)	To final volume	To final volume	To final volume
BIT	20%	20%	–
PDS	10%	10%	10%
PFHM II	5%	5%	5%
GlutaMAX	1×	1×	1×
IL-6	20 ng/ml	–	–
IL-3	20 ng/ml	–	–
EPO	2 U/ml	0.3 U/ml	2 U/ml
SCF	120 ng/ml	–	–
MTG	–	–	0.15 mM
2-ME	55 μM	55 μM	–
Methylcellulose stock solution	55–65%	55–65%	55–65%

Table 2
Media, supplements, and cytokines for the growth of human erythroid progenitor-derived colonies in methylcellulose

Components	BFU-E/CFU-E
IMDM (1×)	To final volume
BIT	20%
PDS	10%
PFHM II	5%
GlutaMAX	1×
IL-6	5 ng/ml
IL-3	40 ng/ml
EPO	2 U/ml
SCF	100 ng/ml
IGF-1	25 ng/ml
2-ME	55 μM
Methylcellulose stock solution	55–65%

2. BIT 9500 Serum Substitute (Stem Cell Technologies): BSA, insulin and transferrin.

2.6 Preparation of Methylcellulose Stock Solution (See Note 6)

1. Autoclave a 2 l Erlenmeyer flask containing a large stir bar and 27 g of methylcellulose powder (Sigma). Use two layers of aluminum foil over top of flask. Do not use a bottle with a screw cap.
2. While the methylcellulose is still warm, in a laminar flow hood, add 500 ml 37 °C pre-warmed sterile tissue culture grade water. While stirring the methylcellulose mix, carefully bring mixture to a boil. Take care not to burn the methylcellulose on the bottom of the flask. Immediately reduce the heat and stir for 2–4 h. Repeat the boiling process once more and then stir for 2–4 h. Do not allow the methylcellulose to boil over thereby compromising sterility.
3. Pre-warm the 500 ml of sterile filtered 2× IMDM/$NaHCO_3$/Pen/Strep (Subheading 2.4, **item 6**) to 37 °C and add it to the warm methylcellulose mix (37 °C) in the laminar flow hood. Stir at room temperature for 1 h. This results in a 2.6% methylcellulose stock solution.
4. Transfer to the cold room and stir overnight. The cloudy solution will be clear by morning.
5. In a laminar flow hood, sterily pour 30–40 ml of the methylcellulose into 50 ml conical tubes and freeze at −20 °C. Methylcellulose must be frozen and thawed before use.
6. Thaw the methylcellulose stock solution in the refrigerator overnight before use. Never thaw methylcellulose at room temperature or 37 °C. Frozen methylcellulose stock solution is good for up to 1 year in a nonfrost free freezer.

2.7 Cell Enumeration and Staining

1. Hemocytometer or cell counter.
2. Trypan blue stain.
3. Benzidine stock: 2 mg/ml benzidine dihydrochloride in 0.5% acetic acid. This stock solution can be stored in the dark at room temperature for several months. Care must be taken when handling benzidine since it is a carcinogen.
4. Working solution for benzidine staining: 1 ml of the benzidine stock with 3 μl of 50% hydrogen peroxide. The final concentration of hydrogen peroxide is 0.15%. This working solution is best made daily but can be stored for up to 1 week in a dark tube at 4 °C.

3 Methods

Studies of the spatial and temporal distribution of erythroid progenitors and their response to various stimuli and perterbations have been carried out most extensively in the mouse [10]. During embryogenesis, erythroid progenitors first emerge in the yolk sac and are subsequently found in the fetal liver. In the adult mouse, the bone marrow is the predominant site of red cell synthesis with the spleen playing a secondary role in steady state and an expanded role following the induction of acute anemia. Specific hematopoietic tissues are isolated and dissociated into single cells, and the single cells are plated in semisolid media with serum and cytokine (s). Hematopoietic progenitors generate colonies of maturing blood cells that are identified and enumerated on an inverted microscope at specific days of in vitro culture.

3.1 Isolation of Bone Marrow Cells

1. Mice are euthanized as per institutional protocols.
2. Dissect the femur and remove the skin and muscle.
3. Cut off both ends of the femur and flush out the marrow using 4 ml of PB2 with a 22 gauge needle, collecting the cells in a 15 ml tube.
4. Since bone marrow cells will be clumped together, pipette up and down several times through a 1 ml pipette tip to generate a single-cell suspension.
5. To determine the total number of erythroid progenitors per femur, collect and count the marrow cells prior to plating a known number of cells.

3.2 Generation of Timed Mouse Pregnancies

1. Mice are mated overnight by placing one male into a cage with 3–4 females. Mating is assumed to occur at the mid-point of the dark cycle. Vaginal plugs are checked the next morning, with noon of that day defined as embryonic day 0.5 (E0.5) if the dark cycle runs from 6 PM to 6 AM.
2. Staged embryos are derived from timed pregnant mice. At specific times, pregnant mice are euthanized by approved animal care protocols.
3. The uteri are removed from the peritoneum and washed with several changes of PB-2.

3.3 Embryo Dissection and Staging

The dissection strategy depends on the age/stage of the mouse embryos [34].

1. For E6.5–E8.5 conceptuses, the uterine muscle is removed using #5 dissecting forceps and the decidual tissue is split into two halves starting from the mesometrial side. Subsequently, a tear is carefully made in Reichert's membrane and the embryos

are freed from the decidual tissue. The conceptuses are transferred with a pipetteman to a fresh 35 mm dish containing 2–3 ml of PB-2 and completely freed of Reichert's membrane.

2. Fetal livers can be dissected away from the remaining embryo proper starting at E10.5.
3. Because the mouse embryo develops extremely rapidly and there can be wide intra- and inter-litter variation, it is important to determine the stage of the dissected embryos using morphologic criteria (*see* **Note 7**).

3.4 Digestion and Single-Cell Preparation

Progenitor assays depend on plating a single-cell suspension in semisolid media to ensure that each colony is clonal in nature. Collagenase or trypsin each can provide effective cell separation of E6.5–E11.5 embryos with equivalent high cell viabilities and hematopoietic progenitor plating efficiencies (*see* **Note 8**). Embryos older than E11.5 must be dissociated with collagenase. However, as the embryo ages and organogenesis proceeds, it becomes increasingly difficult to obtain complete cell dissociation with good viabilities. The volume of trypsin or collagenase depends on the amount of tissue being dissociated. It is important to have as little of the dissection media present during the digestion with trypsin.

3.4.1 Dissociation of Whole Embryos and Yolk Sacs with Trypsin

1. Embryo proper and yolk sac tissues are treated in 0.25% trypsin/1× PBS/1 mM EDTA at 37 °C for 5–7 min (*see* **Note 4**). 5–15 yolk sacs from E7.5 embryos are dissociated in 0.2 ml of trypsin/EDTA.
2. After 3 min the tissue is triturated with a pipette to aid in digestion and checked microscopically to determine if the tissues are completely dissociated.
3. If dissociation is not complete, the dishes are incubated at 37 °C for another 2–5 min and again triturated.
4. Once dissociated, one half volume of IMDM/10% PDS is added to stop the action of the trypsin.
5. The cells are pelleted by centrifugation, resuspended and live cells counted after staining with trypan blue.

3.4.2 Dissociation of Fetal Livers with Trypsin

1. Fetal livers from E10.5–E12.5 mouse embryos are incubated for 2–3 min in 0.25% trypsin/EDTA and then mechanically disrupted with a 1 ml pipet.
2. The fetal livers of E14.5 or later mouse embryos do not require enzymatic digestion, since mechanical trituration in PBS/EDTA with a 1 ml pipet tip is sufficient to generate a single-cell suspension.

3. Once dissociated, one half volume of IMDM/10% PDS is added to stop the action of the trypsin.
4. The cells are pelleted by centrifugation, resuspended, and live counted after staining with trypan blue.

3.4.3 Dissociation with Collagenase

1. Tissues are incubated in 2.5 mg/ml collagenase Type 1/PB2 at 37 °C for 10 min. 5–15 yolk sacs or embryo propers from E7.5 embryos can be dissociated in 1–2 ml of collagenase/PB2.
2. Add 1 ml PBS/1 mM EDTA, pipette 10× with a 1 ml pipette tip and return to the incubator for another 10 min.
3. Pipette 10× with a 1 ml pipette tip.
4. Pellet by centrifugation, resuspend and count cells after staining with trypan blue.

3.5 Methylcellulose Plating

1. Add reagents for BFU-E, CFU-E or EryP-CFC colony assays in the order listed in Table 1 or 2 to a 50 ml tube (i.e., methylcellulose will be added last with a needleless syringe, *see* **step 2**). Vortex on high for 1 min for complete mixing of the reagents with the viscous methylcellulose. Keep on ice or at 4 °C until use. Allow the bubbles to surface for 15–20 min. before plating.
2. For each sample to be plated, aliquot 1.4 ml of the methylcellulose into a 5 ml tube, using a syringe with a 16-gauge needle since the methylcellulose is very viscous. Keep on ice or at 4 °C until use.
3. It is necessary to optimize the number of cells plated per ml of semisolid media (*see* **Note 9**). Plate enough cells to count between 25–200 CFU-E colonies and 20–50 BFU-E colonies per 35 mm dish. Add cells in 0.14 ml of Pb2 or IMDM/PDS at the appropriate concentration of cells per ml (*see* **Note 10**).
4. Vortex vigorously and allow the bubbles to surface for 15–20 min on ice. This ensures even mixing of the cells in the methylcellulose.
5. Use a 3 ml syringe with a 16 gauge needle to plate 1 ml into each 35 mm culture dish.
6. Swirl the dishes enough to distribute the methylcellulose evenly.
7. Place the 35 mm dishes in the 245 × 245 square dishes with a 60 mm dish filled with sterile water. The increase in humidity helps to prevent desiccation of the methylcellulose.
8. Cultures are incubated at 37 °C, 5% CO_2, for 2–14 days for mouse cultures and 7–20 days for human cultures as needed to enumerate colony numbers (*see* Subheading 3.6).

3.6 Colony Identification and Enumeration

1. The identification of hematopoietic colonies in semisolid media is often difficult, though erythroid colonies are easily recognized by the presence of hemoglobinizing cells (*see* **Note 11**). Murine CFU-E-derived colonies are extremely small with faint red color because of the small number of cells comprising the colonies (Fig. 1). Mouse CFU-E-derived colonies are best enumerated at 2–3 days of culture using a 10× objective. Late-stage mouse BFU-E-derived colonies are larger (Fig. 1) and are enumerated at 3 days of culture. Immature mouse BFU-E-derived colonies typically, but not always, have a distinctive burst morphology and are enumerated at 7–10 days of culture with a 4× objective (Fig. 1). Mouse EryP-CFC-derived colonies are composed of larger primitive erythroblasts and have a more orange-red color (Fig. 1). Murine EryP-CFC-derived colonies are enumerated at 5 days of culture.
2. Place the 35 mm dishes in 60 mm gridded dishes to aid in colony counts by the use of the grid lines.
3. Benzidine staining can be used to identify erythroid colonies or the erythroid component of mixed colonies in methylcellulose [35]. This approach is helpful in enumerating CFU-E-derived colonies and for trouble shooting plates that do not contain red colonies (*see* **Note 12**). Layer 0.5 ml of the benzidine working solution onto the methylcellulose dish. Dark blue-brown erythroid colonies are evident after 5–20 min of incubation at room temperature.

4 Notes

1. Strains of mice differ both in hematopoietic progenitor numbers and in progenitor kinetics during embryogenesis. Individual variability in progenitor numbers can be mitigated by pooling tissues from several similarly staged embryos prior to plating. Because the total number of cells in the embryo increases exponentially, we have found it more useful to compare total hematopoietic progenitor numbers per tissue rather than their frequency [24].
2. Non-tissue culture treated 35 mm dishes (Corning 430588). It is necessary to use non-tissue culture treated 35 mm dishes for plating of hematopoietic progenitors in methylcellulose. The plated cells will stick to the bottom of the wells of tissue culture treated dishes and fail to form colonies.
3. PFHM II: In our experience this reagent has a shelf life of 3–4 months after opening the bottle. We have not been able to aliquot and freeze PFHM II due to a persistent yellow precipitate that forms after freezing. It is pH sensitive and

once it has turned pink it should be discarded. We do not aliquot PFHM II into smaller tubes but just access it from the 500 ml original bottle.

4. Each lot of trypsin must be tested to determine the optimum concentration to maximize dissociation and minimize cell death. We test serial dilutions (1:2–1:64) of the 0.25% trypsin in PBS/1 mM EDTA. We have also found that some lots of trypsin can interfere with colony growth, particularly of primitive erythroid progenitors. Optimum concentrations of trypsin provide for effective cell dissociation and produce minimal cell death.
5. PDS will form a white precipitate upon thawing that can be removed by filtration or letting the particulate matter settle after mixing. The particulate matter does not affect the assay.
6. Each batch of methylcellulose must be tested to determine the final concentration for optimal colony growth. We have found that final methylcellulose concentrations between 1.1 and 1.5% (i.e., 50–60% of the stock solution) work best. Growth factors, serum, and other reagents listed below are added to the stock methylcellulose solution. Premade preparations are now readily available and supplied in a 2.2% concentration with IMDM. Premade preparations that contain only methylcellulose and medium without additional cytokines, supplements or reducing agents can be used to facilitate supplementation to support the growth of specific hematopoietic progenitors. If using a premade stock, add 1× Pen/strep to the mix. Stem Cell Technologies has two products that can be used for human and mouse cultures, M3134 and H4100, respectively.
7. There are established gestational "times" at which specific stages of mouse embryos are typically found. However, there is significant variation in the stage of embryos both within a litter and between litters. We use the criteria of Downs and Davies [36] for presomite stage embryos (E6.5–E8.25), somite counts between E8.25–E10.5, and external morphological criteria between E10.5 and birth [36, 37].
8. We have found that mechanical dissociation with syringes and other non-enzymatic approaches, such as the use of glycine, have led to high levels of cell death when dissociating mouse embryos.
9. If plating density is too high, it becomes difficult to distinguish one colony from another and nutrients can be depleted before the colonies are scored. If plating density is too low, statistically meaningful data becomes difficult to generate. The number of cells plated will vary depending on (1) the colony types to be assayed, (2) the stage of the embryo, and (3) the tissue source of the cells. Plating serial dilutions of the cells is helpful in

establishing an optimum cell number per dish. A good starting range is 5 × 10^3 cells/ml to 1 × 10^5 cells/ml. It is often useful to plate more than one concentration and to count the dishes with appropriate numbers of colonies.

10. We have used an alternative plating strategy when cell numbers are limiting. 1 ml of the methylcellulose mix is plated into each 35 mm dish and the cells, resuspended in 0.1 ml of IMDM/10% PDS, are added directly to the dish. The tip of the pipette is used to mix the cells with the methylcellulose. This method leads to an uneven distribution of cells in the dishes, necessitating scoring of all the colonies in the dish. However, this plating method has the advantage of overcoming the loss of 30–40% of the cells because of the viscous nature of methylcellulose.
11. The identity of the cellular components of an individual colony can be confirmed by transferring the colony onto microscope slides to evaluate cell morphology. The colony is plucked out of methylcellulose, transferred into 0.2–0.3 ml of PBS/50% serum, dispersed by pipetting, and cytospun at 400 RPM for 3–5 min on a Shandon cytocentrifuge. After air drying, the cells on the slide are stained with May-Grunwald-Giemsa or with Wright's stain and examined to determine the hematopoietic cell types present. Erythroid colonies should only contain maturing erythroblasts and some reticulocytes.
12. The growth of primitive erythroid and definitive erythroid colonies requires different reducing agents and differs in the need for BIT. In addition, EryP-CFC-derived cultures are particularly sensitive to several factors, including the trypsin used to dissociate the yolk sac, as well as the activity of the reducing agents and of PFHM II. If one or more of these reagents are not optimal then the primitive erythroblasts will not hemoglobinize well in vitro and the colonies will be a pale red color. We have tried several different suppliers of trypsin and find Worthington to work best for us. The standard reducing agent, beta mercaptoethanol does not support primitive erythroid colony growth.

Acknowledgments

We thank Gordon Keller and Marion Kennedy for sharing so generously of their knowledge and expertise of tissue culture and hematopoietic progenitor colony assays. Scott Peslak and Paul Kingsley photographed the erythroid colonies. This work has been supported by funds from the National Institutes of Health and from the Strong Children's Research Center, University of Rochester, Rochester, NY.

References

1. Sun J, Ramos A, Chapman B et al (2014) Clonal dynamics of native hematopoiesis. Nature 514:322–327
2. Busch K, Klapproth K, Barile M et al (2015) Fundamental properties of unperturbed haematopoiesis from stem cells in vivo. Nature 518:542–546
3. Pluznik DH, Sachs L (1965) The cloning of normal "mast" cells in tissue culture. J Cell Comp Physiol 66:319–324
4. Bradley TR, Metcalf D (1966) The growth of mouse bone marrow cells in vitro. Aust J Exp Biol Med Sci 44:287–300
5. Stephenson JR, Axelrad A, McLeod D, Shreeve M (1971) Induction of colonies of hemoglobin-synthesizing cells by erythropoietin in vitro. Proc Natl Acad Sci U S A 68:1542–1546
6. Heath DS, Axelrod AA, McLeod DL, Shreeve MM (1976) Separation of the erythropoietin-responsive BFU-E and CFU-E in mouse bone marrow by unit gravity separation. Blood 47:777–792
7. Iscove NN, Sieber F (1975) Erythroid progenitors in mouse bone marrow detected by macroscopic colony formation in culture. Exp Hematol 3:32–43
8. Koury MJ, Bondurant MC (1990) Erythropoietin retards DNA breakdown and prevents programmed death in erythroid progenitor cells. Science 248:378–381
9. Wojchowski DM, Sathyanarayana P, Dev A (2010) Erythropoietin receptor response circuits. Curr Opin Hematol 17:169–176
10. Koury MJ (2016) Tracking erythroid progenitor cells in times of need and times of plenty. Exp Hematol 44:653–663
11. Cooper MC, Levy J, Cantor LN, Marks PA, Rifkind AR (1974) The effect of erythropoietin on colonial growth of erythroid precursor cells in vitro. Proc Natl Acad Sci U S A 71:1677–1680
12. Gregory CJ, Eaves AC (1977) Human marrow cells capable of erythropoietic differentiation in vitro: definition of three erythroid colony responses. Blood 49:855–864
13. Eliason JF, Van Zant G, Goldwasser E (1979) The relationship of hemoglobin synthesis to erythroid colony and burst formation. Blood 53:935–945
14. Gregory CJ (1976) Erythropoietin sensitivity as a differentiation marker in the hemopoietic system: studies of three erythropoietic colony responses in culture. J Cell Physiol 89:289–301
15. Axelrad AA, McLeod DL, Shreeve MM, Heath DS (1974) Properties of cells that produce erythrocytic colonies in vitro. In: Robinson WA (ed) Proceedings of the Second international workshop on hemopoiesis in culture, DHEW publication no NIH 74-205. US Government Printing Office, Washington, DC, pp p226–p234
16. Sonoda Y, Yang YC, Wong GG et al (1988) Erythroid burst-promoting activity of purified recombinant human GM-CSF and interleukin-3: studies with anti-GM-CSF and anti-IL-3 sera and studies in serum-free cultures. Blood 72:1381–1386
17. Mitjavila MT, Natazawa M, Brignaschi P et al (1989) Effects of five recombinant hematopoietic growth factors on enriched human erythroid progenitors in serum-replaced cultures. J Cell Physiol 138:617–623
18. Sonoda Y, Sakabe H, Ohmisono Y et al (1994) Synergistic actions of stem cell factor and other burst-promoting activities on proliferation of CD34+ highly purified blood progenitors expressing HLA-DR or different levels of c-kit protein. Blood 84:4099–4106
19. Gregory CJ, McCollough EA, Till TA (1973) Erythropoietic progenitors capable of colony formation in culture: state of differentiation. J Cell Physiol 81:411–420
20. Peslak SA, Wenger J, Bemis JC et al (2012) EPO-mediated expansion of late-stage erythroid progenitors in the bone marrow initiates recovery from sublethal radiation stress. Blood 120:2501–2511
21. Clarke BJ, Housman D (1977) Characterization of an erythroid precursor cell of high proliferative capacity in normal human peripheral blood. Proc Natl Acad Sci U S A 74:1105–1109
22. Hara H, Ogawa M (1977) Erythropoietic precursors in murine blood. Exp Hematol 5:161–165
23. Wong PMC, Chung SW, Reicheld SM, Chui DH (1986a) Hemoglobin switching during murine embryonic development: evidence for two populations of embryonic erythropoietic progenitor cells. Blood 67:716–721
24. Palis J, Robertson S, Kennedy M, Wall C, Keller G (1999) Development of erythroid and myeloid progenitors in the yolk sac and embryo proper of the mouse. Development 126:5073–5084
25. Wong PMC, Chung SW, Chui DHK, Eaves CJ (1986b) Properties of the earliest clonogenic hemopoietic precursors to appear in the

developing murine yolk sac. Proc Natl Acad Sci U S A 83:3851–3854

26. Lux C, Yoshimoto M, McGrath KE et al (2008) All primitive and definitive hematopoietic progenitor cells emerging prior to E10 in the mouse embryo are products of the yolk sac. Blood 111:3435–3438
27. McGrath KE, Frame JM, Fromm GJ, Koniski AD, Kingsley PD, Little J, Bulger M, Palis J (2011) A transient definitive erythroid lineage with unique regulation of the beta-globin locus in the mammalian embryo. Blood 117:4600–4608
28. Zimmermann F, Rich IN (1996) The sensitivity of in vitro erythropoietic progenitor cells to different erythropoietin reagents during development and the role of cell death in culture. Exp Hematol 24:330–339
29. Emerson SG, Thomas S, Ferrara JL, Greenstein JL (1989) Developmental regulation of erythropoiesis by hematopoietic growth factors: analysis on populations of BFU-E from bone marrow, peripheral blood, and fetal liver. Blood 74:49–55
30. Valtieri M, Gabbianelli M, Pelosi E et al (1989) Erythropoietin alone induces erythroid burst formation by human embryonic but not adult BFU-E in unicellular serum-free culture. Blood 74:460–470
31. Perry JM, Harandi OF, Paulson RF (2007) BMP4, SCF, and hypoxia cooperatively regulate the expansion of murine stress erythroid progenitors. Blood 15:4494–4502
32. Alter BP (1979) Fetal erythropoiesis in stress hematopoiesis. Exp Hematol 7(Suppl 5):200–209
33. Porayette P, Paulson RF (2008) BMP4/Smad5 dependent stress erythropoiesis is required for the expansion of erythroid progenitors during fetal development. Dev Biol 317:24–35
34. Hogan BLM, Beddington RSP, Constantini F, Lacy E (1994) Manipulating the mouse embryo: a laboratory manual, 2nd edn. Cold Spring Harbor Laboratory Press, Plainview
35. Hara H, Ogawa M (1978) Murine hematopoietic colonies in culture containing normoblasts, macrophages, and megakaryocytes. Am J Hematol 4:23–34
36. Downs KM, Davies T (1993) Staging of gastrulating mouse embryos by morphological landmarks in the dissecting microscope. Development 118:1255–1266
37. Kaufman MH (1992) The atlas of mouse development. Academic, New York

Chapter 8

Analyzing the Formation, Morphology, and Integrity of Erythroblastic Islands

Jia Hao Yeo, Matthew P. Cosgriff, and Stuart T. Fraser

Abstract

The bone marrow is the primary site of erythropoiesis in healthy adult mammals. In the bone marrow, erythroid cells mature within specialized microenvironments termed erythroblastic islands (EBIs). EBIs are multi-cellular clusters comprised of a central macrophage surrounded by red blood cell (erythroid) progenitors. It has been proposed that the central macrophage functions as a "nurse-cell" providing iron, cytokines, and growth factors for the developing erythroid cells. The central macrophage also engulfs and destroys extruded erythroid nuclei. EBIs have recently been shown to play clinically important roles during human hematological disease. The molecular mechanisms regulating this hematopoietic niche are largely unknown. In this chapter, we detail protocols to study isolated EBIs using multiple microscopy platforms. Adhesion molecules regulate cell-cell interactions within the EBI and maintain the integrity of the niche. To improve our understanding of the molecular regulation of erythroid cells in EBIs, we have developed protocols for immuno-gold labeling of erythroid surface antigens to combine with scanning electron microscopy. These protocols have allowed imaging of EBIs at the nanometer scale, offering novel insights into the processes regulating red blood cell production.

Key words Erythroblastic islands, Erythropoiesis, Immuno-gold labeling, Bone marrow hematopoietic niche, Electron microscopy

1 Introduction

1.1 Erythroblastic Islands Are Specialized Niches for Mammalian Erythropoiesis

The bone marrow becomes the dominant site for definitive erythropoiesis shortly before birth in both humans and mice [1]. In the bone marrow, there are specialized niches that are involved in the development of hematopoietic lineage-specific precursor and progenitor cells. Erythroblastic islands (EBIs) are one of these specialized niches found in the bone marrow where mammalian erythroblasts proliferate and differentiate. Discovered by Marcel Bessis in the 1950s, EBIs were described as micro-structures consisting of developing erythroblasts at various stages of differentiation which form rings around a central macrophage, with the outer ring comprised of more mature erythroid progenitors than the inner ring (Fig. 1) [2, 3].

Joyce A. Lloyd (ed.), *Erythropoiesis: Methods and Protocols*, Methods in Molecular Biology, vol. 1698, DOI 10.1007/978-1-4939-7428-3_8, © Springer Science+Business Media LLC 2018

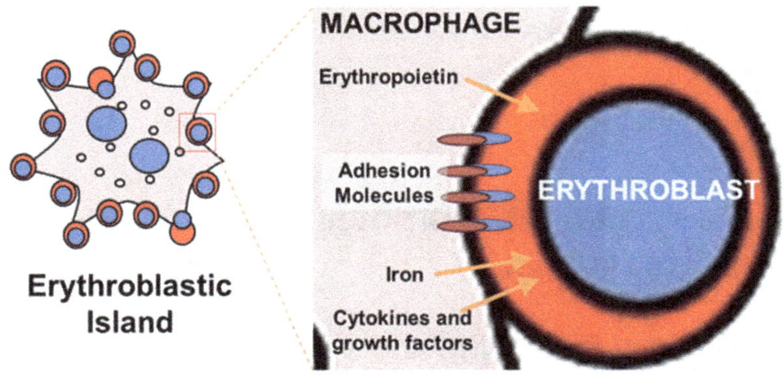

Fig. 1 A schematic diagram illustrating the proposed functions of central macrophages in erythroblastic islands. These macrophages were thought to act as a "nurse cell," providing iron for heme synthesis, erythropoietin, cytokines, and growth factors crucial for proliferation and differentiation

Table 1
Distribution of erythroblastic islands across species

Species (classification)	Tissue	References
Humans (*Homo sapiens*)	Bone marrow Fetal liver Yolk sac	Lee et al. [11], Crocker et al. [14] Crocker et al. [14] Takashina [15]
Monkey (*Macaca mulatta*)	Bone marrow	Switzer [12], Glomski and Pica [13]
Mouse (*Mus musculus*)	E14.5 Yolk sac Fetal liver Bone marrow Spleen	Isern et al. [8] Isern et al. [8] Yeo et al. [10] Sadahira et al. [16]
Rat (*Rattus rattus*)	Bone marrow Spleen	Bessis [2], Berman et al. [17] Mohandas et al. [4] Rassokhin [18]
Guinea pig (*Cavia porcellus*)	Bone marrow Yolk sac	Yeo et al. [10] Sorenson [19]
Dunnart (*Sminthopsis crassicaudata*)	Bone marrow	Yeo et al. [10]
Chicken (*Gallus gallus*)	Bone marrow	Crocker et al. [14]

EBIs have been isolated from a limited number of species. In mice and rats, EBIs have been isolated from adult bone marrow, spleen, and the fetal liver ([2, 4–10]. We have recently reported isolation of EBIs from adult guinea pig (*Cavia porcellus*) bone marrow [10]. EBIs have also been isolated from human bone marrow aspirates [11] and from the bone marrow of Rhesus macaques (*Macaca mulatta*) [12, 13]. We recently asked whether EBIs can be isolated from bone marrow of non-eutherian mammal species. EBIs could be isolated from the marrow of the fat-tailed dunnart (*Sminthopsis crassicaudata*), a carnivorous marsupial [10]. The isolation of EBIs from different species is listed in Table 1.

Earlier studies reported that there are no differences in EBI morphology between humans, rats, and mice [17]. However, the number of erythroblasts surrounding islands in each species is variable. Tissue sections from rat femur reveal about ten cells per island based on 0.5 μm serial sections [4], while isolated macrophages from human bone marrow aspirates range from 5 to 20 erythroblasts per macrophage [11]. We have previously reported that there are two morphologically distinct EBI classes: domed and flat islands [10]. Flat islands are observed to interact with a greater number of reticulocytes compared to domed islands [10].

EBIs were shown to be distributed uniformly throughout the bone marrow as determined by three-dimensional reconstruction of rat bone marrow from 0.5 μm thick serial sections [4]. These EBIs are composed of erythroid cells in various stages of differentiation, ranging from CFU-E to young reticulocytes [4]. Immature myelomonocytic cells can also be found binding to these central bone marrow macrophages [11]. Quantitative microscopic studies of rat bone marrow show that EBIs adjacent to the sinusoids involve more mature erythroblasts as compared to those more distally located EBI [6].

1.2 Proposed Functions of Erythroblastic Islands

The first proposed function of the central macrophage of an EBI was to provide iron to surrounding erythroblasts, acting as a "nurse-cell." Bessis suggested that iron was transferred to the erythroblast by "rhopheocytosis" from the macrophage [20]. However, there is still lack of evidence directly showing this proposed iron transport. Early studies have demonstrated that EBI macrophages contain iron as determined by Perl's staining [3]. These central macrophages also contained heme oxygenase-1 (Hmox-1). No EBIs could be isolated from Hmox1-deficient mouse bone marrow [21]. This evidence suggests a regulatory role of EBI macrophages in iron and heme metabolism.

Erythropoietin (Epo) is a 35 kDa glycoprotein hormone critical for stimulating erythropoiesis. Under hypoxic conditions, an oxygen-sensing mechanism triggers the synthesis of Epo in the kidney. As Epo binds to the Epo receptor on erythroblasts in the bone marrow, intracellular signaling pathways favoring proliferation will be triggered. This prevents apoptosis in erythroblasts resulting in increased erythrocyte production [22–24]. Low amounts of EPO mRNA are detected in the central macrophages of the EBI by in situ hybridization [25]. Cytokines, chemokines, and interleukins also inhibit the proliferation of erythroid progenitors in vitro. It was suggested that these factors may be present at high local concentrations within the erythroid niche to regulate erythropoiesis and may possibly be secreted by the central macrophages [26]. Recent studies show that injection of either human recombinant Fms-like tyrosine kinase 3 ligand (FLT-3-L) or Granulocyte-Colony Stimulating Factor (G-CSF) into mice reduces the phenotypic F4/80$^+$VCAM$^+$CD169$^+$ER-H3$^+$Ly6G$^+$ macrophages in the medullary bone marrow [27, 28].

The most conspicuous function of EBI macrophages is their active phagocytosis of extruded erythroid nuclei within the niche. The expulsion of the nucleus is a characteristic feature of erythropoiesis in mammals, known as enucleation. Orthochromophilic erythroblasts (late erythroblasts) eliminate their highly condensed nucleus by nuclear extrusion. Time-lapse imaging and immunocytochemistry have demonstrated engulfment and destruction of the erythroid nucleus by EBI macrophages [8, 29]. Exposed phosphatidylserine (PS) on the extruded nuclei is detected by T-cell immunoglobulin-and-mucin-domain-containing molecule (Tim-4) present on macrophages [30, 31]. Engulfment is regulated by a MerTK-protein S-dependent mechanism [29].

The central macrophage then destroys the engulfed erythroid nuclei. DNase II is a key enzyme in degrading engulfed nuclei [32]. Engulfed nuclei in DNase II null macrophages, in contrast, are not degraded and accumulate in the macrophage. Undegraded nuclei appeared to be toxic as there were fewer macrophages observed [33]. DNase II null mutations cause severe anemia and embryonic lethality [34]. The transcription factor EKLF/Klf1 regulates DNAse II expression and hence EBI function.

Reticulocytes remain adhered to the macrophages in the bone marrow of rats [35]. These findings coincide with our own observations of EBIs in mouse and guinea pig [10]. Early literature describes distinct reticulocyte morphological changes [3], suggesting that there are different stages of reticulocyte development. It is unclear if there is a correlation between these previously described different stages of reticulocytes and adherence to EBI macrophages, prior to maturing into erythrocytes.

1.3 Interactions Between a Central Macrophage and Developing Erythroblasts

In the bone marrow, proliferation and differentiation of progenitor cells are mediated, not only by soluble factors, but also by adhesion to other cells and extracellular matrix molecules. The formation and integrity of EBIs involves multiple adhesive interactions. To date, only three adhesion molecules have been found to be critical in anchoring erythroid cells to macrophages in the EBI. Blocking any of these interactions disrupts EBI formation in vitro. Unlike the Tim-4-phosphatidylserine proteins involved in engulfing erythroid nuclei described earlier, the molecular mechanism and precise role of these adhesive interactions in regulating erythropoiesis remain largely unclear.

One set of receptors-counter receptors (ligands) identified as mediating cell-cell interactions in EBIs is the α4β1 integrin (also known as Very Late Antigen-4; VLA-4) on erythroblasts binding to VCAM-1 expressed on the EBI central macrophage. Disruption of this interaction by antibodies binding to either α4β1 integrin or VCAM-1 disturbs the formation of EBIs [36]. The importance of this interaction was demonstrated in vivo by injecting anti-VCAM-1 antibodies into mice with transplanted bone marrow. Impaired recovery of erythroid progenitors in the bone marrow was observed in these mice.

The intercellular adhesion molecule-4 (ICAM-4) expressed on erythroid cells interacts with αV integrin on central EBI macrophages. Blocking ICAM-4-αV integrin binding with αV-synthetic peptides resulted in a 70% decrease in islands reconstituted in vitro. ICAM-4-null mice have a 50% decrease in island formation in the bone marrow; however, steady-state erythropoiesis is not adversely affected in adult mice [37]. β3 integrins play a role in mediating this interaction [38].

Erythroblast macrophage protein (EMP) is a 36-kDa transmembrane protein expressed on both erythroblasts and macrophages [39–41]. Homophilic binding of EMP mediates interactions between erythroblasts and macrophages [39, 40]. EMP-null mouse fetuses are severely anemic and die in utero [41]. Erythroblasts cultured in the presence of anti-EMP antibody show a sixfold increase in apoptosis and a marked decrease in erythroid proliferation, maturation, and enucleation [40]. EMP-null erythroblasts fail to enucleate even in the presence of macrophages as EMP regulates the actin cytoskeleton in both macrophages and erythroblasts. EMP co-localizes with concentrated F-actin bundles in wild-type erythroblasts undergoing enucleation [41]. EMP expression is also upregulated on the mononuclear cells of patients who have had a recent bone marrow transplant. Coincidentally, higher numbers of EBIs were also found on bone marrow smears of these patients [42]. While EMP was first identified in EBIs, EMP expression is not exclusive to EBIs [43].

1.4 Clinical Significance of Erythroblastic Islands

The role of EBIs in human health is a poorly defined area of research. Chow and colleagues found that depletion of bone marrow macrophages results in a less severe phenotype in a mouse model of polycythemia vera [44]. Similarly, the severity of a mouse model of β-thalassemia was ameliorated by the depletion of bone marrow macrophages [45]. Very recently, it has been recognized that the poor formation of EBIs is linked to a poorer prognosis in myelodysplastic syndrome patients [46]. Conversely, patients with immune-related pancytopenia and hereditary hemophagocytic histiocytosis present with abnormally activated macrophages and erythrophagocytosis [47].

1.5 Systems for Analyzing EBI at the Micrometer and Nanometer Scales

1.5.1 Light and Fluorescence Microscopy

Early studies of EBIs relied primarily on both light microscopy and stained blood smears (mainly phase contrast) [48]. Observations made under microscopes by relatively simple manipulation of light, such as phase contrast and Normarski microscopy, provided early descriptions of cell dynamics [48]. By overcoming the light diffraction limit, many techniques have evolved from traditional wide-field fluorescence. Currently, wide-field deconvolution microscopy and conventional confocal microscopy are the gold standards in optical bioimaging. One recent development in fluorescence imaging is light sheet microscopy. Light sheet imaging typically visualizes fluorescently labeled tissues cleared by replacing water with a highly refractive index substance in a "wider field of view" compared to other fluorescence techniques. This technique has been extremely valuable in visualizing micro-vasculature in cleared tissues or organs. Light sheet imaging has recently demonstrated non-dividing hematopoietic stem cells residing near perisinusoidal areas in BABB-cleared (Murray's clearing; 1:2 Benzyl Alcohol:Benzyl Benzoate; BABB) mouse tibia [49]. To the best of our knowledge, there is no known light sheet imaging of EBIs in cleared bone marrow.

Live-cell imaging has improved over the years and enables visualization of molecular processes within the niche in real time. The two-photon microscopy enables imaging of living tissue up to 1 mm in depth with reduced out-of-focus fluorescence, providing inherent spatial resolution. It differs from traditional fluorescence microscopy, as it involves simultaneous absorption of two photons (typically a long wavelength such as the infra-red) by the fluorescent dyes to the next excitation state. Multi-photon microscopy is well suited to imaging live cells as it results in less damage than the short-wavelength lasers typically used for single-photon excitation: cells may be observed for longer periods with fewer toxic effects. Super-resolution microscopy can reach a resolution comparable to electron microscopy. The classification of these platforms can vary greatly in the literature such as near-/far-field and "true"/functional resolution. These microscopy techniques are extremely valuable for future studies of the erythroid niche.

1.5.2 Electron Microscopy

Before the development of super-resolution microscopy, electron microscopy served as the sole microscopic technique for imaging tissues or cells with a resolution enabling visualization of an entire cell with its subcellular components. Traditionally, the nomenclature for segregating different hematopoietic cells (mainly in the peripheral blood) is based on individual staining properties. Electron microscopy, particularly the transmission electron microscopy (TEM), revolutionized our understanding of cell biology. TEM can achieve a high resolution for visualizing cells with their subcellular components on ultrathin sections as thin as 70 nm. As a secondary technique developed from TEM, the scanning electron microscope (SEM) provides morphological insights based on ultrastructural topology. SEM has been used to study isolated bone marrow cell clusters [50].

In recent years, novel techniques such as Scanning Transmission Electron Microscopy (STEM), Focused-Ion-Beam SEM, and Serial-Block-Face SEM have been developed. These techniques have reduced the level of difficulty faced in sample preparation for traditional electron microscopy. Specialized image analyses (such as reconstructed three-dimensional volume rendered models) performed on images generated from these techniques have increased our understanding of subcellular components with an appreciation of three-dimensionality of cell biology.

1.5.3 Other Imaging Platforms

In addition to light and electron microscopy, there are also other forms of imaging that have contributed to biology. While these technologies have not been applied to study the erythroid niche, they have been used to analyze haematopoietic cells. Reconstruction from X-ray micro-tomography (MicroCT) contributed to our understanding of the bone marrow endosteal niche [51]. Accurate morphological analyses of bone marrow cells from acute myeloid leukaemia can be obtained from "label-free" Raman-scattering-based imaging [52]. Atomic force microscopy provides topological analyses of human erythrocytes [53–56] and erythrocytes from various animal species [57, 58].

In this chapter, we describe a relatively simple method to study EBI morphology using the light microscope or scanning electron microscope. In the later sections of this chapter, we highlight the major artifacts and how to recognize them, with special reference to our isolation protocol and EBIs.

2 Materials

2.1 Dissecting Equipment and Disposables

1. Dissecting scissors.
2. Forceps.
3. 5 mL syringes.
4. Terumo 25-gauge needles (0.5 mm × 25 mm).
5. 15 mL conical polypropylene tubes.
6. 3 mL sterile transfer pipettes.
7. 12-well tissue culture dish.
8. 15 mm glass coverslips.
9. Microscope Slides 26 × 76 mm with Ground Edges and 20 mm Frosted Ends.

2.2 Specialized Equipment

1. Confocal Microscope (Leica SPE-II or similar).
2. Hitachi Field S4500 Field Emission Guns—Scanning Electron Microscope (or similar).
3. EMITECH K500X Sputter Coater fitted with sputter coater target made of gold (Quorum Technologies Ltd, Lewes, United Kingdom).
4. Zeiss Sigma Variable Pressure Field Emission Guns—Scanning Electron Microscope (VP FEG-SEM) equipped with a sensitive

Gatan backscatter electron detector (Gatan Inc., Pleasanton, CA, USA).

5. EMITECH K950X Carbon Turbo Evaporator fitted properly with carbon rods of 3.0 mm in diameter (Quorum Technologies Ltd).

2.3 EBI Assay: Isolation of Multicellular Clusters from Mammalian Bone Marrow

1. EBI medium—For 1 L: 860 mL of Dulbecco's Modified Eagle's Medium (DMEM), 20 mL of 100× GlutaMAX™ Supplement (Gibco), 10 mL of 100× Penicillin/Streptomycin (100× Pen/Strep contains 10,000 units penicillin and 10 mg streptomycin/mL), 12.6 μL Mono-ThioGlycerol (15.75 mg/L final concentration), 10 mL of 100 mM Sodium Pyruvate (1 mM final concentration), 100 mL of heat-inactivated Fetal Bovine Serum (*see* **Note 1**). Sterilize by filtering through a 0.22 μm filter before use.
2. 70% Ethanol.
3. Fetal Bovine Serum.
4. Phosphate-Buffered Saline (PBS), pH 7.4.

2.4 Immunofluorescent Labeling of Erythroblastic Islands for Confocal Fluorescence Microscopy

1. 4% paraformaldehyde (PFA) in PBS. Mix 4 g paraformaldehyde in PBS at 65 °C until thoroughly dissolved. Cool before use.
2. Blocking buffer: PBS with 0.01% Triton-X-100 (v/v) and 0.05% Bovine Albumin Serum (w/v).
3. Washing buffer: PBS with 0.05% Triton-X-100 (v/v) (PBST).
4. Vectashield with DAPI (Vector Labs).
5. Rat-anti-mouse primary antibody (Ter-119). For alternative markers in the EBI, please *see* Table 2.
6. Goat-anti-rat secondary antibody conjugated to AlexaFluor®576.

Table 2
Markers of erythroblastic islands populations in the laboratory mouse

Cell type	Markers	Cell/structure identified by marker
Erythroblast	Ter-119	Erythroid cells from proerythroblast to mature erythrocyte
	CD71	Transferrin receptor, expressed highly on immature erythroid cells
	α4β1 integrin	Erythroblasts (not restricted to erythroid cells)
Macrophage	F4/80	Macrophage surface marker
	CD169	Macrophage surface marker
	CD11b/Mac1	Myeloid surface marker including macrophages
	Tim4	Phosphatidylserine receptor expressed on macrophage subsets
	Ly6G	EBI macrophages (in combination with other antigens)
	ER-HR3	EBI macrophages (in combination with other antigens)
	VCAM1	Receptor for α4β1integrin
	Forssman Antigen	EBI macrophage marker: surface carbohydrate structure
	Heme Oxygenase-1	Cytoplasmic enzyme present in EBI macrophages
	Phalloidin	Actin (not restricted to macrophages but a useful counterstain)

2.5 Visualizing Erythroblastic Islands by Conventional Scanning Electron Microscopy

1. 2.5% glutaraldehyde: Do a 1:10 dilution of 25% glutaraldehyde in PBS.
2. 1% osmium tetroxide in 0.1 M PBS: Do a 1:1 dilution of 2% osmium tetroxide stock solution with 0.2 M PBS.
3. 30, 50, 70, 95, and 100% Ethanol.
4. Hexamethyldisilazane (HMDS).

2.6 Immunogold-Labeling of Developing Erythroid Cells for Scanning Electron Microscopy

1. 0.5% glutaraldehyde (Dilute 25% glutaraldehyde 1:50 in PBS).
2. PBS with 0.1% (w/v) glycine (add 0.1 g of glycine to 100 mL of PBS).
3. 2.5% glutaraldehyde (dilute 25% glutaraldehyde 1:10 in PBS).
4. HMDS.
5. Colloidal gold-conjugated secondary antibody goat-anti-rat IgG (H&L), EM-grade, 10 nm.
6. Unconjugated rat-anti-mouse primary antibody (Ter-119).
7. 30, 50, 70, 95, and 100% Ethanol.
8. PBS, pH 7.2.

3 Methods

3.1 EBI Assay: Isolation of Multicellular Clusters from Mammalian Bone Marrow

EBIs are multicellular clusters consisting of a central macrophage surrounded by developing erythroid progenitors. Here, we describe an optimized protocol for isolating EBIs from adult laboratory mouse bone marrow (*see* **Note 2** on isolating EBIs from various mouse strains). We have also successfully isolated EBIs from other mammals with this protocol [10]. Figure 2 depicts a schematic diagram of the EBI assay.

1. Prior to performing assay, store 15 mm glass coverslips in absolute or 70% ethanol overnight (*see* **Note 4**).
2. To isolate adult mammalian bone marrow, euthanize animal according to local animal ethics committee guidelines. We typically euthanize mice by cervical dislocation.
3. Soak the animal body with 70% ethanol and transfer to a sterile biosafety cabinet.
4. Make a midline incision and peel back the skin. Expose the upper leg via further dissection.
5. Remove muscle (*quadriceps femoris*) around the femur.
6. Cut the end of the femur adjacent to the pelvis, and then the knee joint (Tip: dislocate the femur-pelvis hip joint by twisting the femur for easy dissection).
7. Cut the two trabecular regions of each end of the femur.

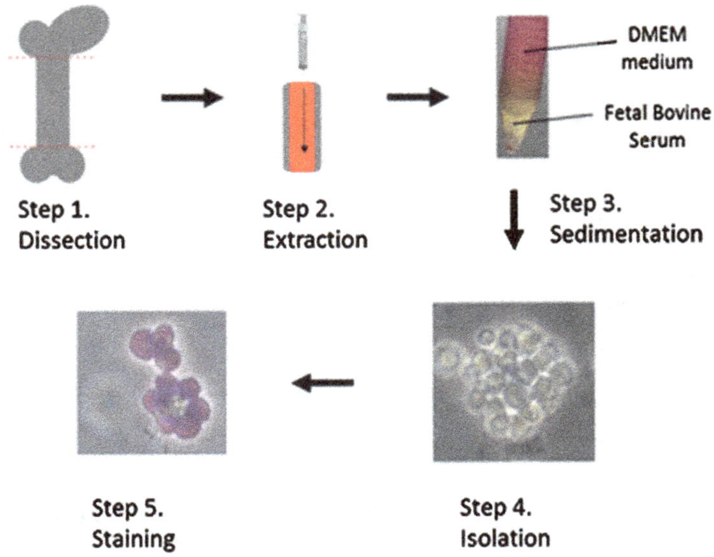

Fig. 2 Schematic diagram of EBI assay. Schematic diagram showing an overview for an EBI assay. Refer to the text for detailed description

8. Gently flush the red bone marrow out of the femur into a 15 mL tube by inserting a 25-gauge needle attached to a 5 mL syringe filled with EBI medium into either end of the femur.
9. Repeat **step 8** until the bone is translucent with no reddish tinge (marrow).
10. In a sterile biosafety cabinet, lightly resuspend marrow tissue using a sterile transfer pipette.
11. Transfer the medium containing marrow cells (approximately 2 drops/s) to a 15 mL conical tube containing ~1.5 mL Fetal Bovine Serum.
12. Rest for 45 min to allow for the multi-cellular clusters to sediment to the bottom of the tube.

    Note: Do not tilt the tube or attempt to resuspend contents within the tube. The serum is a viscous medium to allow cell clusters to sediment.
13. Prepare sterile 15 mm glass coverslips onto each well of a 12-well culture dish (1 coverslip/well). Place coverslips at a 45° angle to allow ethanol to dry.
14. After 45 min, remove serum and medium using a sterile transfer pipette. Be careful not to disturb cells resting at the bottom of the tube (*see* **Note 3**).
15. Resuspend cell clusters in ~250 μL of EBI medium.
16. Gently add the entire volume of cell suspension onto one 15 mm glass coverslip, forming a bubble-like meniscus on the

coverslip. Use a plastic transfer pipette with a wider gauge than a pipette tip to prevent disrupting clusters while plating the cells onto the coverslip. Take care in moving the 12-well dish to the 37 °C, 5% CO_2 incubator to prevent disrupting the meniscus on the coverslip.

17. Allow adherence of marrow cells to coverslip at 37 °C with 5% CO_2 for 90 min.
18. Wash unbound cells with sterile PBS or fresh EBI media gently. Typically, we gently add PBS to each well by touching the wall with the transfer pipette and then releasing the PBS. This prevents the coverslip from receiving the full force of the washing buffer. We have found that dipping the coverslips to wash off unbound cells can result in the loss of cell clusters with results varying greatly between personnel performing this assay. Under a phase contrast microscope, only cell clusters comprised of more than six cells are considered EBIs.

3.2 Cytospin Centrifugation

1. Dissect the mouse and flush the bone marrow as described in Subheading 3.1.
2. Spin the flushed bone marrow cells at 163.8 × *g* for 5 min.
3. Remove the supernatant containing EBI media, and resuspend the pellet containing the bone marrow cells in 2 mL of PBS.
4. Assemble the Cytoclips™, glass slides and the Single White Cytofunnels™ as described by the manufacturer.
5. Aliquot 100 mL of resuspended marrow cells into the Cytofunnel™ (to obtain a cell count of approximately 5000–10,000 cells/mL).
6. Place the assembly into the Cytospin™ 4 instrument and assure the clips are evenly placed.
7. Perform the centrifugation at 200 rpm for 5 min on the Cytospin™ 4 instrument.
8. Disassemble the assembly immediately once the cytospin is done (to obtain the best morphology possible) and leave slides to dry.

3.3 May-Grünwald Giemsa Staining of Multicellular Clusters

1. Dissect the mouse and flush the bone marrow as described in Subheading 3.1.
2. Perform EBI assay described in Subheading 3.1 or Cytospins described in Subheading 3.2.
3. Fix the cells with absolute methanol for 3 min at room temperature.
4. Allow coverslips to dry completely.

5. Stain cells with diluted May-Grünwald Giemsa stain (1:20 concentration; ~2 mL per well in a 12-well dish) for 20 min at room temperature.
6. Wash the cells with tap water (3 × 5 min/wash).
7. Visualize the coverslips in the dish under light microscopes directly under dry conditions. (Alternatively, invert the coverslip and mount the cell clusters onto a glass slide with oil-based mounting medium.)

3.4 Immunofluorescent Labeling of Erythroblastic Islands for Confocal/Fluorescence Microscopy

1. Perform EBI assay as described in Subheading 3.1.
2. EBIs are then fixed with 4% PFA for 20 min, and washed with PBS (3 × 5 min/wash).
3. Block coverslips for 30 min in blocking buffer.
4. Dilute the unconjugated rat-anti-mouse primary antibody (Ter-119) with PBST (1:200) and label the erythroid cells with 400 μL per coverslip for 2 h.
5. Wash the cells with PBST (3 × 5 min/wash).
6. Dilute goat-anti-rat secondary antibody conjugated to Alexa-Fluor®576 (1:500 dilution) in PBST and label the erythroid cells with 400 μL per coverslip for 1 h.
7. Wash the cells with PBST (3 × 5 min/wash).
8. Place a drop of Vectashield mounting medium containing DAPI onto a glass slide. Label the glass slide in a pencil or a solvent-resistant marker pen.
9. Invert the coverslip (the side containing the cell clusters) and place it face-down onto the drop of the DAPI.
10. Place paper towels on the top of the glass slide in a perpendicular orientation. Gently stretch the paper towels sideways (in a direction away from the glass slide) to remove excess DAPI mounting medium. Avoid moving the coverslips to allowing the coverslips to slide along the surface of the slide as this will damage or even destroy cell clusters.
11. Seal the coverslip with nail polish.
12. Store the slides at a 4 °C refrigerator in a dark slide box until imaging on the Leica SPE II confocal microscope. For the best quality and high-resolution images with least amount of background noise, image the samples immediately.

3.5 Visualizing Erythroblastic Islands Under Conventional Scanning Electron Microscopy

The EBI was first described based on careful analysis of transmission electron micrographs of bone marrow sections. They were described as a "reticulum cell surrounded by a ring of erythroblasts, and sometimes two rings are found, the outer being made of more mature cells." Under the scanning electron microscope, the description of EBIs was similar to that first described by Bessis

Unless reconstructed perfectly from serial sections of the EBI under a transmission electron microscope, the SEM offers an advantage of visualizing cell-cell interactions of the EBI at the micro-topographical level. The protocol listed is similar to other SEM sample preparation protocols for monolayer cell cultures. The sample preparation is performed at room temperature.

1. Prepare EBIs as described in Subheading 3.1.
2. Fix the cells with coverslips with 2.5% glutaraldehyde in PBS for 2 h.
3. Wash the cells with PBS pH 7.2 (3 × 5 min per wash).
4. Post-fix the cells with 1% osmium tetroxide for 1 h. Ensure this is done within a chemical fume hood.
5. Wash the cells with PBS pH 7.2 (3 × 5 min/wash). Cell clusters will appear as black particles on the coverslips under the naked eye.
6. Dehydrate cells with graded ethanol (30, 50, 70, 95, 100%; 2× per grade of ethanol; 5 min/change of ethanol). Perform additional change of absolute (100%) ethanol to ensure the cell clusters are fully dehydrated.
7. Dry the cells with HMDS for 2 min (*see* **Note 4**). Air-dry the coverslips in a desiccator (*see* **Note 5**).
8. Mount coverslips onto the aluminum stub for visualizing under the SEM.
9. Coat the coverslips with a 15 nm layer of gold at 25 mA for 2 min using the EMITECH K500X Sputter Coater under vacuum. This layer of gold will cover the surface of the coverslip and the cell clusters to reduce charging artifacts (*see* **Note 6**).
10. Load samples under the Hitachi S4500 FEG-SEM (equipped with a secondary electron detector) at an accelerating voltage of 5.0 kV and a beam current between 6–10 μA. Images can be taken with a Z-height at 15 mm and a working distance of 17–23 mm. Tilt samples at an angle between 30° or 45° for better perspective of EBIs. Note: These settings may vary according to instrument used.

3.6 Immunogold-Labeling of Developing Erythroid Cells for Scanning Electron Microscopy

Developing erythroid cells change their surface phenotype and their surface protein expression as they mature into erythrocytes. This protocol labels erythroid cells with secondary antibody conjugated to 10 nm colloidal gold, allowing correlation of surface protein expression to their surface morphology under both backscatter electron signal and secondary electron signal. Without the use of silver enhancement, we can still visualize labeled erythroid cells within EBIs at a short working distance of 4–5 mm. This protocol is modified from the protocol previously described by de

Harven and colleagues [59] for labeling and visualizing lymphocytes. Sample preparation is performed at [illegible] temperature.

1. Prepare EBIs as described in Subheading 3.[illegible]
2. Briefly fix cells with 0.5% glutaraldehyde in PBS [illegible] This is a mild fixative step).
3. Wash with PBS + 0.1% glycine (3 × 5 min/wash) (Gl[illegible] quenches the fixative; we suggest not using detergents such as Triton-X-100 or Tween-20, as labeling is not intracellular and these detergents may affect cellular morphology).
4. Dilute unconjugated rat-anti-mouse primary antibody (Ter-119) with PBS (1:20) and label the erythroid cells with approximately 400 mL per coverslip for 2 h.
5. Wash cells with PBS (3 × 5 min/wash).
6. Dilute goat-anti-rat secondary antibodies conjugated to 10 nm colloidal gold (1:40 dilution) in PBS and label the erythroid cells with 400 mL per coverslip for 1 h.
7. Wash the cells with PBS (3 × 5 min/wash).
8. Fix the cells in 2.5% glutaraldehyde in PBS for 2 h.
9. Wash the cells with PBS pH 7.2 (3 × 5 min/wash).

   NB: An additional step with 15 min in 1% osmium tetroxide in PBS followed by 3 × 5 min PBS wash can be performed if charging under the electron microscope is problematic.
10. Dehydrate samples with graded ethanol (30, 50, 70, 95, 100%, 2× per grade of ethanol; 5 min/change of ethanol). Perform additional change of absolute (100%) ethanol to ensure the cell cluster is fully dehydrated.
11. Dry the samples with HMDS for 2 min (*see* **Note 5**).
12. Leave the samples to air-dry in a dessicator before imaging. The samples should be dried for a fortnight at least.
13. Mount coverslips onto a metal stub for visualizing under the SEM.
14. Coat the coverslips with 20–25 nm layer of carbon at 25 mA for 2 min using the Emitech K950X Carbon turbo Evaporator under vacuum. Carbon rod evaporation is done at a single pulse (2000 ms evaporating time) to provide 20–25 nm layer of carbon. Coating samples with gold or other metals is not ideal, as it will interfere with the signals generated by the colloidal gold particles.
15. Image the cells using a Zeiss Sigma SEM (equipped with a highly sensitive backscatter electron detector) at an accelerating voltage of 2.0 kV with a working distance of 4–5 mm at high pressure. Note: the key to successful imaging of immunogold labeled cells in this protocol relies on a sensitive backscatter detector, i.e., the GATAN detector.

4 Notes

1. Heat-inactivate FBS in a 65 °C water bath for 1 h. Invert the bottle of FBS approximately ten times every 10 min during heat inactivation. Filter the heat-inactivated serum through a 0.22 μm filter before use. This is to minimize any possible effects from the plasma proteins found in the serum that may affect the protocol such as complement.
2. To date, our protocol has extended to studies involving wild-type mice including inbred BALB/c and C57/BL6 strain mice as well as outbred mouse strains such as ICR and Quackenbush Swiss [8, 10, 21]. Consistent with early studies comparing EBI morphology, we have not identified any major differences in EBIs based on morphology alone [17].
3. We have optimized the protocol by determining which fraction contains the most EBIs. We have found that more EBIs sediment into the bottom fraction than in the serum fraction (data not shown).
4. We do not recommend the use of plastic coverslips due to granulocyte binding to plastic (but not glass) [3].
5. While critical point drying is thought to be more ideal for morphological preservation, and produce less artifacts, HMDS is a cheaper alternative without compromising morphology. HMDS is also an extremely corrosive chemical; read MSDS before proceeding with this protocol.
6. SEM provides invaluable insights into topographical information and cell-cell interactions, which are often more challenging to visualize using other microscopic techniques. Compared to TEM, the possible artifacts for SEM are not as well documented. Here, we provide information on artifacts based on our observations and first-hand experience with reference to our protocol. We recommend readers refer to other sources to better identify artifacts in SEM. In general, we classify the artifacts as handling artifacts, chemical artifacts, and imaging artifacts.

   Handling artifacts is the hardest to describe as they are highly dependent on both the protocol and the experience of laboratory personnel with the sample preparation. In our protocol, observing glass shards on the coverslip under the SEM can be relatively common despite careful handling. One common reason is due to attempts to forcibly remove the coverslip against the fluid tension between the culture dish and the coverslips before mounting the coverslips onto metal stubs. These glass shards can also be generated from poor handling of coverslips during the isolation of EBIs. Prolonged drying of coverslips (>3 months based on experience) prior to mounting onto stubs greatly reduces the fluid tension resulting in fewer glass shards being observed during imaging. Unusual depressions or forcep marks found on cells are also an indication of poor handling of coverslips.

Chemical artifacts can be challenging without a good reference for proper cellular morphology from the literature. Here, we describe the two common artifacts that can be easily identified. Fixation artifacts can be easily recognized by loss of nano-structural integrity, and if severe, the collapse of the specimen integrity resulting in poor morphology of the cells. A good indication of poor fixation is the loss of microvilli on developing erythroid cells. Blebbing of the cell surface can also be an indication of poor fixation. We recommend freshly prepared fixative to reduce this artifact. Drying/dehydration artifacts can range from mild (such as small cracks on cells, usually macrophages) to severe (large cracks, major shrinkage of cells and loss of structural integrity resulting in poor morphology). The least well-described chemical artifact in the literature is crystal formation due to chemical precipitation of the chemicals used for the sample preparation. Based on our protocol, we observed cuboidal-shaped, mesh-like structures that can be mistakenly visualized as erythroid cells in poorly fixed samples. We recommend filtering freshly prepared chemicals prior to sample preparation.

With technological advancements and improvements on acquisition software, many imaging artifacts are no longer present. Here, we describe common imaging artifacts that are present and largely due to the nature of biological samples.

(a) Charging—This is the most commonly observed artifact in biological samples, as samples are not as conductive as material samples (such as metal alloys). This is caused mainly by a buildup of negative charges on the sample that might produce lines or abnormal contrast. With a well gold-coated sample under our prescribed settings, we do not observe any major charging issues at lower magnifications (<10,000 magnification). We suggest reducing the beam intensity or avoiding imaging at a particular spot for a prolonged time if there is a charging issue. Consider coating samples with a thicker coat of gold if necessary. If the problem is severe, the protocol may need to be modified with higher concentrations of osmium tetroxide or cells may need to be impregnated with more heavy metals (e.g., potassium ferrocyanide, uranyl acetate, or lead citrate) prior to dehydration.

(b) Edge effects—This is usually caused when the electron beam is interacting with the edge of a cell (if present, usually observed on erythroid cells when tilting coverslips during imaging). The edges of cells will appear "brighter" than the rest of the cell surface.

(c) Image shifting during imaging—Shifting of the image away from the region of interest can be seen with samples that are not thoroughly dried under the SEM in high vacuum conditions. This can also result in image jerking, whereby the area of interest moves suddenly to a faraway region and back to the region of interest within seconds. Performing quick acquisition by compromising resolution does not rectify this problem; however, imaging under a SEM operating at variable pressure may reduce this artifact. We highly recommend storing the samples in a dry, enclosed desiccated space for at least 1 month prior to coating and imaging to reduce this artifact for routine imaging. We observed obvious shifting of the field of view across the coverslips dried for 7 days. This phenomenon will be very obvious at any magnification higher than 2000× on the SEM.

Acknowledgments

The authors acknowledge the support received from the Bosch Institute Advanced Microscopy Facility, and the expert help of Dr. Louise Cole. The authors also acknowledge the facilities and the scientific and technical assistance of the Australian Microscopy and Microanalysis Research Facility at the Australian Centre for Microscopy and Microanalysis at the University of Sydney. J.H.Y. is supported by The Mrs. Ann & Professor NWG Macintosh PhD Scholarship. This work was supported by funding from the Sydney Medical School New Staff Grant (S.T.F.), The Bosch Institute Translational Grant-in-Aid, the NWG Macintosh Grant 2015 (J. H.Y.) and the NWG Macintosh Grant 2016 (M.P.C). The authors would also like to thank the members of the Laboratory of Blood Cell Development and Andrology Research Group (University of Sydney) for helpful comments.

References

1. Al-Drees MA, Yeo JH, Boumelhem BB et al (2015) Making blood: the haematopoietic niche throughout ontogeny. Stem Cells Int 2015:571893. doi:10.1155/2015/571893
2. Bessis MC (1958) L'ilot erythroblastique. Unite functionelle de la moelle osseuse. Rev Hematol 1958:8–11
3. Bessis MC (1973) Cellules du sang, normal et pathologique. Springer, Berlin
4. Mohandas N, Prenant M (1978) Three-dimensional model of bone marrow. Blood 51:633–643
5. Sadahira Y, Yasuda T, Kimoto T (1991) Regulation of Forssman antigen expression during maturation of mouse stromal macrophages in haematopoietic foci. Immunology 73:498–504
6. Yokoyama T, Etoh T, Kitagawa H et al (2003) Migration of erythroblastic islands toward the

sinusoid as erythroid maturation proceeds in rat bone marrow. J Vet Med Sci 65:449–452
7. Rhodes MM, Kopsombut P, Bondurant MC et al (2007) Adherence to macrophages in erythroblastic islands enhances erythroblast proliferation and increases erythrocyte production by a different mechanism than erythropoietin. Blood 111:1700–1708. doi:10.1182/blood-2007-06-098178
8. Isern J, Fraser ST, He Z, Baron MH (2008) The fetal liver is a niche for maturation of primitive erythroid cells. Proc Natl Acad Sci U S A 105:6662–6667. doi:10.1073/pnas.0802032105
9. Kusakabe M, Hasegawa K, Hamada M et al (2011) c-Maf plays a crucial role for the definitive erythropoiesis that accompanies erythroblastic island formation in the fetal liver. Blood 118:1374–1385. doi:10.1182/blood-2010-08-300400
10. Yeo JH, McAllan BM, Fraser ST (2016) Scanning electron microscopy reveals two distinct classes of erythroblastic island isolated from adult mammalian bone marrow. Microsc Microanal 22:368–378. doi:10.1017/S1431927616000155
11. Lee SH, Crocker PR, Westaby S et al (1988) Isolation and immunocytochemical characterization of human bone marrow stromal macrophages in hemopoietic clusters. J Exp Med 168:1193–1198
12. Switzer JW (1967) Bone marrow composition in the adult rhesus monkey (Macaca mulatta). J Am Vet Med Assoc 151:823–829
13. Glomski CA, Alessandra Pica JF (2015) Erythrocytes of the rhesus and cynomolgus monkeys. CRC Press, Boca Raton, FL
14. Crocker PR, Morris L, Gordon S (1988) Novel cell surface adhesion receptors involved in interactions between stromal macrophages and haematopoietic cells. J Cell Sci Suppl 9:185–206
15. Takashina T (1987) Haemopoiesis in the human yolk sac. J Anat 151:125–135. doi:10.1111/(ISSN)1469-7580
16. Sadahira Y, Mori M, Kimoto T (1990) Isolation and short-term culture of mouse splenic erythroblastic islands. Cell Struct Funct 15:59–65
17. Berman I (1967) The ultrastructure of erythroblastic islands and reticular cells in mouse bone marrow. J Ultrastruct Res 17:291–313
18. Rassokhin AG (2002) Regularities of hemopoietic development in bone marrow erythroblastic islands during early ontogenesis. Vestn Akad Med Nauk:7–12
19. Sorenson G (1961) An electron microscopic study of hematopoiesis in the yolk sac. Lab Investig 10:178–193
20. Bessis MC, Breton-Gorius J (1962) Iron metabolism in the bone marrow as seen by electron microscopy: a critical review. Blood 19:635–663
21. Fraser ST, Midwinter RG, Coupland LA et al (2015) Heme oxygenase-1 deficiency alters erythroblastic island formation, steady-state erythropoiesis and red blood cell lifespan in mice. Haematologica. doi:10.3324/haematol.2014.116368
22. Keerthivasan G, Wickrema A, Crispino JD (2011) Erythroblast enucleation. Stem Cells Int 2011:139851–139859. doi:10.4061/2011/139851
23. Koulnis M, Liu Y, Hallstrom K, Socolovsky M (2011) Negative autoregulation by Fas stabilizes adult erythropoiesis and accelerates its stress response. PLoS One 6:e21192–e21114. doi:10.1371/journal.pone.0021192
24. Ingley E (2012) Integrating novel signaling pathways involved in erythropoiesis. IUBMB Life 64:402–410. doi:10.1002/iub.1024
25. Rich IN, Vogt C, Pentz S (1988) Erythropoietin gene expression in macrophages detected by in situ hybridization. Behring Inst Mitt:202–206
26. Chasis JA, Mohandas N (2008) Erythroblastic islands: niches for erythropoiesis. Blood 112:470–478. doi:10.1182/blood-2008-03-077883
27. Jacobsen RN, Forristal CE, Raggatt LJ et al (2014) Mobilization with granulocyte colony-stimulating factor blocks medullar erythropoiesis by depleting F4/80+VCAM1+CD169+ER-HR3+Ly6G+ erythroid island macrophages in the mouse. Exp Hematol 42:547–561.e4. doi:10.1016/j.exphem.2014.03.009
28. Jacobsen RN, Nowlan B, Brunck ME et al (2016) Fms-like tyrosine kinase 3 (Flt3) ligand depletes erythroid island macrophages and blocks medullar erythropoiesis in the mouse. Exp Hematol 44:207–212.e4. doi:10.1016/j.exphem.2015.11.004
29. Toda S, Segawa K, Nagata S (2014) MerTK-mediated engulfment of pyrenocytes by central macrophages in erythroblastic islands. Blood 123:3963–3971. doi:10.1182/blood-2014-01-547976
30. Miyanishi M, Tada K, Koike M et al (2007) Identification of Tim4 as a phosphatidylserine receptor. Nature 450:435–439. doi:10.1038/nature06307

31. Savill J, Gregory C (2007) Apoptotic PS to phagocyte TIM-4: eat me. Immunity 27:830–832. doi:10.1016/j.immuni.2007.12.002
32. Nagata S (2007) Autoimmune diseases caused by defects in clearing dead cells and nuclei expelled from erythroid precursors. Immunol Rev 220:237–250. doi:10.1111/j.1600-065X.2007.00571.x
33. Yoshida H, Kawane K, Koike M et al (2005) Phosphatidylserine-dependent engulfment by macrophages of nuclei from erythroid precursor cells. Nat Cell Biol 437:754–758. doi:10.1038/nature03964
34. Kawane K, Fukuyama H, Kondoh G et al (2001) Requirement of DNase II for definitive erythropoiesis in the mouse fetal liver. Science 292:1546–1549. doi:10.1126/science.292.5521.1546
35. Liu J, Guo X, Mohandas N et al (2010) Membrane remodeling during reticulocyte maturation. Blood 115:2021–2027. doi:10.1182/blood-2009-08-241182
36. Sadahira Y, Yoshino T, Monobe Y (1995) Very late activation antigen 4-vascular cell adhesion molecule 1 interaction is involved in the formation of erythroblastic islands. J Exp Med 181:411–415
37. Lee G, Lo A, Short SA et al (2006) Targeted gene deletion demonstrates that the cell adhesion molecule ICAM-4 is critical for erythroblastic island formation. Blood 108:2064–2071. doi:10.1182/blood-2006-03-006759
38. Wang Z, Vogel O, Kuhn G et al (2013) Decreased stability of erythroblastic islands in integrin β3-deficient mice. Physiol Rep 1. doi:10.1002/phy2.18
39. Hanspal M, Hanspal JS (1994) The association of erythroblasts with macrophages promotes erythroid proliferation and maturation: a 30-kD heparin-binding protein is involved in this contact. Blood 84:3494–3504
40. Hanspal M, Smockova Y, Uong Q (1998) Molecular identification and functional characterization of a novel protein that mediates the attachment of erythroblasts to macrophages. Blood 92:2940–2950
41. Soni S, Bala S, Gwynn B et al (2006) Absence of erythroblast macrophage protein (Emp) leads to failure of erythroblast nuclear extrusion. J Biol Chem 281:20181–20189. doi:10.1074/jbc.M603226200
42. Mao X, Shi X, Liu F et al (2013) Evaluation of erythroblast macrophage protein related to erythroblastic islands in patients with hematopoietic stem cell transplantation. Eur J Med Res 18:9. doi:10.1186/2047-783X-18-9
43. Bala S, Kumar A, Soni S et al (2006) Emp is a component of the nuclear matrix of mammalian cells and undergoes dynamic rearrangements during cell division. Biochem Biophys Res Commun 342:1040–1048. doi:10.1016/j.bbrc.2006.02.060
44. Chow A, Huggins M, Ahmed J et al (2013) CD169+ macrophages provide a niche promoting erythropoiesis under homeostasis and stress. Nat Med. doi:10.1038/nm.3057
45. Ramos P, Casu C, Gardenghi S et al (2013) Macrophages support pathological erythropoiesis in polycythemia vera and β-thalassemia. Nat Med:1–11. doi:10.1038/nm.3126
46. Buesche G, Teoman H, Giagounidis A et al (2016) Impaired formation of erythroblastic islands is associated with erythroid failure and poor prognosis in a significant proportion of patients with myelodysplastic syndromes. Haematologica 101:e177–e181. doi:10.3324/haematol.2015.129015
47. Wang Y-H, Fu R, Dong S-W et al (2014) Erythroblastic islands in the bone marrow of patients with immune-related pancytopenia. PLoS One 9:e95143. doi:10.1371/journal.pone.0095143
48. Bessis M (1955) Phase contrast microscopy and electron microscopy applied to the blood cells: general review. Blood 10:272–286
49. Acar M, Kocherlakota KS, Murphy MM et al (2015) Deep imaging of bone marrow shows non-dividing stem cells are mainly perisinusoidal. Nature 526:126–130. doi:10.1038/nature15250
50. Fujii Y, Terada N, Ueda H et al (1997) Electron microscopic study of erythroblastic islands obtained by "tissue-stamp culture" method. J Electron Microsc 46:477–484
51. Ellis SL, Grassinger J, Jones A et al (2011) The relationship between bone, hemopoietic stem cells, and vasculature. Blood 118:1516–1524. doi:10.1182/blood-2010-08-303800
52. Vanna R, Ronchi P, Lenferink ATM et al (2015) Label-free imaging and identification of typical cells of acute myeloid leukaemia and myelodysplastic syndrome by Raman microspectroscopy. Analyst 140:1054–1064. doi:10.1039/C4AN02127D
53. Asghari-Khiavi M, Wood BR, Mechler A et al (2010) Correlation of atomic force microscopy and Raman micro-spectroscopy to study the effects of ex vivo treatment procedures on human red blood cells. Analyst 135:525–526. doi:10.1039/b919245j

54. Mukherjee R, Saha M, Routray A, Chakraborty C (2015) Nanoscale surface characterization of human erythrocytes by atomic force microscopy: a critical review. IEEE Trans Nanobiosci 14:625–633. doi:10.1109/TNB.2015.2424674
55. Maciaszek JL, Andemariam B, Huber G, Lykotrafitis G (2012) Epinephrine modulates BCAM/Lu and ICAM-4 expression on the sickle cell trait red blood cell membrane. Biophys J 102:1137–1143. doi:10.1016/j.bpj.2012.01.050
56. Krause MA, Diakite SAS, Lopera-Mesa TM et al (2012) α-Thalassemia impairs the cytoadherence of Plasmodium falciparum-infected erythrocytes. PLoS One 7:e37214–e37217. doi:10.1371/journal.pone.0037214
57. Tian Y, Cai M, Xu H et al (2014) Atomic force microscopy of asymmetric membranes from turtle erythrocytes. Mol Cells 37:592–597. doi:10.14348/molcells.2014.0115
58. Tian Y-M, Cai M-J, Zhao W-D et al (2014) The asymmetric membrane structure of erythrocytes from Crucian carp studied by atomic force microscopy. Chin Sci Bull 59:2582–2587. doi:10.1007/s11434-014-0375-6
59. de Harven E (1984) A novel approach for scanning electron microscopy of colloidal gold-labeled cell surfaces. J Cell Biol 99:53–57. doi:10.1083/jcb.99.1.53

Chapter 9

Flow Cytometry (FCM) Analysis and Fluorescence-Activated Cell Sorting (FACS) of Erythroid Cells

Xiuli An and Lixiang Chen

Abstract

To study the process of erythropoiesis, it is important to be able to isolate erythroid progenitors and erythroblasts at distinct stages of development. During the past decade, considerable progress has been made on the development of flow cytometry (FCM) and fluorescence-activated cell sorting (FACS) methods for the analysis and isolation of both murine and human erythroid cells at distinct stages of erythropoiesis, based on changes in the expression of cell surface markers. A method for the identification of murine BFU-E and CFU-E cells was reported by Flygare et al., by negative selection for Ter119, B220, Mac-1, CD3, Grl, Sca-1, CD16/CD32, CD41, and CD34 cells, followed by separation based on the expression levels of CD71. We developed an alternative method in which Ter119 is used as an erythroid lineage marker, and in conjunction with CD44 and cell size as differentiation markers, it is possible to unambiguously distinguish erythroblasts at each developmental stage during murine terminal erythroid differentiation. We also developed methods for the analysis and isolation of human erythroid cells at all developmental stages. BFU-E and CFU-E are characterized by $CD45^{+}GPA^{-}IL\text{-}3R^{-}CD34^{+}CD36^{-}CD71^{low}$ and $CD45^{+}GPA^{-}IL\text{-}3R^{-}CD34^{-}CD36^{+}CD71^{high}$ phenotypes, respectively; the combination of GPA, band 3 and α4-integrin are used to isolate erythroid cells at all of the terminal stages of human erythropoiesis, including proerythroblasts, early basophilic, late basophilic, polychromatic and orthochromatic erythroblasts.

Key words Erythropoiesis, Erythroid progenitor, Erythroblasts, Flow cytometry, Flow cytometry-activated cell sorting (FACS), Surface marker

1 Introduction

Flow cytometry (FCM) is a method for measuring the chemical and/or physical characteristics of cells passing through a detection device in a fluid stream [1, 2]. Fluorescence-activated cell sorting (FACS) is a specialized type of FCM. It provides a method for sorting a heterogeneous mixture of biological cells into two or more containers, one cell at a time, based upon the specific light scattering and fluorescent characteristics of each cell [3, 4]. FCM and FACS are vital strategies in biological research and clinical applications, as they provide fast, objective, and quantitative

Joyce A. Lloyd (ed.), *Erythropoiesis: Methods and Protocols*, Methods in Molecular Biology, vol. 1698,
DOI 10.1007/978-1-4939-7428-3_9, © Springer Science+Business Media LLC 2018

recording of fluorescent signals from individual cells as well as physical separation of cells of particular interest [5, 6].

Erythropoiesis is the process by which hematopoietic stem cells (HSCs) proliferate and differentiate to produce mature red blood cells. It is a tightly regulated process that can be divided into two stages, early and terminal. Early erythropoiesis refers to the process by which multi-potential hematopoietic stem cells proliferate and differentiate into committed erythroid progenitors: erythroid burst-forming unit (BFU-E) and then erythroid colony-forming unit (CFU-E) cells that then differentiate into proerythroblasts [7–10]. During the terminal stage, morphologically recognizable proerythroblasts undergo 4–5 mitotic cell divisions, to produce basophilic, polychromatic, and orthochromatic erythroblasts. Orthochromatic erythroblasts then expel the nuclei to generate reticulocytes [8].

To study the process of erythropoiesis, it is important to be able to identify and isolate erythroid progenitors and erythroblasts at distinct stages of development. During the past decade, considerable progress has been made in analyzing and isolating erythroid cells at each developmental stage during murine and human erythroid differentiation. Here, we describe a method for the identification of mouse BFU-E and CFU-E cells using negative selection for Ter119, B220, Mac-1, CD3, Gr1, Sca-1, CD16/CD32, CD41, and CD34 cells, followed by separation based on the level of expression of CD71 [11]. We also describe FCM and FACS strategies in which Ter119 is used as an erythroid lineage marker, in conjunction with CD44 and cell size as differentiation markers, to unambiguously distinguish erythroblasts at each developmental stage during murine terminal erythroid differentiation [12, 13].

It has long been known that there are fundamental differences between human and murine erythropoiesis. Recently, our genome-wide-based studies provide strong evidence to support this hypothesis [14–16]. Based on advances in the understanding of human erythropoiesis, we developed novel methods for the identification and isolation of human erythroid cells at each development stage. Erythroid progenitors, BFU-E and CFU-E, are characterized by $CD45^{+}GPA^{-}IL\text{-}3R^{-}CD34^{+}CD36^{-}CD71^{low}$ and $CD45^{+}$ $GPA^{-}IL\text{-}3R^{-}CD34^{-}CD36^{+}$- $CD71^{high}$ phenotypes, respectively [14]. Erythroid cells at successive terminal stages of human erythropoiesis were identified by using the combination of GPA, band 3, and α4-integrin to distinguish proerythroblasts, early basophilic, late basophilic, polychromatic and orthochromatic erythroblasts [15].

2 Materials

2.1 Murine BFU-E and CFU-E Cells

1. Pregnant mice at 14.5 days post conception.
2. 1× Phosphate-Buffered Saline (PBS).

3. PBS-BSA buffer: 1× PBS, supplemented with 0.5% BSA and 0.1% sodium azide.
4. Biotin mouse lineage panel (contains biotin-conjugated antibodies against mouse Ter119, B220, Mac-1, CD3, and Gr-1).
5. Biotin rat anti-mouse-Sca-1.
6. Biotin rat anti-mouse-CD16/32 (FcgR).
7. Biotin rat anti-mouse-CD41.
8. Biotin rat anti-mouse-CD34.
9. PE streptavidin.
10. FITC rat anti-mouse-CD71.
11. APC rat anti-mouse-CD117.
12. 7-AAD.
13. 70 μm cell strainer.
14. Streptavidin particles plus-DM.
15. Cell separation magnet.
16. 1.5 mL Eppendorf tubes.
17. 15 mL, 50 mL corning polypropylene conical centrifuge tubes.
18. Eppendorf refrigerated centrifuge 5810R with swinging bucket rotor.
19. BD FACSCanto II cell analyzer.
20. Beckman-Coulter MOFLO high-speed cell sorter.

2.2 Murine Erythroblasts

1. Mice at the age of 2 months or older.
2. 1× Phosphate-Buffered Saline (PBS).
3. PBS-BSA buffer: 1× Phosphate-Buffered Saline supplemented with 0.5% BSA and 0.1% sodium azide.
4. Rat anti-mouse CD16/CD32.
5. FITC rat anti-mouse TER-119.
6. APC-Cy7 rat anti-mouse CD11b.
7. APC rat anti-mouse CD44.
8. APC-Cy7-conjugated CD45.
9. 7-AAD.
10. 1.5 mL Eppendorf tubes.
11. 15 and 50 ml Corning polypropylene conical centrifuge tubes.
12. Eppendorf refrigerated centrifuge 5810R with swinging bucket rotor.
13. BD FACSCanto II cell analyzer.
14. Beckman-Coulter MOFLO high-speed cell sorter.

2.3 Human BFU-E and CFU-E Cells from Cultured CD34+ Cells

1. Umbilical cord blood or peripheral blood from volunteer donors.
2. Ficoll-Paque Plus.
3. Buffer I: 1× PBS with 2% FBS and 2 mM EDTA.
4. Buffer II: 1× PBS with 0.5% BSA and 2 mM EDTA.
5. PBS-BSA buffer: 1× PBS with 0.5% BSA.
6. Blocking buffer: 1× PBS with 4% human AB Serum.
7. Erythroid Culture Medium:

   Phase I (Day 0–6): Iscove's Modified Dulbecco's Medium (IMDM), 2% human peripheral blood plasma, 3% human AB serum, 200 μg/ml Holo-human transferrin, 3 IU/ml heparin, 10 μg/ml insulin, 10 ng/ml stem cell factor (SCF), 1 ng/ml IL-3, and 3 IU/ml erythropoietin (EPO).
8. Human CD34 Isolation Kit (Miltenyi Biotec).
9. PE mouse anti human CD34 (BD Bioscience).
10. FITC mouse anti human CD36 (BD Bioscience).
11. APC mouse anti human GPA (BD Bioscience).
12. PE-Cy7 mouse anti human IL-3R (CD123) (eBioscience).
13. 7-AAD (BD Bioscience).
14. 70 μm Cell Strainer.
15. Magnetic Cell Separation system (Miltenyi Biotec).
16. 1.5 mL Eppendorf Tubes, 15 and 50 ml Corning Polypropylene Conical Centrifuge Tubes.
17. Refrigerated Centrifuge with Swinging Bucket Rotor.
18. BD FACSCanto II cell analyzer.
19. Beckman-Coulter MOFLO high-speed cell sorter.

2.4 Primary Human BFU-E and CFU-E Cells from Bone Marrow

1. Bone marrow from volunteer donors.
2. Ficoll-Paque Plus.
3. Buffer I: 1× PBS with 2% FBS and 2 mM EDTA.
4. Buffer II: 1× PBS with 0.5% BSA and 2 mM EDTA.
5. PBS-BSA buffer: 1× PBS with 0.5% BSA.
6. Blocking buffer: 1× PBS with 4% human AB Serum.
7. Human CD45 Isolation Kit (Miltenyi Biotec).
8. PE mouse anti human CD34 (BD Bioscience).
9. FITC mouse anti human CD36 (BD Bioscience).
10. APC mouse anti human GPA (BD Bioscience).
11. PE-Cy7 mouse anti human IL-3R (CD123) (eBioscience).
12. V450 mouse anti human CD45 (BD Bioscience).

13. APC mouse anti human CD41 (BD Bioscience).
14. PerCp mouse anti human CD14 (BD Bioscience).
15. PerCp mouse anti human CD3 (BD Bioscience).
16. PerCp mouse anti human CD4 (BD Bioscience).
17. PerCp mouse anti human CD19 (BD Bioscience).
18. APC-H7 mouse anti human CD71 (BD Bioscience).
19. 7-AAD (BD Bioscience).
20. 70 μm Cell Strainer.
21. Magnetic Cell Separation system (Miltenyi Biotec).
22. 1.5 mL Eppendorf Tubes, 15 and 50 ml Corning Polypropylene Conical Centrifuge Tubes.
23. Refrigerated Centrifuge with Swinging Bucket Rotor.
24. BD FACSCanto II cell analyzer.
25. Beckman-Coulter MOFLO high-speed cell sorter.

2.5 Human Erythroblasts from Cultured CD34$^+$ Cells

1. Umbilical cord blood or peripheral blood from volunteer donors.
2. Ficoll-Paque Plus.
3. Buffer I: 1× PBS with 2% FBS and 2 mM EDTA.
4. Buffer II: 1× PBS with 0.5% BSA and 2 mM EDTA.
5. PBS-BSA buffer: 1× PBS with 0.5% BSA.
6. Blocking buffer: 1× PBS with 4% human AB Serum.
7. Erythroid Culture Medium:

   Phase I (Day 0–6): Iscove's Modified Dulbecco's Medium (IMDM), 2% human peripheral blood plasma, 3% human AB serum, 200 μg/ml Holo-human transferrin, 3 IU/ml heparin, 10 μg/ml insulin, 10 ng/ml stem cell factor (SCF), 1 ng/ml IL-3, and 3 IU/ml erythropoietin (EPO).

   Phase II (Day 7–11): IMDM, 2% human peripheral blood plasma, 3% human AB serum, 200 μg/ml Holo-human transferrin, 3 IU/ml heparin, 10 μg/ml insulin, 10 ng/ml SCF, and 1 IU/ml EPO.

   Phase III (Day 11–18): IMDM, 2% human peripheral blood plasma, 3% human AB serum, 1 mg/ml Holo-human transferrin, 3 IU/ml heparin, 10 μg/ml insulin, and 1 IU/ml EPO.
8. Human CD34 Isolation Kit (Miltenyi Biotec).
9. PE Mouse anti human CD235a (BD Bioscience).
10. APC mouse anti human CD49d (α4-integrin) (Miltenyi Biotec).
11. DyLight 488-conjugated mouse anti human Band3.

12. 7-AAD (BD Bioscience).
13. 70 μm Cell Strainer.
14. Magnetic Cell Separation system (Miltenyi Biotec).
15. 1.5 mL Eppendorf Tubes, 15 and 50 ml Corning Polypropylene Conical Centrifuge Tubes.
16. Refrigerated Centrifuge with Swinging Bucket Rotor.
17. BD FACSCanto II cell analyzer.
18. Beckman-Coulter MOFLO high-speed cell sorter.

2.6 Primary Human Erythroblasts from Bone Marrow

1. Bone marrow from volunteer donors.
2. Ficoll-Paque Plus.
3. Buffer I: 1× PBS with 2% FBS and 2 mM EDTA.
4. Buffer II: 1× PBS with 0.5% BSA and 2 mM EDTA.
5. PBS-BSA buffer: 1× PBS with 0.5% BSA.
6. Blocking buffer: 1× PBS with 4% human AB Serum.
7. Human CD45 Isolation Kit (Miltenyi Biotec).
8. PE Mouse anti human CD235a (BD Bioscience).
9. APC mouse anti human CD49d (α4-integrin) (Miltenyi Biotec).
10. DyLight 488-conjugated mouse anti human Band3.
11. 7-AAD (BD Bioscience).
12. 70 μm Cell Strainer.
13. Magnetic Cell Separation system (Miltenyi Biotec).
14. 1.5 mL Eppendorf Tubes, 15 and 50 ml Corning Polypropylene Conical Centrifuge Tubes.
15. Refrigerated Centrifuge with Swinging Bucket Rotor.
16. BD FACSCanto II cell analyzer.
17. Beckman-Coulter MOFLO high-speed cell sorter.

3 Methods

3.1 Identification and Isolation of Murine BFU-E and CFU-E Cells

3.1.1 Preparation of Fetal Liver Cells

1. Sacrifice the pregnant mice at 14.5 days post conception.
2. Dissect the uterus and wash in 1× PBS.
3. Move fetuses into fresh 1× PBS.
4. Dissect the fetal livers and place them in fresh 1× PBS.
5. Pool ~8 fetal livers into a 1.5 mL Eppendorf tube in 1 mL 1× PBS.
6. Disrupt the fetal livers with a 1000 μl Rainin Pipette (~10 passes).

7. Pool all material and filter the cell suspension through a 70 μm cell strainer into a 15 mL conical tube.
8. Pellet the cells at 300 × *g* for 5 min.
9. Aspirate the supernatant. Resuspend the cells in 1 mL PBS-BSA buffer.

3.1.2 Enrichment of Fetal Liver Erythroid Progenitors

1. Count the cells, and resuspend them in PBS-BSA buffer at a final density of 100 × 10^6/ml.
2. According to the manufacturer's instruction, add the biotinylated antibodies from the Mouse Lineage Panel, using 2 μl of each antibody per 10^6 cells, and incubate on ice for 15 min.
3. Wash the stained cells twice with PBS-BSA buffer and centrifuge for 5 min at 300 × *g* at 4 °C, carefully aspirate all the supernatant.
4. Resuspend the cells at a final density of 20–100 × 10^6 cells/ml.
5. Vortex the streptavidin particles thoroughly, and add 50 μl of particles for every 1 × 10^7 cells.
6. Mix thoroughly, incubate for 30 min at 4 °C.
7. Bring the cell density to 20–80 × 10^6 cells/ml with PBS-BSA buffer.
8. Transfer the labeled cells to a round-bottom test tube, maximum volume added not to exceed 1.0 mL. Place this positive-fraction tube on the Cell Separation Magnet (horizontal position) for 6–8 min.
9. With the tube on the Cell Separation Magnet and using a glass Pasteur pipette, carefully aspirate the supernatant (negative-fraction) and place in a new tube.
10. Remove the positive-fraction tube from the Cell Separation Magnet, and add PBS-BSA buffer to the same volume as in **step** 7.
11. Resuspend the positive fraction well by pipetting up and down 10–15 times, and place the tube back on the Cell Separation Magnet for 8 min.
12. Using a new sterile Pasteur pipette, carefully aspirate the supernatant and combine with the negative-fraction from **step 9** above.
13. Place the tube containing the combined depleted fraction on the Cell Separation Magnet for a final 8 min.
14. Carefully aspirate the supernatant and place in a new sterile tube.
15. Pellet the cells at 300 × *g* for 5 min.
16. Repeat **steps 1–14** to further deplete Sca, CD16/32 (FcgR), CD41 and CD34 positive cells.

17. Collect Ter119, B220, Mac-1, CD3, Grl, Sca-1, CD16/CD32, CD41, and CD34 negative cells for the following cell sorting.

3.1.3 Antibody Staining of Murine Fetal Liver Erythroid Progenitor Cells

1. Count the depleted cells, resuspend in PBS-BSA buffer at a final density of 10^6/80 μl for the antibody staining.
2. Stain the cells as below:

   PE Streptavidin (0.2 μg/10^6 cells)

   APC rat anti-mouse CD117 (0.2 μg/10^6 cells)

   FITC rat anti-mouse CD71 (0.2 μg/10^6 cells)
3. Incubate on ice for 30 min in the dark.
4. Wash two times with PBS-BSA buffer.
5. Stain with the viability marker PI (0.25 μg/10^6 cells) on ice for 10 min in the dark.
6. Spin at 300 × *g* or 5 min at 4 °C.
7. Resuspend cell pellet in PBS-BSA.
8. For samples to be analyzed, resuspend at 1 × 10^6 cells/ml. For sorting samples, resuspend the cells at 10–20 × 10^6 cells/ml.

3.1.4 Flow Cytometry Analysis of Murine Fetal Liver Erythroid Progenitor Cells

1. Run the stained samples on BD FACSCanto II.
2. Gate out cellular debris and PI^+ cells.
3. Appropriate cytometry gates are determined according to Fig. 1.
4. For PI^- population, gate on PE-$Biotin^-$ population.
5. Show CD117 and CD71 expression of PE negative cells.
6. For PE-$Biotin^-$ population, gate on two populations, BFU-E and CFU-E.
7. The BFU-E fraction is the 10% lowest CD71-expressing part of the cKit (CD117)+ fraction; the CFU-E fraction is the 20% highest CD71-expressing part of the cKit+ fraction.

3.1.5 Isolation of Murine Fetal Liver BFU-E and CFU-E Cells by FACS

1. Wash the cells twice with 40 mL PBS-BSA buffer and resuspend the cells at the concentration of 20 × 10^6 cells/ml.
2. Stain the cells as described in 3.1.3.
3. Follow the manufacturer's protocol for cell sorter and the gating strategy described in 3.1.4.
4. Collect populations of interest in FBS.
5. Wash cells with PBS and centrifuge at 300 × *g* for 10 min at 4 °C.
6. Aspirate the supernatant and the purified cells are ready for downstream applications.

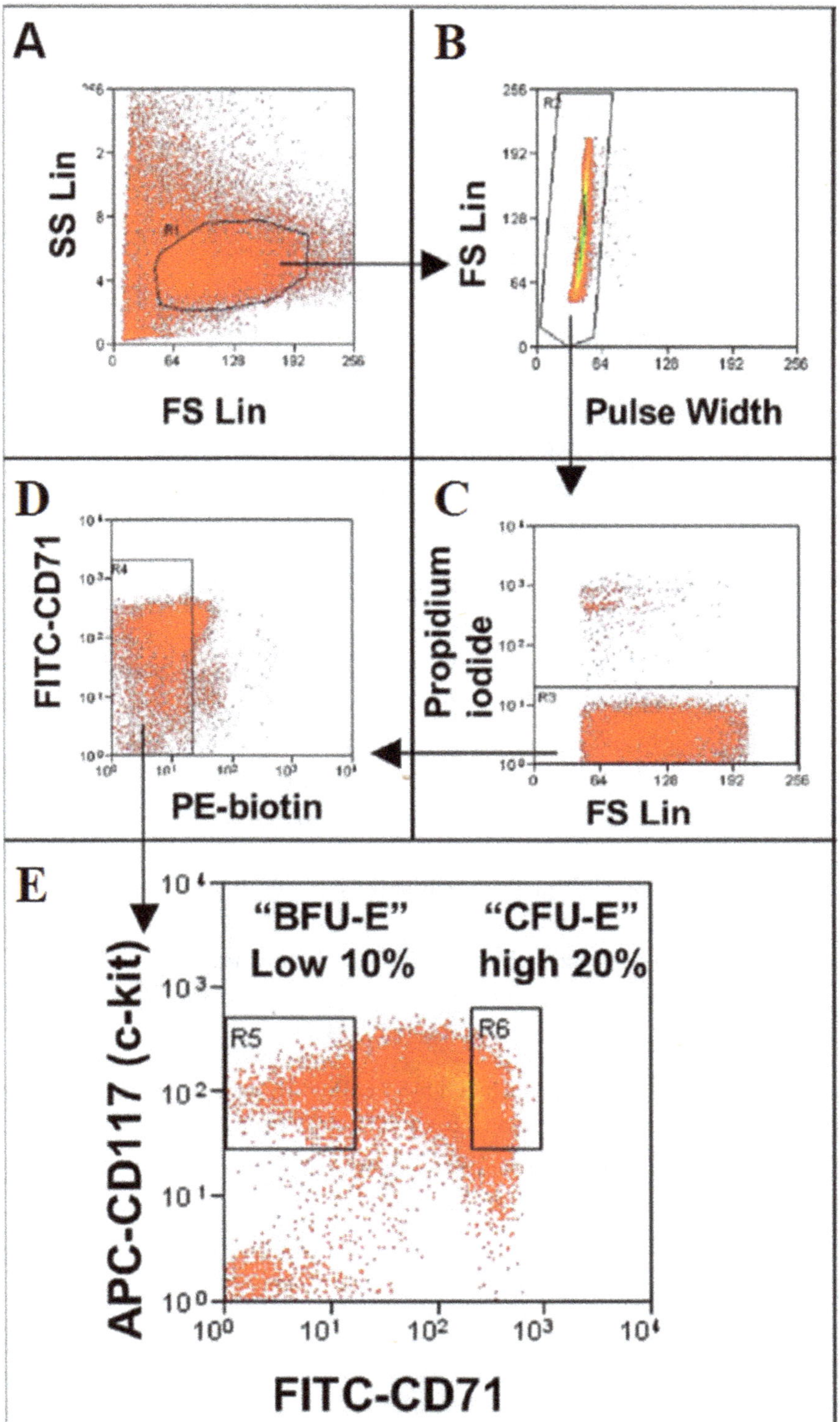

Fig. 1 Identification and isolation of murine BFU-E and CFU-E cells by flow cytometric cell sorting. Day 14.5–15.5 mouse fetal liver cells stained with biotin-conjugated antibodies against murine Ter119, B220,

3.2 Identification and Isolation of Murine Erythroblasts

3.2.1 Preparation of Murine Bone Marrow Cells

1. Sacrifice mice by cervical dislocation.
2. Remove the femur and tibia and cut off the ends of the bones.
3. Flush out the bone marrow into a 50 mL Falcon tube by inserting a 26-gauge needle attached to a 20 mL syringe filled with PBS-BSA. Pass the flushing buffer through the bone until the color of the bone turns from red to white.
4. Disperse cells by gently passing through a 26-gauge needle three to four times.
5. Centrifuge cells at 300 × *g* for 5 min at 4°˜C.
6. Discard the supernatant and resuspend cells in PBS-BSA to the final density of 10^7 cells/90 μl.
7. Add 10 μl of mouse CD45 MicroBeads (Miltenyi Biotec) per 10^7 cells, and incubate on ice for 15 min.
8. Wash cells twice with PBS-BSA buffer and centrifuge at 300 × *g* for 10 min at 4°˜C, carefully aspirate all the supernatant.
9. Add an additional 1 mL of PBS-BSA buffer and mix thoroughly.
10. Centrifuge cells at 300 × *g* for 10 min at 4°˜C, aspirate the supernatant.
11. Resuspend up to 10^8 cells/ml in of PBS-BSA buffer.
12. Apply the cell suspension onto the LS column that has been placed on the magnetic separator and rinsed with PBS-BSA buffer.
13. Collect the unlabeled cells which pass through and wash column with 3 mL PBS-BSA buffer for three times.
14. Collect the total effluent, centrifuge to pellet the $CD45^-$ cells. Resuspend the cells at the density of 10^6/80 μl in PBS-BSA buffer.

3.2.2 Antibody Staining of Murine Bone Marrow Cells

1. Incubate the $CD45^-$ cells with rat anti-mouse CD16/CD32 at the dose of 2.5 μg/10^6 cells on ice for 15 min.
2. Stain the cells as below:

   FITC rat anti-mouse TER-119 (1 μg/10^6 cells)

   APC rat anti-mouse CD44 (1 μg/10^6 cells)

Fig. 1 (continued) Mac-1, CD3, Gr-1, CD32/CD16, Sca-1, CD41, and CD34 were first depleted by magnetic beads. (**a**) Plot of FSC versus SSC of all the labeled cells. (**b**) Plot of FSC versus Pulse Width of the gated cells. (**c**) Plot of FSC versus PI of the Pulse Width negative cells. (**d**) Plot of Biotin versus CD71 of the biotin negative cells. (**e**) Plot of CD117 versus CD71 of the CD117 positive cells. BFU-E fraction is the 10% lowest CD71-expressing part of the $CD117^+$ fraction; CFU-E fraction is the 20% highest CD71-expressing part of the $CD117^+$ fraction

APC-Cy7 rat anti-mouse CD45 (0.1 μg/10^6 cells)
APC-Cy7 rat anti-mouse CD11b (0.1 μg/10^6 cells)
APC-Cy7 rat anti-mouse GR1 (0.1 μg/10^6 cells)

3. Incubate on ice for 30 min in the dark.
4. Stain with the viability marker 7-AAD (0.25 μg/10^6 cells) on ice for 10 min in the dark.
5. Run the stained samples on BD FACSCanto II.

3.2.3 Flow Cytometry Analysis of Murine Bone Marrow Cells

1. Gate out cellular debris and 7-AAD^+ cells.
2. For 7-AAD^- cells, use APC-Cy7-conjugated CD45, GR1, and CD11b to gate out the residual leukocytes, monocytes, and granulocytes.
3. Show Ter119 expression of $CD45^-GR1^-CD11b^-$ cells.
4. For $Ter119^+$ cells, plot in two ways:
   (a) CD44/Ter119 plot, proerythroblasts (population I) is the $CD44^{hi}TER119^{low}$ fraction;
   (b) CD44/FSC plot, populations II through population VI are basophilic erythroblasts, polychromatic erythroblasts, orthochromatic erythroblasts, reticulocytes and mature red cells, respectively (Fig. 2).

3.2.4 Isolation of Murine Bone Marrow Erythroblasts by FACS

1. Wash the cells twice with 40 mL PBS-BSA buffer and resuspend the cells at the concentration of 20×10^6 cells/ml.
2. Stain the cells as described in Subheading 3.2.2.
3. Follow the manufacturer's protocol for cell sorter and the gating strategy described in Subheading 3.2.3.
4. Collect populations of interest in FBS.
5. Wash the cells with PBS and centrifuge at $300 \times g$ for 10 min at 4 °C.
6. Aspirate the supernatant and the purified cells are ready for downstream applications.
7. Take 5×10^4 cells from sorted populations to prepare cytospin, and use Giemsa staining to confirm the purity by cell morphology (Fig. 3).

3.3 Flow Cytometry Analysis and Isolation of Human BFU-E and CFU-E Cells from Cultured CD34+ Cells

3.3.1 Purification of Human $CD34^+$ Cells

1. Obtain umbilical cord blood or peripheral blood from volunteer donors after signing informed consent and following the institutional guidelines.
2. Dilute blood sample 2.5 times with buffer I at room temperature.
3. Pour 15 mL Ficoll-Paque Plus into 50 mL Conical Centrifuge tube and carefully add 35 mL of the diluted blood on the Ficoll.

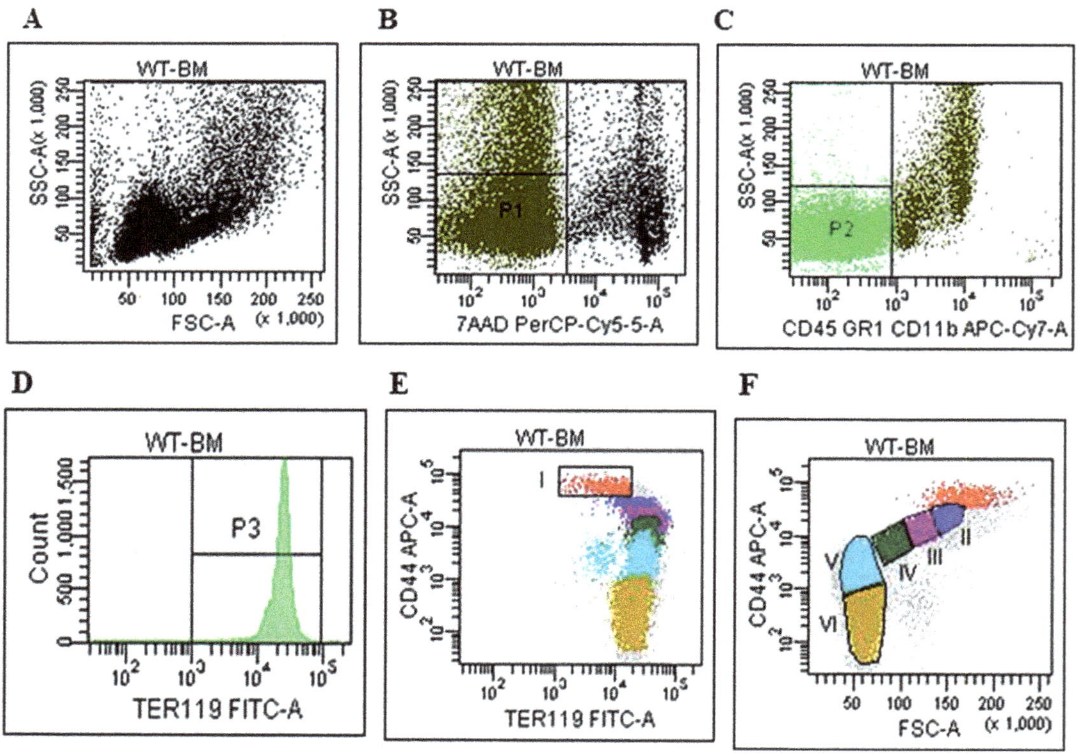

Fig. 2 Flow cytometry analysis of murine erythroblasts. (**a**) Plot of FSC versus SSC of all the cells. (**b**) Plot of 7AAD versus SSC of all the cells. (**c**) Plot of CD45CD11bGR1 versus SSC of all the cells. (**d**) Histogram of TER119 of $CD45^{-}CD11b^{-}GR1^{-}$cells. (**e**) Plot of CD44 versus TER119 of the TER119 positive cells with the gating on $CD44^{hi}TER^{low}$ population I. (**f**) Plot of CD44 versus FSC of the TER119 positive cells with gating the population II, III, IV, V and VI

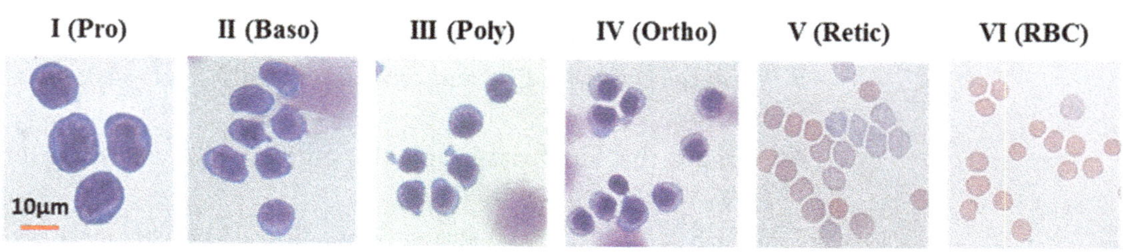

Fig. 3 Morphology of murine erythroblasts at distinct stage of development

4. Centrifuge at 400 × *g* for 30 min with no brake at room temperature.
5. Collect the buffy coat (middle layer) containing mononuclear cells with a transfer pipette into a 50 mL tube (up to 25 mL/tube), and fill the tube up with buffer I and invert the tubes two to three times to wash the cells.
6. Centrifuge at 300 × *g* for 10 min with low break at room temperature.

7. Discard the supernatant and pool the pellets into one tube, add buffer I to 50 mL and centrifuge for at 200 × *g* 10 min with low break at room temperature.
8. Resuspend the pellet in cold buffer II, count the cells at proper dilution.
9. Wash the cells with cold buffer II and centrifuge at 200 × *g* for 10 min with low break at 4 °C.
10. Resuspend the cells at the density of up to 1 × 10^8 cells/300 μl cold buffer II.
11. Add 100 μl of FcR Blocking Reagent and 100 μl of CD34 microbeads for up to 1 × 10^8 cells. Mix well and incubate in the refrigerator (2–8 °C) for 30 min.
12. Wash the cells with buffer II as described in **step 9**.
13. Resuspend up to 10^8 cells in 500 μl of buffer II and apply to 70 μm Cell Strainer to obtain uniform single-cell suspensions.
14. Place column in the magnetic field of a suitable MACS Separator and prepare MACS column by rinsing with buffer II.
15. Apply single-cell suspensions onto the column.
16. Wash the column two to three times with 2–5 ml buffer II.
17. Remove column from the separator and place it on a suitable collection tube.
18. Add 2–5 ml buffer II to the column and immediately flush out the magnetically labeled cells by firmly pushing the plunger into the column.
19. (Optional) To increase the purity of $CD34^+$ cells, the eluted fraction can be enriched over a second column, repeat the magnetic separation procedure (**steps 14–18**).
20. Count the cells at proper dilution, centrifuge to remove buffer II and culture the cells in a fresh medium.

3.3.2 In Vitro Culture of $CD34^+$ Cells

1. Resuspend $CD34^+$ cells with the Phase I Erythroid Culture Medium at the cell density of 10^5 cells/ml.
2. Culture the cells at 37 °C in a humidified atmosphere with 5% CO_2 in air.
3. Refresh the Phase I medium at day 4, adjust the cell density to 10^5 cells/ml.
4. Culture the cells until day 7.

3.3.3 Antibody Staining

1. Collect 1 × 10^5 cells in a 1.5 mL Eppendorf tube, wash with PBS-BSA buffer.
2. Centrifuge at 300 × *g* for 5 min at 4 °C, aspirate the supernatant.

3. Resuspend cells in 22.5 μl of PBS-BSA buffer, add 2.5 μl blocking buffer and incubate at 4 °C for 10 min.
4. Add antibodies as follows:

   PE mouse anti human CD34: 2 μl.

   PE-Cy7 mouse anti human IL-3R (CD123): 2 μl.

   FITC mouse anti human CD36: 1 μl.

   APC mouse anti human GPA (1:100 diluted): 1 μl.
5. Mix well and incubate at room temperature for 15 min, keep in the dark.
6. Add 1 mL PBS-BSA buffer to wash the cells as described in **step 2**.
7. Resuspend cells in 200–300 μl of PBS-BSA buffer and transfer into flow tube pre-loaded with 5 μl of 7-AAD.
8. Run the stained samples on BD FACSCanto II.

3.3.4 Flow Cytometry Analysis

1. Conduct flow cytometry analysis using BD FACSDiva.
2. Gate out cellular debris in FSC-SSC plot.
3. Gate 7-AAD$^-$ cells as live cells.
4. Show the expression of IL-3R and GPA of live cells by plotting IL-3R versus GPA, gate IL-3R$^-$GPA$^-$ cells.
5. Show the expression of CD34 and CD36 in IL-3R$^-$GPA$^-$ cells by plotting CD34 versus CD36.
6. The BFU-E and CFU-E populations are defined as IL-3R$^-$GPA$^-$CD34$^+$CD36$^-$ and IL-3R$^-$GPA$^-$CD34$^-$CD36$^+$, respectively (Fig. 4).

3.3.5 Isolation of BFU-E and CFU-E Cells by FACS

1. Collect the erythroid cells at day 5 of culture. Wash with PBS-BSA buffer and centrifuge at 200 × *g* for 10 min at 4 °C.
2. Resuspend up to 10 × 10^6 cells in 800 μl of PBS-BSA buffer, add 100 μl blocking buffer and incubate at 4 °C for 10 min.

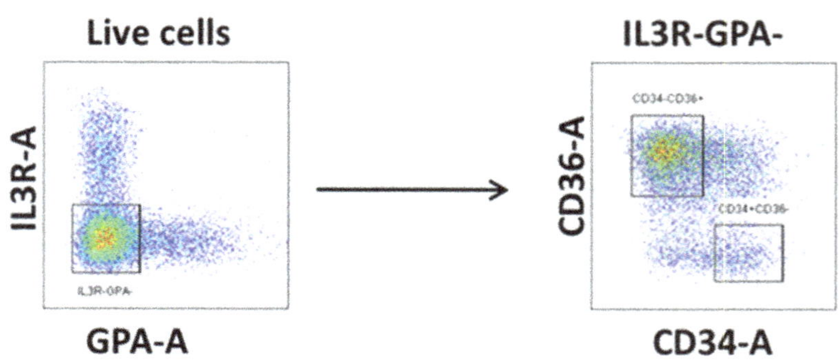

Fig. 4 Flow cytometry analysis of human BFU-E and CFU-E from in vitro culture

3. Stain the cells as follows (for 10×10^6 cells):

   PE mouse anti human CD34: 40 μl.

   PE-Cy7 mouse anti human IL-3R (CD123): 40 μl.

   FITC mouse anti human CD36: 20 μl.

   APC mouse anti human GPA: 0.4 μl.

4. Mix well and incubate at room temperature for 15 min, keep in the dark.
5. Wash the cells twice with 40 mL PBS-BSA buffer and resuspend cells in PBS-BSA buffer at density of 15×10^6 cells/ml.
6. Stain with the viability dye 7-AAD (15 μl/10×10^6 cells) on ice for 10 min in the dark.
7. Follow the manufacturer's protocol for cell sorter and the gating strategy described in Subheading 3.3.4.
8. Collect populations of interest in FBS.
9. Wash cells with PBS and centrifuge at $300 \times g$ for 10 min at 4 °C.
10. Aspirate the supernatant and the purified BFU-E and CFU-E cells are ready for downstream applications.

3.4 Isolation of Primary Human BFU-E and CFU-E Cells from Bone Marrow

3.4.1 Preparation of Human Bone Marrow Cells

1. Dilute bone marrow sample with equal volume of buffer II and following the procedure described in Subheading 3.3.1 to isolate mononuclear cells.
2. Resuspend 10×10^6 cells in 80 μl of ice-cold buffer II, add 20 μl CD45 microbeads, and incubate at 4 °C for 15 min.
3. Wash and apply cell suspension to MACS column as described in Subheading 3.3.1.
4. Collect the flow-through containing $CD45^-$ cells for the isolation of erythroblasts at different stages.
5. Flush out the $CD45^+$ cells with 2 ml of buffer II for isolation of BFU-E and CFU-E.

3.4.2 Antibody Staining

1. Resuspend 2×10^5 $CD45^+$ cells in 22.5 μl of PBS-BSA buffer, add 2.5 μl blocking buffer, and incubate at 4 °C for 10 min.
2. Stain the cells as below:

   PE mouse anti human CD34: 2 μl.

   FITC mouse anti human CD36: 1 μl.

   APC mouse anti human GPA (1:40 diluted): 2 μl.

   APC mouse anti human CD41(1:10 diluted): 2.5 μl.

   V450 mouse anti human CD45(1:10 diluted): 2.5 μl.

   PE-Cy7 mouse anti human IL-3R (CD123): 2 μl.

   PerCp mouse anti human CD3: 2 μl.

PerCp mouse anti human CD4: 2 μl.

PerCp mouse anti human CD14: 2 μl.

PerCp mouse anti human CD19: 2 μl.

APC-H7 mouse anti human CD71: 0.5 μl.

3. Mix well and incubate at room temperature for 30 min in the dark.
4. Wash and stain with the viability dye as described in Subheading 3.3.3.
5. Run the stained samples on BD FACSCanto II.

3.4.3 Flow Cytometry Analysis

1. Conduct flow cytometry analysis using BD FACSDiva.
2. Plot SSC vs CD45 expression of all cells to gate $CD45^+$ cells as P1.
3. Plot SSC vs CD3, CD4, CD14, CD19, and 7AAD of $CD45^+$ population, gate the negative population as P2.
4. Plot SSC vs CD41 and GPA of the P2 population to gate the negative population as P3.
5. Plot SSC vs IL-3R of the P3 population to gate $IL\text{-}3R^-$ population.
6. Plot SSC vs CD36 expression of $IL\text{-}3R^-$ population, resulting in $CD36^-$ and $CD36^+$ populations.
7. Plot CD34 vs SSC of the $CD36^-$ population to gate the $CD34^+$ population as BFU-E, which gave rise to BFU-E colonies.
8. Plot CD71 vs SSC of the $CD36^+$ population to gate the $CD71^+$ population as CFU-E, which gave rise to CFU-E colonies (Fig. 5).

3.4.4 Isolation of Human Bone Marrow BFU-E and CFU-E Cells by FACS

1. Resuspend up to 10×10^6 $CD45^+$ cells in 800 μl of PBS-BSA buffer, add 100 μl blocking buffer and incubate at 4 °C for 10 min.
2. Stain the cells as below (for 10×10^6 cells):

PE mouse anti human CD34: 40 μl.

FITC mouse anti human CD36: 20 μl.

APC mouse anti human GPA: 2 μl.

APC mouse anti human CD41: 10 μl.

V450 mouse anti human CD45: 10 μl.

PE-Cy7 mouse anti human IL-3R (CD123): 40 μl.

PerCp mouse anti human CD3: 40 μl.

PerCp mouse anti human CD4: 40 μl.

PerCp mouse anti human CD14: 40 μl.

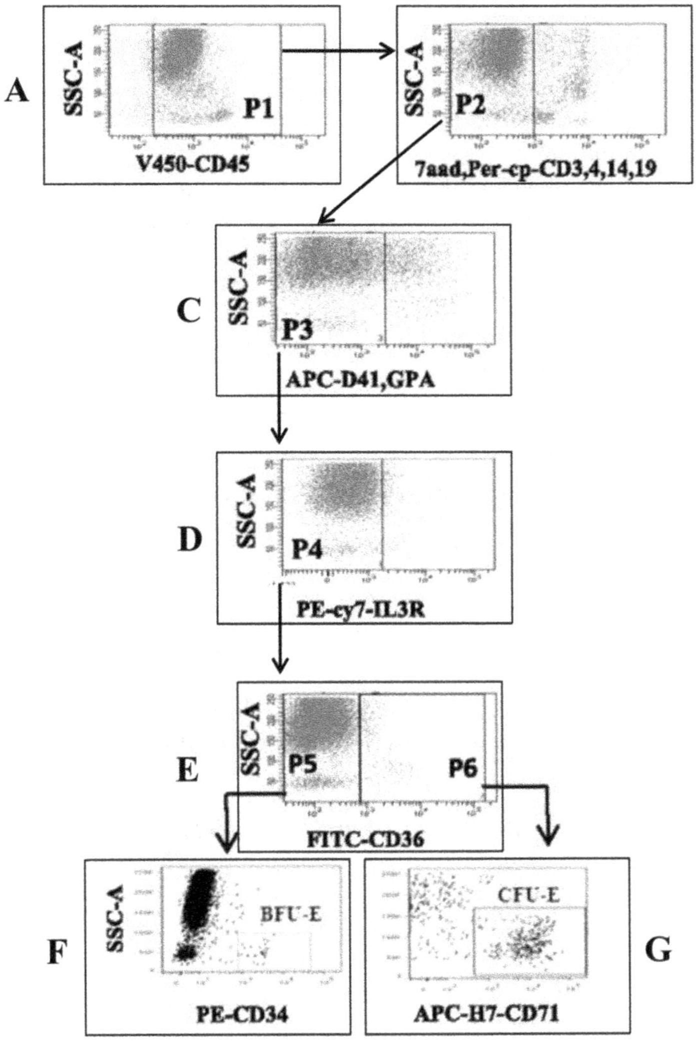

Fig. 5 Flow cytometry analysis and isolation of human BFU-E and CFU-E cells from primary samples. (**a**) Plot of SSC vs CD45 expression of all cells being analyzed; (**b**) SSC vs CD3, CD4, CD14, CD19, and 7AAD of $CD45^+$ population (P1); (**c**) Plot of SSC versus CD41 and GPA of the P2 population, namely P3; (**d**) Plot of SSC versus IL-3R expression of the P3 population; (**e**) Plot of SSC versus CD36 expression of $IL\text{-}3R^-$ population, resulting in $CD36^-$ population (P5) and $CD36^+$ populations (P6); (**f**) Plot of CD34 versus SSC of the $CD36^-$ population revealed $CD34^+$, which gave rise to BFU-E colonies; (**g**) Plot of CD71 versus SSC of the $CD36^+$ cells revealed $CD71^+$ population, which gave rise to CFU-E colonies

PerCp mouse anti human CD19: 40 μl.

APC-H7 mouse anti human CD71: 10 μl.

3. Wash cells and stain with the viability dye as described in Subheading 3.3.5.

4. Follow the manufacturer's protocol for cell sorter and the gating strategy described in Subheading 3.4.4.
5. Collect the populations of interest and wash as described in Subheading 3.3.5.

3.5 Isolation of Human Erythroblasts from Cultured CD34⁺ Cells

3.5.1 In Vitro Culture of CD34⁺ Cells

1. Resuspend $CD34^+$ cells with the Phase I Erythroid Culture Medium at the cell density of 10^5 cells/ml.
2. Culture the cells at 37 °C in a humidified atmosphere with 5% CO_2 in air.
3. Refresh the Phase I medium at day 4, adjust the cell density to 10^5 cells/ml.
4. Change to the Phase II medium at day 7, keep the cell density of 10^5 cells/ml.
5. Change to the Phase III medium at day 11, adjust the cell density to 10^6 cells/ml.
6. Refresh the Phase III medium at day 15, adjust the cell density to 5×10^6 cells/ml.
7. Complete the erythroid cell culture at day 18.

3.5.2 Antibody Staining

1. Collect 1×10^5 cells at different day of culture, wash and blocking as described in Subheading 3.3.3.
2. Stain cells as below:

   PE Mouse anti human CD235a (1:50 diluted): 1 μl

   APC mouse anti human α4-integrin: 1.3 μl

   DyLight 488-conjugated mouse anti human Band3 (1:10 diluted): 2.5 μl

3. Wash and stain with the viability dye as described in Subheading 3.3.3.
4. Run the stained samples on BD FACSCanto II.

3.5.3 Flow Cytometry Analysis

1. Conduct flow cytometry analysis using BD FACSDiva.
2. Gate out cellular debris and $7AAD^+$ cells.
3. Gate GPA^+ cells to show the expression of α4-integrin and band 3.
5. According to Fig. 6, proerythroblasts (I), early baso erythroblasts (II) and late baso erythroblasts (III) are defined as α4-$integrin^{hi}$/$Band3^{neg}$, α4-$integrin^{hi}$/$Band3^{low}$ and α4-$integrin^{hi}$/$Band3^{med}$ populations, respectively. Polychromatic erythroblasts (IV) and orthochromatic erythroblasts (V) are defined as α4-$integrin^{med}$/band 3^{med} and α4-$integrin^{low}$/band 3^{hi} populations, respectively.

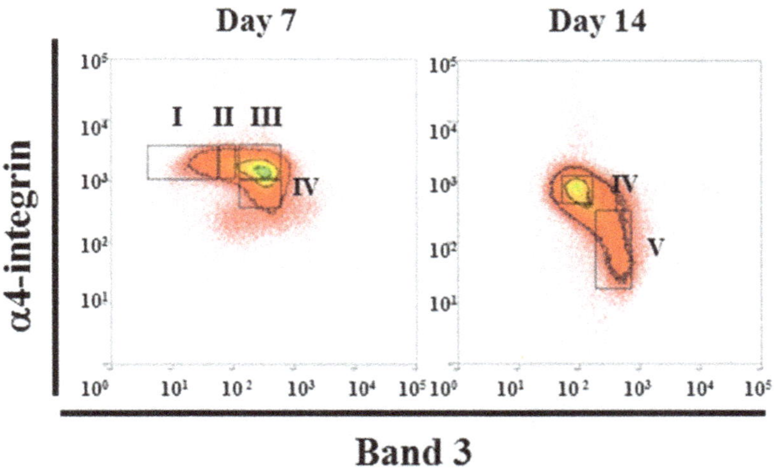

Fig. 6 Flow cytometry analysis of human erythroblasts at distinct stages of development from in vitro culture. The expression levels of α4 integrin of all GPA^+ cells were plotted against the expression levels of band 3. Band 3 negative cells are gated as population I. Population III is represented by the cluster expressing medium level of band 3 and high level of α4 integrin. The region between I and III is gated as population II. Two distinct populations (IV and V) are clearly separated on cells cultured for 14 days

3.5.4 Isolation of Cultured Human Erythroblasts by FACS

1. To sort proerythroblasts, early baso erythroblasts and late baso erythroblasts, collect cells at day 7 of culture; to sort polychromatic erythroblasts and orthochromatic erythroblasts, collect cells at day 14 of culture.
2. Prepare the cells as described in Subheading 3.3.5.
3. Stain the cells as below (for 10×10^6 cells):

   PE Mouse anti human CD235a: 0.4 μl

   APC mouse anti human α4-integrin: 60 μl

   DyLight 488-conjugated mouse anti human band 3: 10 μl
4. Wash cells and stain with the viability dye as described in Subheading 3.3.5.
5. Follow the manufacturer's protocol for cell sorter and the gating strategy described in Subheading 3.5.3.
6. Collect the populations of interest and wash as described in Subheading 3.3.5.
7. Take 5×10^4 cells from sorted populations to prepare cytospin, and perform Giemsa staining to confirm the purity by cell morphology (Fig. 7).

3.6 Isolation of Primary Human Erythroblasts from Bone Marrow

3.6.1 Antibody Staining

1. Following the procedures described in Subheading 3.4.1 to isolate $CD45^-$ cells.
2. Resuspend 2×10^5 $CD45^-$ cells in 22.5 μl of PBS-BSA buffer, add 2.5 μl blocking buffer and incubate at 4 °C for 10 min.
3. Stain cells as described in 3.5.2.

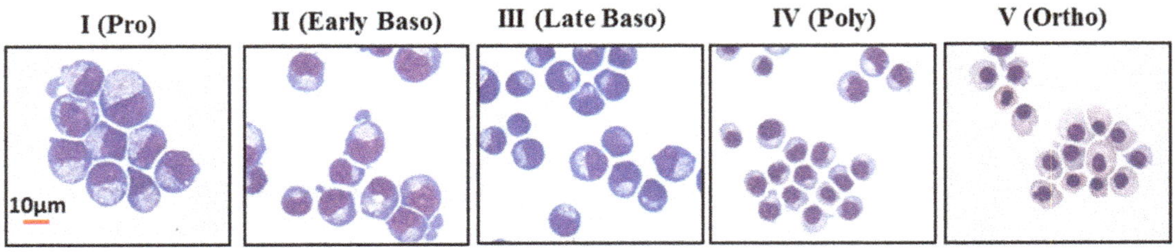

Fig. 7 Morphology of cultured human erythroblasts at distinct stage of development

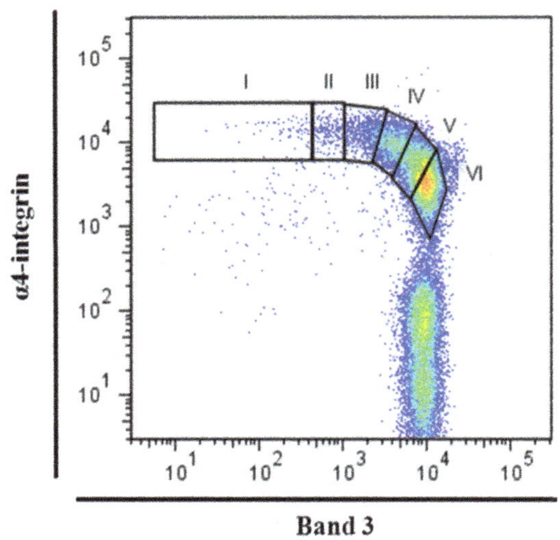

Fig. 8 Flow cytometry analysis of primary human bone marrow erythroblasts at distinct stage of development

4. Wash and stain with the viability dye as described in Subheading 3.3.3.
5. Run the stained samples on BD FACSCanto II.

3.6.2 FACS Analysis

1. Conduct FACS analysis using BD FACSDiva.
2. Gate out cellular debris and $7AAD^+$ cells.
3. Gate GPA^+ cells to show the expression of α4-integrin and Band3.
4. Proerythroblasts (I), early baso erythroblasts (II), late baso erythroblasts (III), Polychromatic erythroblasts (IV) and orthochromatic erythroblasts (V) are gated according to Fig. 8.

3.6.3 Isolation of Primary Human Erythroblasts from Bone Marrow

1. Follow the procedure described in Subheading 3.4.1 to isolate $CD45^-$ cells.
2. Resuspend up to 10×10^6 $CD45^-$ cells in 800 μl of PBS-BSA buffer, add 100 μl blocking buffer and incubate at 4 °C for 10 min.

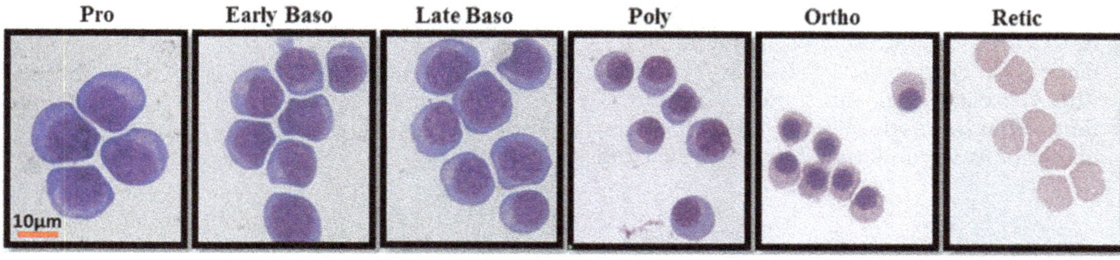

Fig. 9 Morphology of sorted bone marrow erythroblasts at distinct stage of development

3. Stain the cells as described in Subheading 3.5.3 and sort erythroblasts at distinct stage following the gating strategy described in Subheading 3.6.2.
4. Collect the populations of interest and wash as described in Subheading 3.3.5.
5. Take 5×10^4 cells from sorted populations to prepare cytospin, and perform Giemsa staining to confirm the purity by cell morphology (Fig. 9).

References

1. Melamed MR, Mullaney PF, Mendelsohn ML (eds) (1979) Flow cytometry and sorting. John Wiley & Sons, New York, NY
2. Muirhead KA, Horan PK, Poste G (1985) Flow cytometry: present and future. Nat Biotechnol 3:337–356
3. Herzenberg LA, Sweet RG, Herzenberg LA (1976) Fluorescence-activated cell sorting. Sci Am 234:108–117
4. Ibrahim SF, van den Engh G (2003) High-speed cell sorting: fundamentals and recent advances. Curr Opin Biotechnol 14(1):5–12
5. Mattanovich D, Borth N (2006) Applications of cell sorting in biotechnology. Microb Cell Fact 21(5):12
6. Herzenberg LA, Parks D, Sahaf B, Perez O, Roederer M, Herzenberg LA (2002) The history and future of the fluorescence activated cell sorter and flow cytometry: a view from Stanford. Clin Chem 48:1819–1827
7. Stephenson JR, Axelrad AA, McLeod DL, Shreeve MM (1971) Induction of colonies of hemoglobin synthesizing cells by erythropoietin in vitro. Proc Natl Acad Sci U S A 68:1542–1546
8. Palis J (2008) Ontogeny of erythropoiesis. Curr Opin Hematol 15(3):155–161
9. McLeod DL, Shreeve MM, Axelrad AA (1974) Improved plasma culture system for production of erythrocytic colonies in vitro: quantitative assay method for CFU-E. Blood 44:517–534
10. Iscove NN, Sieber F, Winterhalter KH (1974) Erythroid colony formation in cultures of mouse and human bone marrow: analysis of the requirement for erythropoietin by gel filtration and affinity chromatography on agarose-concanavalin A. J Cell Physiol 83:309–320
11. Flygare J, Rayon Estrada V, Shin C, Gupta S, Lodish HF (2011) HIF1alpha synergizes with glucocorticoids to promote BFU-E progenitor self renewal. Blood 117(12):3435–3444
12. Chen K, Liu J, Heck S, Chasis JA, An X, Mohandas N (2009) Resolving the distinct stages in erythroid differentiation based on dynamic changes in membrane protein expression during erythropoiesis. Proc Natl Acad Sci U S A 106(41):17413–17418
13. Liu J, Zhang J, Ginzburg Y, Li H, Xue F, De Franceschi L, Chasis JA, Mohandas N, An X (2013) Quantitative analysis of murine terminal erythroid differentiation in vivo: novel method to study normal and disordered erythropoiesis. Blood 121(8):e43–e49
14. Li J, Hale J, Bhagia P, Xue F, Chen L, Jaffray J, Yan H, Lane J, Gallagher PG, Mohandas N, Liu J, An X (2014) Isolation and transcriptome analyses of human erythroid progenitors: BFU-E and CFU-E. Blood 124(24):3636–3645

15. Hu J, Liu J, Xue F, Halverson G, Reid M, Guo A, Chen L, Raza A, Galili N, Jaffray J, Lane J, Chasis JA, Taylor N, Mohandas N, An X (2013) Isolation and functional characterization of human erythroblasts at distinct stages: implications for understanding of normal and disordered erythropoiesis in vivo. Blood 121 (16):3246–3253
16. An X, Schulz VP, Mohandas N, Gallagher PG (2015) Human and murine erythropoiesis. Curr Opin Hematol 22(3):206–211

Chapter 10

Analysis of Erythropoiesis Using Imaging Flow Cytometry

Theodosia Kalfa and Kathleen E. McGrath

Abstract

Erythroid maturation has been classically defined based on the remarkable changes visualized through microscopy. These involve the decrease in cell size, nuclear condensation and organelle loss, and include the final unique asymmetric division creating the short-lived nucleated pyrenocyte and the enucleate reticulocyte that matures into the red blood cell. Understanding the regulation of these processes has been challenging due to the difficulty in obtaining sufficient numbers of cells, particularly of rare intermediates, to study by microscopy. While flow cytometry can provide quantitative analysis of high cell numbers as well as critical tools for assaying processes like cell cycle, apoptosis and cell signaling, it cannot analyze or categorize cells based on morphology. Imaging flow cytometry (IFC) combines microscopy and flow cytometry by capturing brightfield and fluorescent images of large numbers of cells, which can be quantitated for both morphometric and fluorescent characteristics. Over the past 10 years, this approach has been increasingly used to study aspects of erythropoiesis. This chapter describes how to utilize IFC to enumerate multiple specific stages of erythropoiesis from primary tissue, as well as how to culture primary progenitors to enrich for the rare late stage enucleating cells in order to examine intracellular proteins involved in enucleation. These methods demonstrate the approaches and strength of IFC as a tool to bridge the power of microscopy and flow cytometry to more fully interrogate erythropoiesis.

Key words Erythropoiesis, Erythroblast, Enucleation, Bone marrow, ImageStream, Imaging flow cytometry

1 Introduction

The analysis of intermediate stages of erythropoiesis has greatly improved our understanding of the regulation of erythroid cell production. Quantitative colony-forming assays have been valuable for the study of the early progenitor phase of erythropoiesis. However, the later precursor stages have been more challenging to study due, in part, to the need to consistently quantify morphological characteristics important in their maturation, as well as the dynamic and short-lived nature of some critical stages, such as enucleation. Imaging flow cytometry (IFC) combines the power of flow cytometry for single-cell analysis of large numbers of cells with the acquisition of morphological data for each cell. Brightfield and

Joyce A. Lloyd (ed.), *Erythropoiesis: Methods and Protocols*, Methods in Molecular Biology, vol. 1698,
DOI 10.1007/978-1-4939-7428-3_10, © Springer Science+Business Media LLC 2018

multiple fluorescent images are captured that can be analyzed not only for intensity, but also for the shape, size, and patterns of that intensity. Not surprisingly, one of the earliest uses for this technology was delineating erythroblast stages, and IFC continues to be utilized in an expanding variety of applications to analyze shape changes, heterochromatin, cell cycle, apoptosis, cell signaling, and enucleation of erythroid cells [1–8]. In this paper, methods are presented for the IFC analysis of erythroid intermediates in primary tissue, as well for the qualitative and quantitative evaluation of intracellular structures in cultured erythroid cells. Erythroid precursor populations in bone marrow are distinguished utilizing a combination of immunophenotype and morphometric criteria that correlate with classically defined stages of erythropoiesis (Fig. 1) [1, 9, 10]. Characteristics of size, shape, texture, and distribution of the fluorescent stains and brightfield images are used to remove artifacts with similar immunophenotypes, which cannot be distinguished from real events with regular flow cytometry [11]. In order

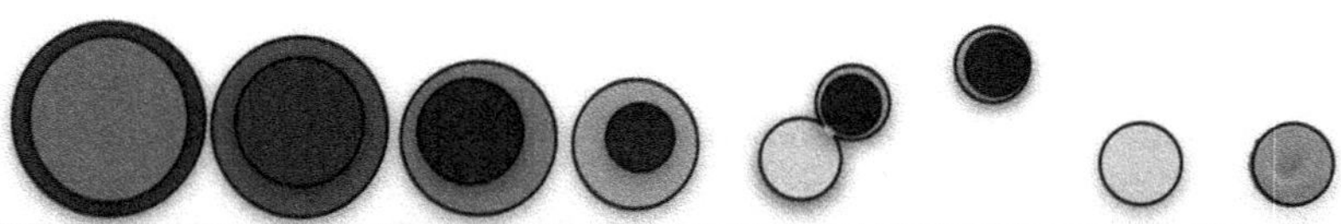

	ProE	BasoE	PolyE	OrthoE	Enucl.	Pyren.	Retic	RBC
Cell Size/Shape	Large	→	→	Small	Elongated	Small	Small	Small
Nuclear Size	Large	→	→	Small	Small Off Center	Small	-	-
Kit	Lo	-	-	-	-	-	-	-
Ter119 (Total)	Lo	hi	hi	hi	hi/lo	Lo	hi	hi
CD71	+	+	+	+	+	+	+	-
Ter119 (Median)	Dim	→	→	Bright	Bright/Dim	Dim	Bright	Bright

Fig. 1 Maturation characteristics of erythropoiesis. The earliest precursor, the proerythroblast (ProE), is defined in Giemsa stained histological samples as a large cell with highly basophilic cytoplasm and a very large nucleus. As maturation proceeds, successive erythroblasts (basophilic (BasoE), polychromatophilic (PolyE), and orthochromatic (OrthoE)) decrease in cell size and cytoplasmic basophilic staining as well as nuclear size as the nucleus condenses. Immunophenotypically, ProE still express a low level of kit as they are turning off its expression, which marked earlier progenitor stages. All erythroblasts express the erythroid-specific marker Ter119, which begins at a low level in ProE, increasing to high in BasoE and then remaining at a similar level throughout the rest of erythropoiesis. However, due to the decreasing size of the cell, the median Ter119 intensity is increasing with maturation and this can be detected by IFC. OrthoE will then enucleate (Enucl.) through an asymmetric cytokinesis process that produces an enucleate reticulocyte (Retic) and a nucleated pyrenocyte (Pyren.). During this process, Ter119 concentrates on the reticulocyte part of the enucleating cell leaving the pryenocyte dim for this marker. The retic continues to mature by changing shape, increasing in hemoglobin concentration and loosing expression of CD71

to study the rare, transient process of enucleation, erythroid cells are first cultured to enrich for terminal maturing forms. Then they are analyzed by IFC, providing a population in which the intracellular structures involved in this process can reliably be investigated, by gating enucleating cells based on their morphologic characteristics [5].

2 Materials

2.1 Isolation of Murine Bone Marrow

1. 7–9-week-old mouse (*see* **Note 1**).
2. PB2 solution: 1 L packet of Dulbecco's phosphate-buffered saline (DPBS) w/calcium chloride, Calcium chloride (provided w/PBS), 1.0 g of Glucose, 3.0 g of BSA, tissue culture grade water to 1 L. Sterilize, passing through a 0.2 μm sterile filter, and store at 4 °C.
3. PB2 + heparin: Supplement PB2 solution with 25 μg/ml heparin (*see* **Note 2**).
4. 5 mL syringes.
5. 22 gauge (22-G) needle (length = 25 mm).
6. 20 gauge (20-G) needle (length = 40 mm).
7. Scalpel.
8. Dissecting scissors.
9. Forceps.
10. 15 mL conical tubes.
11. 40 μm nylon cell strainers.
12. Hemocytometer.
13. Microscope.

2.2 Erythropoiesis Culture from Low-Density Bone Marrow (LDBM) Cells

1. IMDM with 2% Fetal Bovine Serum (FBS).
2. Density gradient cell separation medium (Histopaque) 1.083 mg/ml (Sigma).
3. Erythroblast growth medium (EGM): StemPro-34 medium with 2.6% StemPro-34 medium supplement (Gibco Invitrogen), 20% BIT-9500 (Stemcell Technologies), 900 ng/ml ferrous sulfate, 90 ng/ml ferric nitrate, 100 U/ml penicillin/streptomycin, 2 mM L-glutamine, 1 μM hydrocortisone and freshly added cytokines to a final concentration 100 ng/ml SCF, 5 ng/ml IL-3, and 3 IU/ml Erythropoietin.
4. 6-well cell-culture plates.
5. Cell-culture incubator (37 °C/5% CO_2).
6. MS5 cells (murine stromal cell line).

7. Culture medium for MS5: α-MEM containing 20% FBS, 100 U/ml penicillin/streptomycin, and 2 mM L-glutamine.
8. Hemocytometer and microscope or alternatively an automated cell counter.

2.3 Cell Staining and Preparation for IFC

1. Blocking buffer: 10% normal rat serum prepared in PB2 (*see* **Note 3**).
2. 5 mL round-bottom FACS tubes.
3. Swinging bucket centrifuge.
4. Anti-mouse antibodies: PE-CD71 and unconjugated CD71 (clone R17217) PE-CF594 CD117 (e.g., Kit, clone 2B8), FITC-Ter119 (clone TER-119) from eBioscience (*see* **Note 4**).
5. 1.7 mL microfuge tubes.
6. DRAQ5.
7. Compensation beads that bind antibodies (such as BD Comp-Bead Anti-Rat Ig, BDBioscience).

2.4 Fixation and Permeabilization of Cells, IFC Data Acquisition, and Analysis

1. Stains for intracellular proteins/structures, such as F-actin antibody with AF488-phalloidin, or β-tubulin antibody with AF488 (Invitrogen) and Ter119-PECy7.
2. 3.7% formaldehyde in PBS.
3. Ice-cold 100% acetone. Keep in a −20 °C freezer.
4. Ice-cold 50% acetone. Dilute 100% acetone 1:1 with ddH_2O.
5. 1.7 mL microfuge tubes.
6. Bench mini microcentrifuge.
7. Imaging flow cytometer with Inspire collection software (MilliporeSigma) (*see* **Note 5**).
8. IDEAS analysis software (MilliporeSigma).

3 Methods

These methods describe two overlapping approaches. The first is quantifying erythropoietic intermediates in live bone marrow samples. The second is the analysis of intracellular structures in fixed and permeabilized enucleating erythroid cells, produced in ex vivo in expanded erythroid cultures. These techniques can be adapted to alternative starting materials and expanded to other morphometric and flow cytometric analyses.

3.1 Isolation of Bone Marrow Cells

1. Mice are euthanized as per institutional protocol (e.g., CO_2 inhalation, followed by cervical dislocation).
2. Remove hind limbs. Dissect and clean femurs (*see* **Note 6**).

3. Open marrow cavity by cutting both ends of the bone with the scalpel.
4. Flush in a distal to proximal direction (knee to hip) with 4 mL of PB2 + heparin in 5 mL syringe with 22-G needle into a 15 mL conical tube.
5. Dissociate clumps of marrow by drawing suspension in and out of a 5 mL syringe with 20-G needle three to four times.
6. Pass cell suspension through a 40 μm nylon cell strainer on a 5 mL round-bottom tube.
7. Count single-cell suspension using a hemocytometer and determine total cells per flushed femur. This number can be used to determine the total number of erythroid precursors per femur (*see* **Note 7**).

3.2 In Vitro Erythropoiesis Culture

To produce an erythroblast population enriched in enucleating cells, we use a 3-phase erythroid differentiation culture protocol, based on Giarratana et al. [12]. As described here, it is modified and adapted for mouse cells. To obtain the cells for the cultures, low-density bone marrow cells (LDBM) are prepared by density gradient centrifugation of the bone marrow cell suspension obtained in Subheading 3.1.

1. Carefully layer 5 mL of bone marrow cell suspension on 5 mL of 1.083 g/ml density gradient cell separation medium (Histopaque) in a 15 mL tube and spin at 750 × *g* for 25 min at room temperature (RT) with no brake/low acceleration.
2. Transfer the supernatant that contains the buffy coat (collect everything up to about the 2 mL mark of the 15 mL tube) to a new 15 mL tube and perform one more wash with IMDM + 2% FBS at 525 × *g* for 5 min at RT.
3. Aspirate the supernatant. Suspend the pellet in 3 mL RBC lysis buffer and incubate for 5 min at RT to lyse any remaining red blood cells (RBC).
4. Add 7 mL IMDM + 2% FBS to rinse and spin at 525 × *g* for 5 min at RT.
5. Aspirate the supernatant and resuspend the pelleted cells in 2 mL erythroblast growth medium (EGM), with freshly added cytokines.
6. Count the cells using a hemocytometer or an automated cell counter.
7. Adjust concentration of cells in EGM to 2×10^5 LDBM cells/ml.
8. Plate 2.5 mL of cells/well (5×10^5 cells/well) in a 6-well cell-culture plate and incubate at 37 °C/5% CO_2. This is considered day #0 of culture.

9. Count cells on day #2, day #3, and day #4 and change the medium on each of these days with fresh medium containing all three cytokines. Resuspend cells such that the cell concentration is $<5 \times 10^5$ cells/well, and use the correspondingly appropriate number of wells containing 2.5 mL. This will optimize the proliferative phase. Typically, cells need to be spilt into additional wells every 2–3 days.
10. On day 4, plate MS5 cells (murine stromal cell line) to wells of a new 6-well cell-culture plate in a MS5 cell-culture medium.
11. On day 5, lift the erythroid cells from each well of the erythroid cultures, count number of cells (which are now significantly enriched in erythroblasts), and transfer to 15 mL conical tubes.
12. Spin down at $525 \times g$ for 5 min at RT, aspirate the supernatant, and dissolve pellets with fresh EGM containing only Epo (3 IU/ml) (no SCF or IL3) at a concentration of 2×10^5 cells/ml.
13. Aspirate the supernatant from the wells with the MS-5 cells (should be at about 70–80% confluency at this time) and add the erythroid cells in EGM + Epo.
14. On day #6, change medium adding fresh EGM + Epo.
15. On day #7, change the medium to EGM with no cytokines.
16. On days 7–9, test samples daily for enucleation with flow cytometry after staining with anti-Ter119 and a nuclear stain (e.g., Syto-16) and proceed to staining samples for IFC (*see* **Note 8**).

3.3 Bone Marrow Live-Cell Staining and Preparation for IFC

1. Pellet 5×10^6–1×10^7 cells in a 5 mL round-bottom tube for 5 min at $200 \times g$ at 4 °C.
2. Resuspend the cell pellet in 50 μl of blocking buffer. Incubate for 20 min on ice.
3. Dilute CD71-PE with 2 parts unstained CD71, due to its abundant number of targets.
4. To make the erythroblast panel antibodies, make a solution in PB2 containing 1:50 dilutions of FITC-Ter119, PE-Cy594-CD117-Kit and the PE/unstained CD71 from **step 3**.
5. Add 50 μl of erythroblast panel antibodies prepared in PB2 to the cell suspension from **step 2**. Incubate on ice in the dark for 30 min.
6. Wash cells with 4 mL of PB2 and pellet for 5 min at $200 \times g$ at 4 °C.
7. Remove all but approximately 100 μl of supernatant, resuspend the pellet and transfer to a 1.7 mL microfuge tube. Keep samples on ice and in the dark.

8. Add 2.5 μM DRAQ5 to each sample 5–10 min prior to data collection on an imaging flow cytometer.
9. While cells are staining, make single-stained controls for compensation. For DRAQ5- have one 1.7 mL tube with one million cells in 100 μl PB2 and add DRAQ5 1–10 min prior to data collection. For each antibody, add 0.5 μl antibody to one drop of positive compensation beads (CompBeads) in a 1.7 mL microfuge tube. After 5 min, wash the antibody single-stained controls with 1 mL PB2 and pellet at 5 min at 200 × *g* at 4 °C. Remove all but approximately 100 μl of supernatant. Keep samples on ice and in the dark.

3.4 Fixation, Permeabilization, and Staining of Cultured Erythroid Cells

The following steps are used to visualize intracellular proteins [13] (*see* **Note 9**).

1. Wash cells in PBS, spin 525 × *g* for 5 min at RT, and aspirate the supernatant.
2. Resuspend the cell pellet by gentle pipetting in 500 μl of 3.7% formaldehyde in PBS, transfer to a 1.7 mL plastic centrifuge tube, and incubate for 15 min at RT (*see* **Note 10**).
3. Spin at 2000 × *g* for 20 s in a microcentrifuge, aspirate the supernatant, and perform one wash by adding 500 μl PBS to each tube and pipetting gently. In all subsequent steps in this heading, to spin indicates 2000 × *g* for 20 s in microcentrifuge.
4. Spin, aspirate supernatant except the last 50 μl. Gently resuspend pellet in remaining PBS and keep tubes on ice for at least 15 min.
5. Take acetone solutions out from −20 °C freezer and put on ice. Permeabilize cells by resuspending cell pellet in 500 μl ice-cold 50% acetone (1:1 with dH_2O) with gentle pipetting.
6. Spin, aspirate the supernatant and resuspend cell pellet in 500 μl ice-cold 100% acetone with gentle pipetting.
7. Spin, aspirate supernatant and resuspend cell pellet in 500 μl ice-cold 50% acetone with gentle pipetting.
8. Spin, aspirate the supernatant and resuspend cells in cold FACS buffer (PBS + 0.5% BSA) for blocking. Incubate for 20 min on ice. Divide into four aliquots in different 1.7 mL tubes to have the experimental sample plus three samples for single-color stain (intracellular molecule of interest, Ter119, and DRAQ5).
9. Prepare labeling cocktail with antibodies and markers for the molecules of interest: 1 μl/100 μl Ter119-PECy7 for erythroid cell staining and 0.1 U/100 μl AF488-phalloidin for F-actin staining or AF488-anti-β-tubulin antibody (1:50) to visualize microtubules (*see* **Note 11**).

10. Spin cells, aspirate the supernatant, and resuspend cell pellets in 100 μl of the antibody/marker cocktail or single-color stain (except DRAQ5), pipette gently and incubate for 30 min at RT.
11. Prepare FACS buffer containing 2.5 μM of the nuclear stain DRAQ5.
12. Wash cells in FACS buffer, spin, aspirate the supernatant, and resuspend the experimental sample and the single-color DRAQ5 calibration sample in 60 μl FACS buffer containing DRAQ5. Resuspend the other two single-color stain controls in 60 μl of FACS buffer.

3.5 IFC Setup and Acquisition

1. Activate channels and turn on lasers based on stains used and manufacturer's instructions for the imaging flow cytometer. Use the 40× objective for collection of live cells and the 60× objective for imaging of intracellular structures (*see* **Note 12**).
2. Begin acquisition with a fully stained bright sample and confirm that laser settings do not cause saturation of any of the stains (*see* **Note 13**). For each sample of live bone marrow, collect one or more files of 50,000 events, depending on the rarity of the events being analyzed. For standard quantitative analysis of live bone marrow, two files are sufficient. For qualitative analysis of intracellular structures in enucleating cells starting from cultured fixed and permeabilized cells, the cells lost during permeabilization may limit the number of cells available to be collected. We aim to collect at least 10,000 events, typically having at least 100 enucleating cells for analysis.
3. Collect single-stained controls for spectral compensation as per manufacturer's instructions.

3.6 IFC Gating of Erythropoietic Intermediates

See Fig. 1 for reference to stages and gate according to the example shown in Fig. 2 (*see* **Note 14**).

1. Compensate the raw image files using the macro "wizard" in IDEAS software.
2. Open up the compensated image file for the control sample of your data set (*see* **Note 15**) in IDEAS and optimize color and image sensitivity using the "begin analysis" wizard.
3. Create all new masks and features not included in the default program as specified in Table 1. *Single cells are gated out of All* (Fig. 2a, i) using aspect ratio (elongatedness) and area of brightfield images and gated to include the large proerythroblasts (like cell #2) and elongated erythrocytes (#14) and enucleating orthochromatic erythroblasts (OrthoE) (#10) while still excluding large clumps of cells (like events # 17, 18, and

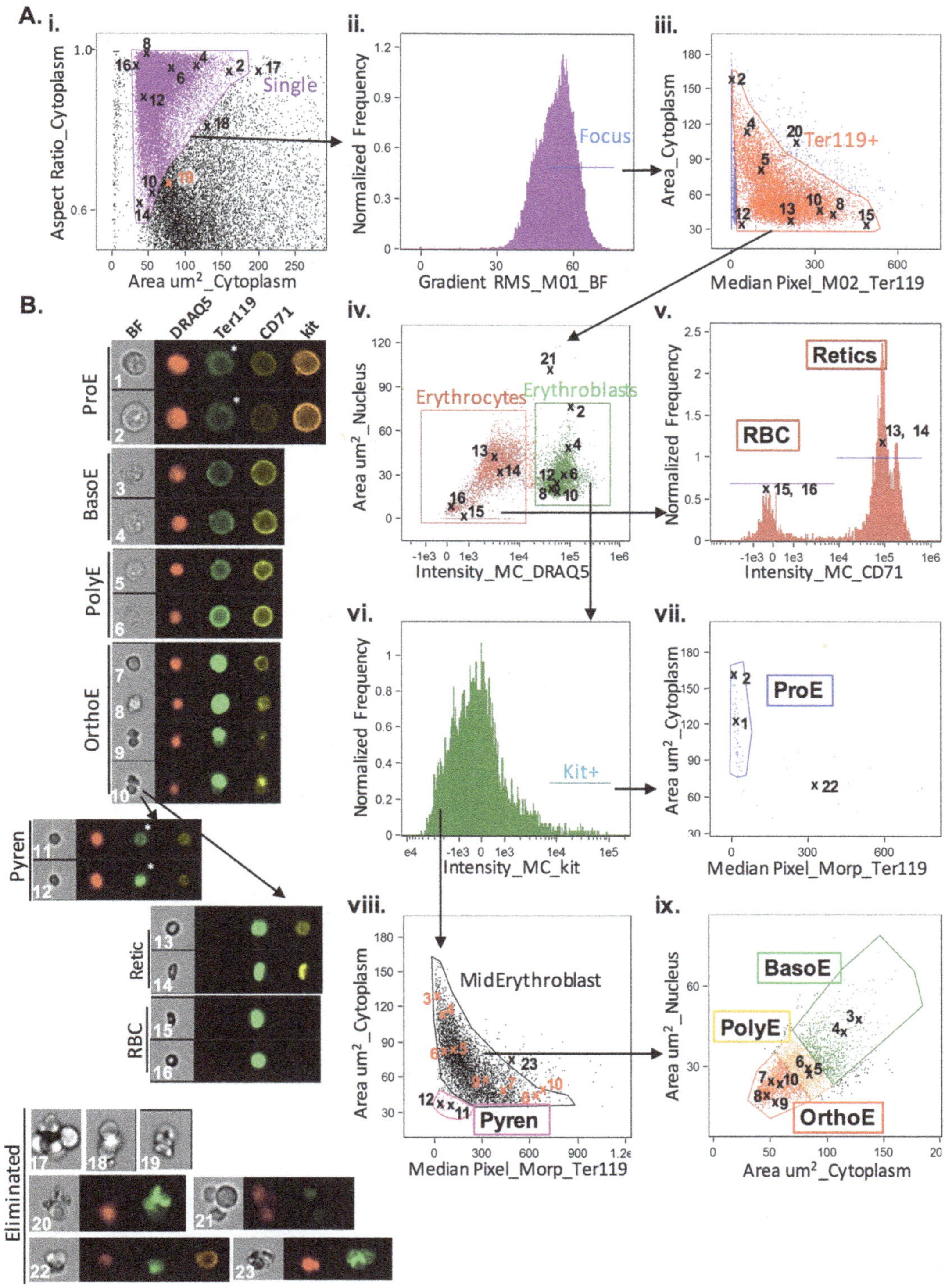

Fig. 2 Initial gating of erythroid intermediates: (**a**) Gating strategy for identifying seven intermediates of erythropoiesis. Table 1 contains a full description of the features on the axes. (**b**) Images of cells in the plots in (**a**), numbers correlate to the *x* that marks the cell-position in the graph. Cells #*1* through *16* represent examples of

19, Fig. 2b). Backgated populations can be used to assure this gate is set for yield to capture the full range of erythroid intermediates. Further clumps (#20–23 for example) will be removed on successive gating.

4. *Focused cells are gated out of Single* (Fig. 2a, ii) based on gradient RMS of brightfield images. Setting of 50 or above selects cells sufficiently focused to "mask" the cytoplasm and nucleus for this analysis.
5. *Erythroid (Ter119 positive) cells are gated out of focused* (Fig. 2a, iii) using median intensity of Ter119 and cell size to retain the dim large proerythroblasts (#2) and small dim pyrenocytes (#12) while excluding bright larger clumps of mature erythrocytes (#20).
6. *Nucleated erythroblasts* (#1 through 12) and *enucleated erythrocytes* (#13–16) are gated *out of Ter119 positive cells* (Fig. 2a, iv) based on DNA intensity with elimination of clumps with multiple nuclei by nuclear area (#21).
7. *Reticulocytes* (Retic #13, 14) *and mature RBC* (#15 and 16) *are gated out of erythocytes* (Fig. 2a, v) based on CD71 expression.
8. *Kit positive cells are gated out of erythroblasts* (Fig. 2a, vi) (*see* **Note 16**).
9. *Proerythroblasts* (ProE, #1, 2) *are gated out of Kit positive* (Fig. 2a, vii) based on the further criteria of being large and dim for Ter119 (*see* **Note 12**) unlike small artifact due to doublets (#22).
10. A combined population of erythroblasts not gated as ProE ("erythroblast and not ProE" under population manager in IDEAS) is plotted by cell area and median pixel Ter119 (Fig. 2a, viii) (*see* **Note 17**). *Pyrenocytes* (Pyren #11, 12) are identified as small dim cells while the rest of *mid-erythroblasts* are gated away from small circular clumps of more mature cells that are too bright in Ter119 for their size (#23) relative to the normal continuum found (#3 through 6).
11. The final discrimination of the remaining erythroblasts (Basophilic (BasoE #3, 4) Polychromatophilic (PolyE #5, 6) and orthochromatic (OrthoE#7–10) is made by ranking these cells by cell and nuclear size (Fig. 2a, ix). Standard proportions of each are determined for each strain and then used to gate the control sample for each experimental set (*see* **Note 18**).

Fig. 2 (continued) the erythroid intermediates gated and *#17* through *23* represent clumps of cells that are removed. Images are brightfield (*gray*), DNA stained by DRAQ5 (*red*), Ter119-FITC (*green*), CD71-PE (*yellow*) and kit- PE-Cy594 (*gold*). All the images are to the same scale and exposure except for Ter119 images indicated by *asterisk* which have had the sensitivity for Ter119 increased due to their dim expression

Table 1
Masks and features used in analysis of erythropoiesis by IFC

Axis Label	Feature	Mask
Intensity_DRAQ5	[a]Intensity_MC_DRAQ5	[a]MC
Intensity_CD71	[a]Intensity_MC_CD71	
Intensity_Kit	[a]Intensity_MC_kit	
Gradient RMS_M01_BF	[a]Gradient RMS_M01_BF	[a]M01
Aspect Ratio_BF	[a]Aspect Ratio_M01_BF	
Median Pixel_M02_Ter119	[a]Median Pixel_M02_Ter119	[a]M02
Area um^2_Nucleus	Area_Morphology(M11, DRAQ5)	Morphology (M11, DRAQ5)
Median Pixel_Morph_Ter119	Median Pixel_Morphology(M02, Ter119) _Ter119	Morphology (M02, Ter119)
Area um^2-Cytoplasm	Area_Cytoplasm	"cytoplasm" = (Morphology (M02, Ter119) or AdaptiveErode (M01, BF, 81) or Morphology (M03, CD71))
Aspect Ratio Cytoplasm	Aspect Ratio_cytoplasm	
Delta Centroid Ter119_DRAQ5	Delta Centroid XY_M06_Ter119_IW_M11_DRAQ5_IW (Intensity Weighted)	[a]M06, [a]M11
Delta Centroid BF-DRAQ5	Delta Centroid XY_M01_1_BF_M11_11_DRAQ (not Intensity Weighted)	[a]M01, [a]M11

[a]Masks and feature automatically made by IDEAS, all others need to be made according to the default names. MC = combined mask, M01–M11 = MChannel#, BF = brightfield

12. The gating strategy created for the control sample can now be applied along with the compensation matrix to all of the experimental files for consistent gating. This is done within the "open" function in IDEAS using either a template file created from the bone marrow file or by referring to the data analysis file for the control bone marrow.
13. For future experiments, instead of starting over from **step 3**, apply the previous controls template to the new control as a starting point and only minor adjustments should be needed.

3.7 IFC Gating and Analysis of Enucleating Erythroblasts (Fig. 3a, b)

1. Enucleating erythroblasts are gated out of the orthochromatic erythroblasts or, in a simpler and more inclusive fashion, out of the middle stage erythropoietic intermediates (Fig. 2a, viii) in a scatter plot of "Aspect Ratio_BF" (defined as minor axis/major axis of the cell and decreasing as the cell becomes more

elongated) versus "Delta Centroid Ter119_DRAQ5"(defined by the distance between the centroids of the Ter119$^+$ area and the DRAQ5$^+$ area). The enucleating cells are characterized by low Aspect Ratio and increasing Delta Centroid Ter119_-DRAQ5, as the Ter119$^+$ reticulocyte is separating from the DRAQ5$^+$ nucleus (Fig. 3a).

2. Within the gate of enucleating cells, different stages of enucleation can be visualized in a progressive fashion with decreasing Aspect Ratio_BF and increasing Delta Centroid Ter119_-DRAQ5 (cells #1–5, Fig. 3b), and F-actin organized in a contractile actin ring can be followed in images like the ones numbered 2–5.

3.8 IFC to Evaluate of Polarity of Orthochromatic Erythroblasts

The polarity/asymmetry of orthochromatic erythroblasts with and without inhibition of microtubule polymerization with colchicine can be evaluated quantitatively by IFC analysis (Fig. 3c, d) [13].

1. Orthochromatic erythroblasts (OrthoE) are gated as shown in Fig. 2a, ix.
2. In control WT OrthoE (stained with anti-β-tubulin–AF488 and the nuclear stain DRAQ5), a unipolar microtubule formation is visible (Fig. 3c).
3. In WT erythroblasts incubated with 5 μM colchicine for 6 h in a fast in vitro enucleation assay (*see* **Note 19**), β-tubulin is diffusely stained [5, 13] (Fig. 3c).
4. Delta Centroid BF_DRAQ5 (the distance between the centroids of the BF area and the DRAQ5$^+$ area) is used to evaluate quantitatively the polarization of the nucleus within the OrthoE population, which likely precedes enucleation. Delta Centroid BF_DRAQ5 decreases with colchicine incubation, indicating a decrease in establishment of polarity (Fig. 3d).

4 Notes

1. The data on live bone marrow provided here is from 7 to 9-week-old CD-1 (Charles River) outbred mice; the data on enucleation is from 2 to 6 month old C57Bl/6 mice (Jackson Laboratories). Other strains of mice, genetically modified mouse models, and alternative tissues, e.g., spleen enriched in erythroblasts after induction of stress erythropoiesis [5], have been used with similar approach. However, the proportions of stages of erythroblasts may differ depending on age and strain.
2. PB2 buffer is used throughout processing of live cells. PB2 was optimized for the preservation of erythroblast precursors for flow cytometric analyses [14]. Other buffers will function as

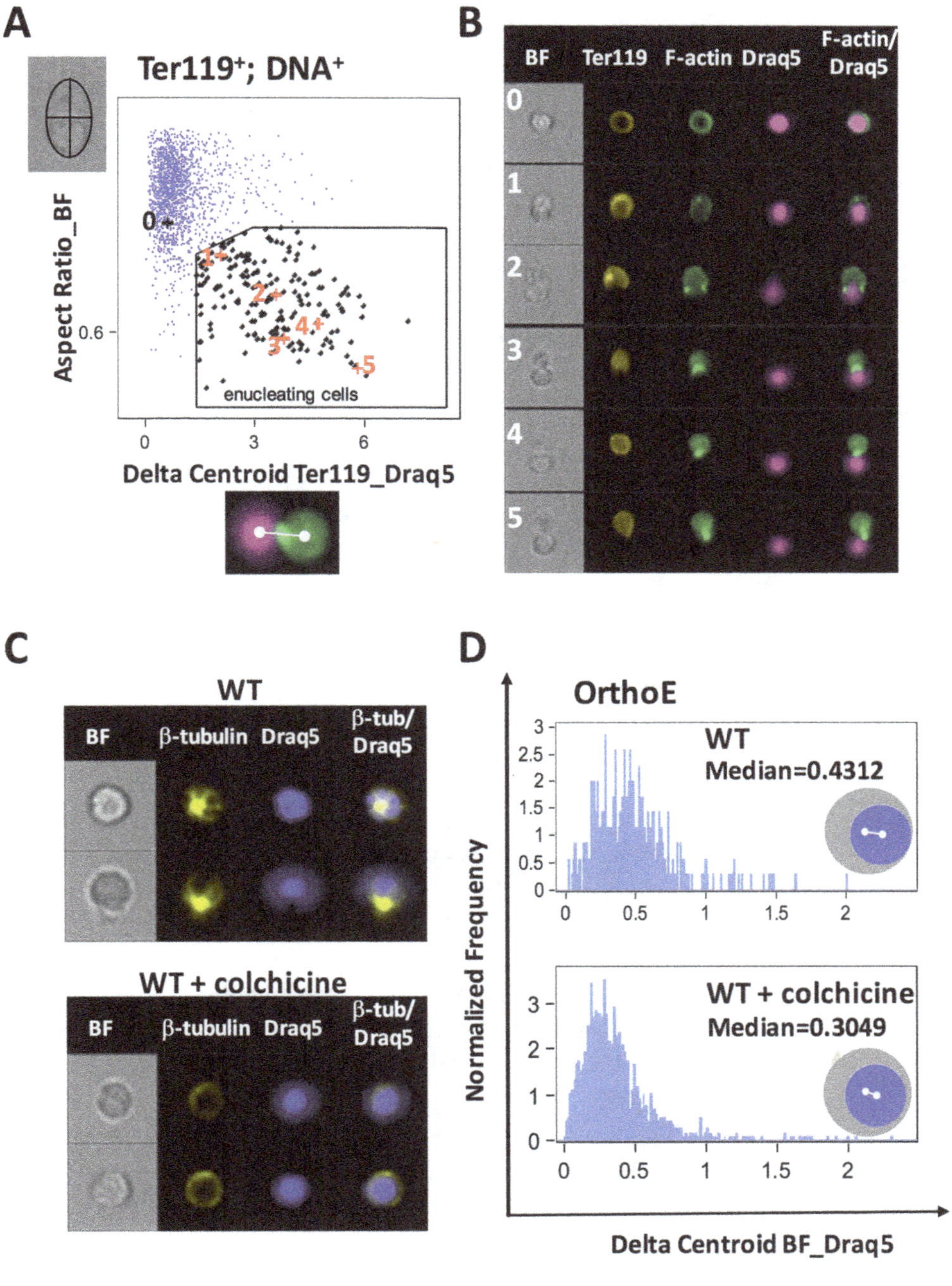

Fig. 3 Analysis of intracellular structures during enucleation: (**a**) Gating of enucleating cells based on their characteristics of decreased aspect ratio in brightfield image (elongated cells) and increased delta centroid (distance between centroids) of the Ter119^{+} reticulocyte and DRAQ5^{+} nucleus. (**b**) Representative images of enucleating WT mouse orthochromatic erythroblasts, produced in ex vivo erythropoiesis culture and stained for Ter119 (*yellow*), F-actin (*green*), and DRAQ5 (*magenda*). Cells are shown at progressive stages of enucleation, as the aspect ratio is decreasing and the delta centroid Ter119_DRAQ5 is increasing (*red-plus sign next to the #1-5* within the "enucleating cells" gate shows the position of the cells #*1–5* imaged on the right. (**c**) A unipolar microtubule assembly can be imaged with IFC in WT orthochromatic erythroblasts, presumably preceding enucleation. This structure is inhibited with colchicine inhibition of microtubule polymerization. (**d**) IFC analysis demonstrates evidence of polarization of the nucleus within the WT orthochromatic erythroblast by increasing distance between the center of the cytoplasm area (Brightfield) and the center of the nucleus (DRAQ5-positive area). This distance and polarization decreased with colchicine inhibition

well when preparing to culture cells, such as phosphate-buffered saline or IMDM + 2% FBS.

3. Normal rat serum is used here because these antibodies are produced in rat. Again, PBS can be substituted for PB2.
4. The antibodies and DNA stain (DRAQ5) listed here will function well, however other fluorescent conjugates or clones of these antibodies or other vital DNA stains such as vibrant violet (Invitrogen) may also work as well and can be used if additional stains for other populations or aspects of erythropoiesis are needed.
5. For these analyses a 2-camera ImageStreamX (MilliporeSigma) imaging flow cytometer with INSPIRE acquisition software and IDEAS analysis software (version 6.2) was used. The ImageStream MarkII would function equivalently. The FlowCyte with Quantitative Imaging upgrade could also be used; however, lower resolution images will require the adjustment of masks and may diminish discrimination of subcellular asymmetry as in Fig. 2.
6. Depending on strain and age, single femur can generate 20–25 million cells, sufficient for live-cell staining. However, a combination of marrow from both femurs, tibiae and pelvic bones may be needed if fixation and permeabilization is to be done, which causes cell loss, or if LDBM cells are going to be derived for culture.
7. Ultimately if numbers of erythroid intermediates are to be compared, it is critical to know the denominator of cell numbers in total sample. Some conditions that affect erythropoiesis will also change numbers of total cells in the marrow as well as the number of circulating RBCs that are included in the bone marrow samples. To calculate number of erythroid cells in the original material, the proportion of erythroid cells per focused gated single cells can be multiplied by total cell number of the sample. To account for loss of RBCs if fixation is used, run an un-fixed sample stained only with DRAQ5 to determine the % nucleated cells and this can be used to determine the total nucleated cells per sample. This then can be multiplied by proportion of erythroid cells per nucleated single cells as determined by IFC to return the total erythorid cell number in the sample.
8. Before plating the LDBM cells and on day 2, 4, 6, and 7 of culture, monitor the cultured cells for erythroblast differentiation with flow cytometry, by assessing surface markers CD45, CD44, and Ter119 versus light scatter (FSC) [15, 16] or CD71, Ter119, and FSC [17, 18].
9. With the permeabilization steps of cultured mature erythroblasts and reticulocytes there may be a loss of up to 50–80% of

cells due to leakage of hemoglobin and transformation to "ghosts" which are excluded from the initial gating where we take under account size and shape characteristics. Therefore, we aim to start with fixation of at least 2×10^6 cultured cells to have enough of the relatively small percentage of actively enucleating cells for imaging flow cytometry. Since we likely have preferential loss of the most mature cells, we do not use this method for quantitative calculations of enucleation efficiency, but only for imaging analysis to determine distribution of, e.g., F-actin or microtubules within the cells. Permeabilization Subheadings 3.4, **steps 5–8** should be done gently, quickly, and at low temperature. Following spinning/aspiration at RT, cell pellets should immediately be put back on ice, in order for cells to better retain their integrity.

10. Fixation time may need to be altered/optimized depending on the antigen targeted and the antibody used.
11. Alternative or additional staining can also be done using AF594-conjugated cholera toxin subunit B to label lipid rafts (1:200), anti-pMRLC (Ser19) primary antibody for the phosphorylated myosin regulatory light chain (1:50), followed by anti-rabbit AF488-conjugated secondary antibody (1:400), and anti-γ-tubulin primary rabbit antibody (1:100) followed by anti-rabbit AF555-conjugated secondary antibody (1:300) [13].
12. Machine setup will vary depending on model of imaging flow cytometer used as well as what stains are utilized. For collection, a cell classifier is used to limit cells numbers collected. In the ImageStreamX, a standard classifier is 20 as lower area in brightfield which captures all cells, including RBC. In the ImageStream MarkII, gated populations can be specified for collection, and typically a "single" gate as in Fig. 2a, i is used. In each case, the classifier is set broadly to capture all relevant cells as contaminating objects can be removed during analysis.
13. Typical laser settings for live cells with the described stains are 100 mW 488 nm, 20 mW 658 nm. Stains were read in the following channels: Brightfield channels 1 and 9, FITC or AF488 channel 2, PE channel 3, PECF594 channel 4, DRAQ5 channel 11. Note that detection of stains cannot be individually adjusted as in standard flow cytometry. Thus an abundant bright stain can limit the laser setting and the detection of dimmer stains off of that same laser. However, the bright antibody can be diluted with unconjugated antibody, maintaining same total antibody concentration, for better stain balance. This is why CD71 is diluted in Subheading 3.3, **step 3**.

14. This gating strategy is designed to give robust quantitation of stages of erythropoiesis in live samples. However, if analysis only involves a subset of erythroid cells the staining and gating can be adjusted. For example, kit or CD71 is not needed if not gating ProE or RBC, respectively. Or in the case of examining subcellular structures in enucleating cells (Subheading 3.7), the initial gates of single, focus, Ter119+ and DNA+ from Fig. 2 can be used and then specific gating described in Subheading 3.7 can be used. Conversely, the final gating that identifies enucleating cells in Subheading 3.7 can be applied to the orthoE cells identified here to quantitate proportion of enucleating cells if permeabilization is not done [8].
15. A critical component of any quantitative analysis of erythroid intermediates is a control for gating [19]. This would be a standard available sample where consistent proportions of the cells of interest would be expected. For fluorescent intensity, standard controls such as FMO can be done, but also morphometric features must be gated consistently, which is even more challenging as many are on a continuum. In this case, normal bone marrow can be run with each experiment and used to set the initial gates and adjusted to obtain the consistent proportions of cells ranked by maturation characteristics. This gated file can be used as a template to open the experimental sample compensated files without further changes (Subheading 3.6, **step 12**). This template can also be used similarly as a starting point for new control samples for other experiments with the same stain, adjusting to obtain consistent populations. We have found that little adjustment should need to be made over time, particularly in measurements of morphology. In this way, all the cells within an experiment are all ranked and classified using consistent metrics and differences are meaningful and repeatable.
16. The kit positive gate can be set based on FMO controls. We also use as a guide, cells that have kit pixel intensities (seen under image analysis software in IDEAS) above background.
17. The use of two different masks in detecting median intensity Ter119 (e.g., Fig. 2a, iii vs. vii and viii) is used because they have been found to each discriminate different subsets of background cells [11].
18. While it is clear that plotting the mid-erythroblasts on a continuum of nuclear and cell size will rank them by maturation, the exact boundaries of the gates to divide this continuum into the three middle erythroblasts are less clear. What is most important is that it is done consistently so that differences in maturation can be compared between samples. The boundaries set in this example were based values for erythroblasts

determined by the historical criteria of morphology on stained samples of bone marrow from this strain (20%, 35%, and 45% respectively for BasoE, PolyE, and OrthoE) [1]. In further support for accuracy of the identification of OrthoE, enucleating cells within this group (#9 and 10), their nuclear size is similar to pyrenocytes and there is an expected sharp drop in S-phase in this group [11].

19. Inhibition assays of enucleation in vitro require a system where erythroblasts are fairly synchronized in differentiation stage and enucleation proceeds fast within 6–8 h. This way, enucleation can be efficiently inhibited when samples are incubated with an inhibitor for just 6 h (a short time is preferable to avoid toxicity). Such a system is the short-term erythropoiesis culture starting from mouse splenocytes enriched in erythroblasts after induction of stress erythropoiesis [5, 20].

Acknowledgments

The authors thank Seana Catherman for technical support and the Flow Cytometry Core Facility at the University of Rochester Medical Center and the Research Flow Cytometry Core at Cincinnati Children's Hospital Research Foundation National Institutes of Health (NHLBI) grant R01HL116352 (T.A.K.) and Michael Napoleone Memorial Foundation and the NIH/NHLBI (R01 HL130670; R01 HL116364 (K.E.M))

References

1. McGrath KE, Bushnell TP, Palis J (2008) Multispectral imaging of hematopoietic cells: where flow meets morphology. J Immunol Methods 336(2):91–97. doi:10.1016/j.jim.2008.04.012
2. McGrath KE, Kingsley PD, Koniski AD, Porter RL, Bushnell TP, Palis J (2008) Enucleation of primitive erythroid cells generates a transient population of "pyrenocytes" in the mammalian fetus. Blood 111(4):2409–2417. doi:10.1182/blood-2007-08-107581
3. Peslak SA, Wenger J, Bemis JC, Kingsley PD, Koniski AD, McGrath KE, Palis J (2012) EPO-mediated expansion of late-stage erythroid progenitors in the bone marrow initiates recovery from sublethal radiation stress. Blood 120(12):2501–2511. doi:10.1182/blood-2011-11-394304
4. Malik J, Kim AR, Tyre KA, Cherukuri AR, Palis J (2013) Erythropoietin critically regulates the terminal maturation of murine and human primitive erythroblasts. Haematologica 98(11):1778–1787. doi:10.3324/haematol.2013.087361
5. Konstantinidis DG, Pushkaran S, Giger K, Manganaris S, Zheng Y, Kalfa TA (2014) Identification of a murine erythroblast subpopulation enriched in enucleating events by multispectral imaging flow cytometry. J Vis Exp 88. doi:10.3791/50990
6. Liang R, Camprecios G, Kou Y, McGrath K, Nowak R, Catherman S, Bigarella CL, Rimmele P, Zhang X, Gnanapragasam MN, Bieker JJ, Papatsenko D, Ma'ayan A, Bresnick E, Fowler V, Palis J, Ghaffari S (2015) A systems approach identifies essential FOXO3 functions at key steps of terminal erythropoiesis. PLoS Genet 11(10):e1005526. doi:10.1371/journal.pgen.1005526
7. Malik J, Getman M, Steiner LA (2015) Histone methyltransferase Setd8 represses Gata2 expression and regulates erythroid maturation. Mol Cell Biol 35(12):2059–2072. doi:10.1128/MCB.01413-14

8. Gnanapragasam MN, McGrath KE, Catherman S, Xue L, Palis J, Bieker JJ (2016) EKLF/KLF1-regulated cell cycle exit is essential for erythroblast enucleation. Blood 128:1631. doi:10.1182/blood-2016-03-706671
9. Bessis M (1973) Living blood cells and their ultrastructure. Springer, Berlin
10. Palis J (2014) Primitive and definitive erythropoiesis in mammals. Front Physiol 5:3. doi:10.3389/fphys.2014.00003
11. McGrath KE, Catherman SC, Palis J (2016) Delineating stages of erythropoiesis using imaging flow cytometry. Methods 112:68. doi:10.1016/j.ymeth.2016.08.012
12. Giarratana MC, Kobari L, Lapillonne H, Chalmers D, Kiger L, Cynober T, Marden MC, Wajcman H, Douay L (2005) Ex vivo generation of fully mature human red blood cells from hematopoietic stem cells. Nat Biotechnol 23(1):69–74. doi:10.1038/nbt1047
13. Konstantinidis DG, Pushkaran S, Johnson JF, Cancelas JA, Manganaris S, Harris CE, Williams DA, Zheng Y, Kalfa TA (2012) Signaling and cytoskeletal requirements in erythroblast enucleation. Blood 119(25):6118–6127. doi:10.1182/blood-2011-09-379263
14. England SJ, McGrath KE, Frame JM, Palis J (2011) Immature erythroblasts with extensive ex vivo self-renewal capacity emerge from the early mammalian fetus. Blood 117(9):2708–2717. doi:10.1182/blood-2010-07-299743
15. Chen K, Liu J, Heck S, Chasis JA, An X, Mohandas N (2009) Resolving the distinct stages in erythroid differentiation based on dynamic changes in membrane protein expression during erythropoiesis. Proc Natl Acad Sci U S A 106(41):17413–17418. doi:10.1073/pnas.0909296106
16. Liu J, Zhang J, Ginzburg Y, Li H, Xue F, De Franceschi L, Chasis JA, Mohandas N, An X (2013) Quantitative analysis of murine terminal erythroid differentiation in vivo: novel method to study normal and disordered erythropoiesis. Blood 121(8):e43–e49. doi:10.1182/blood-2012-09-456079
17. Kalfa TA, Pushkaran S, Zhang X, Johnson JF, Pan D, Daria D, Geiger H, Cancelas JA, Williams DA, Zheng Y (2010) Rac1 and Rac2 GTPases are necessary for early erythropoietic expansion in the bone marrow but not in the spleen. Haematologica 95(1):27–35. doi:10.3324/haematol.2009.006239
18. Koulnis M, Pop R, Porpiglia E, Shearstone JR, Hidalgo D, Socolovsky M (2011) Identification and analysis of mouse erythroid progenitors using the CD71/TER119 flow-cytometric assay. J Vis Exp 54. doi:10.3791/2809
19. Filby A, Davies D (2012) Reporting imaging flow cytometry data for publication: why mask the detail? Cytometry A 81(8):637–642. doi:10.1002/cyto.a.22091
20. Yoshida H, Kawane K, Koike M, Mori Y, Uchiyama Y, Nagata S (2005) Phosphatidylserine-dependent engulfment by macrophages of nuclei from erythroid precursor cells. Nature 437:754–758. doi:10.1038/nature03964

Chapter 11

Flow Cytometric Analysis of Erythroblast Enucleation

Xiuli An and Lixiang Chen

Abstract

Enucleation is the final step in mammalian erythropoiesis. In this process, the nucleus is extruded by budding off from the erythroblast, forming the reticulocyte. Herein, we describe the flow cytometry-based assays for enucleation assessment. The separation of nucleated erythroblasts, reticulocytes, and extruded nuclei by flow cytometry is based on DNA staining, surface expression of erythrocyte specific markers, or forward scatter (FSC). The enucleation of murine erythroblasts is assessed by the surface expression of murine erythrocyte marker Ter119 and DNA staining. Three discrete populations that represent nucleated erythroblasts, reticulocytes, and extruded nuclei are defined as Hoechst^{med}TER119^{high}, Hoechst^{low-}TER119^{high}, and Hoechst^{high}TER119^{med}, respectively. Another nuclei acid staining dye, SYTO16, is used for the assessment of human enucleation in combination with FSC. For human cells, the three populations that represent nucleated erythroblasts, reticulocyte, and extruded nuclei are identified as FSC^{high} SYTO16^{+}, FSC^{high} SYTO16^{-}, FSC^{low}SYTO16^{+}, respectively.

Key words Enucleation, Flow cytometry, Glycophorin A, Hoechst 33342, SYTO16

1 Introduction

Enucleation is the last step in terminal differentiation during mammalian erythropoiesis. After exiting the final cell cycle, the nuclei of orthochromatic erythroblasts are polarized to one side of the cell. Eventually, the nucleus is extruded by budding off from the erythroblast to form the reticulocyte and the "pyrenocyte," which has a thin rim of cytoplasm surrounded by a plasma membrane [1]. Enucleation has critical physiological and evolutionary significance in that it allows elevation of hemoglobin levels in the blood and also gives red cells their flexible biconcave shape. Enucleation also is one of the critical rate-limiting steps of ex vivo red cell production. An improved understanding of enucleation would not only provide insight into the process of erythropoiesis, but also has significant implications for the development of highly efficient ex vivo red blood cell production systems [2].

Enucleation is accompanied by a sorting process, which directs membrane proteins to enter either the reticulocyte or the

Joyce A. Lloyd (ed.), *Erythropoiesis: Methods and Protocols*, Methods in Molecular Biology, vol. 1698,
DOI 10.1007/978-1-4939-7428-3_11, © Springer Science+Business Media LLC 2018

pyrenocyte [3]. Studies have shown that many key red membrane surface proteins, including band 3, glycophorin A (GPA), glycophorin C (GPC), and Rh-associated glycoprotein (RhAG), are predominantly segregated to the nascent reticulocytes following enucleation of erythroblasts [4, 5], while several other membrane proteins are selectively lost, such as β1 integrin, the vitamin C transporter SVCT2, and erythroblast macrophage protein [6–8].

Yoshida et al. first described the method to monitor the enucleation process with two different DNA-staining dyes, SYTO16 and SYTOX blue, which are cell permeable and cell-impermeable fluorescent dyes, respectively [9]. We adopted the method to assess human erythroblast enucleation. By combining SYTO16 with FSC, nucleated erythroblasts, reticulocytes, and extruded nuclei are identified as FSC^{high} $SYTO16^{+}$, FSC^{high} $SYTO16^{-}$, $FSC^{low}SYTO16^{+}$, respectively. Ter119, the murine erythroid-specific surface marker, was used together with Hoechst 33342 nucleic acid dye to monitor the enucleation of murine erythroblasts. The murine nucleated erythroblasts, reticulocytes, and extruded nuclei are defined as $Hoechst^{med}TER119^{high}$, $Hoechst^{low}TER119^{high}$, and $Hoechst^{high}TER119^{med}$, respectively [10, 11].

2 Materials

2.1 Murine Erythroblast Enucleation

1. Pregnant C57BL/6 mice at 14.5 days post conception.
2. 1× Phosphate-Buffered Saline (PBS).
3. PBS-BSA: PBS with 0.5% bovine serum albumin.
4. Biotin mouse lineage panel (contains biotin-conjugated antibodies against mouse Ter119, B220, Mac-1, CD3 and Gr-1).
5. Biotin rat anti-mouse-Sca-1.
6. Biotin rat anti-mouse-CD16/32 (FcgR).
7. Biotin rat anti-mouse-CD41.
8. Biotin rat anti-mouse-CD34.
9. PE streptavidin.
10. FITC rat anti-mouse-CD71.
11. APC rat anti-mouse-CD117.
12. 7-AAD.
13. Streptavidin particles plus-DM.
14. 70 μm cell strainer.
15. Murine erythroblast culture medium: Iscove's Modified Dulbecco's Medium (IMDM) with 15% FBS, 1% BSA, 200 μg/ml holo-transferrin, 10 μg/ml recombinant human insulin, 2 mM L-glutamine, 10^{-4} M β-mercaptoethanol, and 2 U/ml erythropoietin (EPO).

16. Erythroid-differentiation medium (EDM): IMDM with 20% FBS, 2 mM L-glutamine, and 10^{-4} M β-mercaptoethanol.
17. PE Rat anti-mouse TER-119 (BD Pharmingen).
18. Hoechst 33342 dye.
19. Cell Separation Magnet.
20. Fibronectin-coated cell culture dish.
21. 1.5 mL Eppendorf Tubes.
22. 15 and 50 ml polypropylene conical centrifuge tubes.
23. Refrigerated microcentrifuge with swinging bucket rotor.
24. BD LSRFortessa flow cytometry analyzer or similar.
25. Beckman-Coulter MOFLO high-speed cell sorter or similar.

2.2 Human Erythroblast Enucleation

1. Umbilical cord blood or peripheral blood from volunteer donors.
2. Ficoll-Paque Plus.
3. Buffer I: 1× PBS with 2% FBS and 2 mM EDTA.
4. Buffer II: 1× PBS with 0.5% BSA and 2 mM EDTA.
5. PBS-BSA: Phosphate-Buffered Saline with 0.5% BSA.
6. Blocking buffer: 1× PBS with 4% human AB Serum.
7. Staining Buffer: PBS with 5 mM $MgCl_2$.
8. SYTO® 16 Green Fluorescent Nucleic Acid Stain (Invitrogen): 1 mM solution in DMSO.
9. Human Erythroid Culture Medium:

   Phase I (Day 0–6): Iscove's Modified Dulbecco's Medium (IMDM) with 2% human peripheral blood plasma, 3% human AB serum, 200 μg/ml holo-human transferrin, 3 IU/ml heparin, 10 μg/ml insulin, 10 ng/ml stem cell factor (SCF), 1 ng/ml IL-3, and 3 IU/ml erythropoietin (EPO).

   Phase II (Day 7–11): IMDM with 2% human peripheral blood plasma, 3% human AB serum, 200 μg/ml holo-human transferrin, 3 IU/ml heparin, 10 μg/ml insulin, 10 ng/ml SCF, and 1 IU/ml EPO.

   Phase III (Day 11–18): IMDM with 2% human peripheral blood plasma, 3% human AB serum, 1 mg/ml holo-human transferrin, 3 IU/ml heparin, 10 μg/ml insulin, and 1 IU/ml EPO.
10. Human CD34 Isolation Kit.
11. Cell Separation Magnet.
12. PE Mouse anti human CD235a.
13. APC mouse anti human CD49d (α4-integrin).

14. DyLight 488-conjugated mouse anti human Band3.
15. 7-AAD (BD Bioscience).
16. 70 μm Cell Strainer.
17. Fibronectin-coated cell culture dish.
18. 1.5 mL Eppendorf Tubes.
19. 15 and 50 ml polypropylene conical centrifuge tubes.
20. Refrigerated microcentrifuge with swinging bucket rotor.
21. BD LSRFortessa flow cytometry analyzer or similar.
22. Beckman-Coulter MOFLO high-speed cell sorter or similar.

3 Methods

3.1 Measurement of Murine Erythroblast Enucleation

3.1.1 Preparation of Fetal Liver Cells

1. Sacrifice the pregnant mice at 14.5 days post conception.
2. Dissect uterus and wash in 1× PBS.
3. Move fetuses into fresh 1× PBS.
4. Dissect fetal livers and place them in fresh 1× PBS.
5. Pool ~8 fetal livers into a 1.5 mL Eppendorf tube in 1 mL 1× PBS.
6. Disrupt fetal livers with a 1000 μl Rainin Pipette (~10 passes).
7. Pool all material and filter the cell suspension through a 70 μm cell strainer into a 15 mL conical tube.
8. Pellet the cells at 300 × *g* for 5 min.
9. Aspirate the supernatant. Resuspend the cells in 1 mL PBS-BSA buffer.

3.1.2 Preparation of Murine Erythroid Progenitors

1. Count the cells, and resuspend them in PBS-BSA buffer at a final density of 100 × 10^6/ml.
2. According to the manufacturer's instruction, add the biotinylated antibodies from the Mouse Lineage Panel, using 2 μl of each antibody per 10^6 cells, and incubate on ice for 15 min.
3. Wash the stained cells twice with PBS-BSA buffer and centrifuge for 5 min at 300 × *g* at 4 °C, carefully aspirate all the supernatant.
4. Resuspend the cells at a final density of 20–100 × 10^6 cells/ml.
5. Vortex the streptavidin particles thoroughly, and add 50 μl of particles for every 1 × 10^7 cells.
6. Mix thoroughly, incubate for 30 min at 4 °C.
7. Bring the cell density to 20–80 × 10^6 cells/ml with PBS-BSA buffer.

8. Transfer the labeled cells to a round-bottom test tube, maximum volume added not to exceed 1.0 mL. Place this positive-fraction tube on the Cell Separation Magnet (horizontal position) for 6–8 min.
9. With the tube on the Cell Separation Magnet and using a glass Pasteur pipette, carefully aspirate the supernatant (negative-fraction), and place in a new tube.
10. Remove the positive-fraction tube from the Cell Separation Magnet, and add PBS-BSA buffer to the same volume as in **step 7**.
11. Resuspend the positive fraction well by pipetting up and down 10–15 times, and place the tube back on the Cell Separation Magnet for 8 min.
12. Using a new sterile Pasteur pipette, carefully aspirate the supernatant, and combine with the negative-fraction from **step 9** above.
13. Place the tube containing the combined depleted fraction on the Cell Separation Magnet for a final 8 min.
14. Carefully aspirate the supernatant and place in a new sterile tube.
15. Pellet the cells at 300 × *g* for 5 min.
16. Repeat **steps 1–14** to further deplete Sca, CD16/32 (FcgR), CD41 and CD34 positive cells.
17. Collect Ter119, B220, Mac-1, CD3, Grl, Sca-1, CD16/CD32, CD41, and CD34 negative cells for the following cell staining.
18. Count the depleted cells, resuspend in PBS-BSA buffer at a final density of 10^6 cells/80 μl for the antibody staining.
19. Stain the cells as below:

    PE Streptavidin (0.2 μg/10^6 cells)

    APC rat anti-mouse CD117 (0.2 μg/10^6 cells)

    FITC rat anti-mouse CD71 (0.2 μg/10^6 cells)
20. Incubate on ice for 30 min in the dark.
21. Wash two times with PBS-BSA buffer.
22. Stain with the viability marker PI (0.25 μg/10^6 cells) on ice for 10 min in the dark.
23. Spin at 300 × *g* or 5 min at 4 °C.
24. Resuspend the cell pellet in PBS-BSA at 10–20 × 10^6 cells/ml.
25. Run the stained samples on the cell sorter following the manufacturer's protocol.
26. Gate out cellular debris and PI^+ cells.

27. For PI^- population, gate on PE-Biotin- population.
28. Show CD117 and CD71 expression of PE negative cells.
29. For $PE\text{-}Biotin^-$ population, gate on two populations, BFU-E and CFU-E.
30. The BFU-E fraction is the 10% lowest CD71-expressing part of the cKit $(CD117)^+$ fraction.
31. Collect BFU-E cells in FBS.
32. Wash the cells with PBS and centrifuge at 300 × *g* for 10 min at 4 °C.
33. Aspirate the supernatant and the purified cells are ready for downstream applications.

3.1.3 In Vitro Culture of Erythroblasts

1. Seed the cells in fibronectin-coated wells with murine erythroblast culture medium at the density of 1×10^5 cells/ml on day 1.
2. Change to erythroid-differentiation medium on day 2.
3. Collect the differentiated cells on day 3, centrifuge at 300 × *g* for 5 min.
4. Resuspend the cell pellet in PBS-BSA at the density of 1×10^6 cells/ml.

3.1.4 Antibody Staining

1. Stain the cells with 10 μg/ml Hoechst 33342 and PE-conjugated antibody against TER119 (1 μg/10^6cells), for 15 min at room temperature.
2. Wash twice with PBS-BSA.
3. Centrifuge at 300 × *g* for 5 min, resuspend the cell pellet in PBS-BSA at the density of 1×10^6 cells/ml.

3.1.5 Flow Cytometry Analysis

1. Follow the manufacturer's protocol for FCM analysis using the analyzer.
2. Gate on three distinct populations: extruded nuclei, nucleated erythroblasts, and reticulocytes are defined as $Hoechst^{high}TER119^{med}$, $Hoechst^{med}TER119^{high}$, and $Hoechst^{low}TER119^{high}$, respectively (Fig. 1a).

3.1.6 Isolation of Extruded Nuclei, Nucleated Erythroblasts, Reticulocytes

1. Wash the cells twice with PBS-BSA buffer and resuspend the cells at the concentration of 20×10^6 cells/ml.
2. Stain the cells as described in 3.1.4.
3. Follow the manufacturer's protocol for cell sorter and the gating strategy described in 3.1.5.
4. Collect populations of interest in FBS.
5. Wash cells with PBS and centrifuge at 300 × *g* for 10 min at 4 °C.

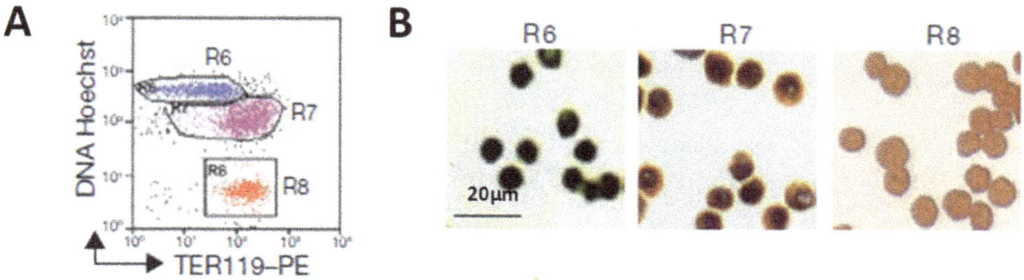

Fig. 1 Analysis of enucleation of murine erythroid cells during in vitro culture. (**a**) Three distinct populations of cells are defined: R6 (*blue*; extruded nuclei; Hoechst^{high}TER119^{med}), R7 (*magenta*; nucleated erythroblasts; Hoechst^{med}TER119^{high}), and R8 (*red*; reticulocytes; Hoechst^{low}TER119^{high}). (**b**) R6 to R8 cells are analysed by Benzidine–Giemsa staining

6. Take 5×10^4 cells from sorted populations to prepare cytospin, and perform Benzidine–Giemsa staining to examine cell morphology (Fig. 1b).

3.2 Measurement of Human Erythroblast Enucleation

3.2.1 Purification of Human CD34$^+$ Cells

1. Obtain umbilical cord blood or peripheral blood from volunteer donors after signing informed consent and following the institutional guidelines.
2. Dilute blood sample 2.5 times with buffer I at room temperature.
3. Pour 15 mL Ficoll-Paque Plus into a 50 mL Conical Centrifuge tube and carefully add 35 mL of the diluted blood on the Ficoll.
4. Centrifuge at $400 \times g$ for 30 min with no brake at room temperature.
5. Collect the buffy coat (middle layer) containing mononuclear cells with a transfer pipette into a 50 mL tube (up to 25 mL/tube), and fill the tube up with buffer I and invert the tubes two to three times to wash the cells.
6. Centrifuge at $300 \times g$ for 10 min with low break at room temperature.
7. Discard the supernatant and pool the pellets into one tube, add buffer I to 50 mL and centrifuge at $200 \times g$ for 10 min with low break at room temperature.
8. Resuspend the pellet in cold buffer II, count the cells at proper dilution.
9. Wash the cells with cold buffer II and centrifuge at $200 \times g$ for 10 min with low break at 4 °C.
10. Resuspend the cells at the density of up to 1×10^8 cells/300 µl cold buffer II.
11. Add 100 µl of FcR Blocking Reagent and 100 µl of CD34 microbeads for up to 1×10^8 cells. Mix well and incubate in the refrigerator (2–8 °C) for 30 min.

12. Wash the cells with buffer II as described in **step 9**.
13. Resuspend up to 10^8 cells in 500 μl of buffer II and apply to 70 μm Cell Strainer to obtain uniform single-cell suspensions.
14. Place the column in the magnetic field of a suitable MACS Separator and prepare MACS column by rinsing with buffer II.
15. Apply single-cell suspensions onto the column.
16. Wash the column two to three times with 2–5 mL buffer II.
17. Remove the column from the separator and place it on a suitable collection tube.
18. Add 2–5 mL buffer II to the column and immediately flush out the magnetically labeled cells by firmly pushing the plunger into the column.
19. (Optional) To increase the purity of CD34+ cells, the eluted fraction can be enriched over a second column, repeat the magnetic separation procedure (**steps 14–18**).
20. Count cells at proper dilution, centrifuge to remove buffer II, and culture cells in a fresh medium.

3.2.2 *In Vitro Culture of $CD34^+$ Cells*

1. Resuspend $CD34^+$ cells with the Phase I Erythroid Culture Medium at a cell density of 10^5 cells/ml.
2. Culture the cells at 37 °C in a humidified atmosphere with 5% CO_2.
3. Centrifuge at 300 × *g* for 5 min, resuspend the cell pellet in fresh Phase I medium at day 4, and adjust the cell density to 10^5 cells/ml.
4. Centrifuge at 300 × *g* for 5 min, resuspend the cell pellet in fresh Phase II medium at day 7, and adjust the cell density to 10^5 cells/ml.
5. Centrifuge at 300 × *g* for 5 min, resuspend cell pellet in fresh Phase III medium at day 11, adjust the cell density to 10^6 cells/ml.
6. Culture the cells until at least day 15.

3.2.3 *Antibody Staining*

1. Collect 1 × 10^5 cells on day 15 (or later) in a 1.5 ml Eppendorf tube, wash with PBS-BSA. Centrifuge at 300 × *g* for 5 min at 4 °C.
2. Resuspend cells in 30 μl of staining buffer.
3. Add 0.6 μl diluted SYTO16 (1:200 diluted with staining buffer) to get a working concentration of 100 nM.
4. Mix well and incubate at room temperature for 15 min, keep in the dark.
5. Wash stained cells with PBS-BSA buffer. Centrifuge at 300 × *g* for 5 min at 4 °C.

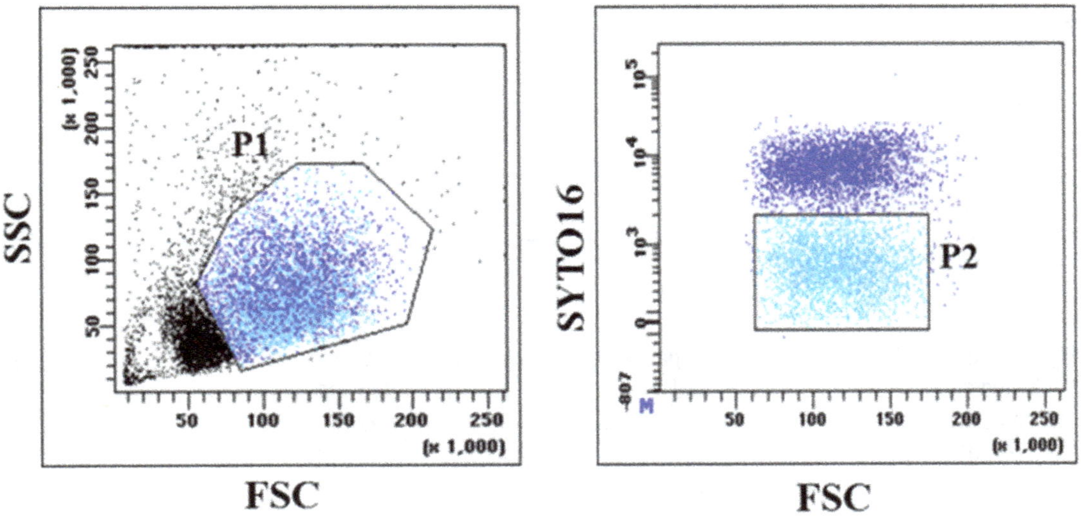

Fig. 2 Flow Cytometry analysis of human enucleation. Enucleation percentage is defined as the percentage of SYTO16⁻ cells (reticulocytes, P2) in the total cells except debris and nuclei (P1)

6. Resuspend cells in 200 μl of PBS-BSA buffer and transfer into a flow tube.
7. Run the stained samples on BD LSRFortessa.

3.2.4 Flow Cytometry Analysis

1. Conduct flow cytometry analysis using BD FACSDiva.
2. Gate out cellular debris and nuclei (black dots) in FSC-SSC plot.
3. Plot the SYTO16 versus FSC of P1 population: $SYTO16^+$ cells are nucleated erythroblasts, $SYTO16^-$ cells are reticulocytes.
4. Calculate the enucleation rate: $SYTO16^-$ cells (reticulocytes: P2) divided by the total cells excluding cell debris and nuclei (P1) (Fig. 2).

3.2.5 Isolation of Nucleated Erythroblasts, Reticulocytes, Extruded Nuclei by FACS

1. Collect the cells on day 15 (or later) of culture. Wash the cells with PBS-BSA, centrifuge at 300 × g for 10 min at 4 °C.
2. Resuspend up to 10×10^6 cells in 2.5 mL of staining buffer.
3. Add SYTO16 at a working concentration of 160 nM for up to 10×10^6 cells.
4. Mix well and incubate at room temperature for 15 min, keep in the dark.
5. Wash the cells twice with 40 mL PBS-BSA and resuspend cells in PBS-BSA at a density of 15×10^6 cells/ml.
6. Follow the manufacturer's protocol for cell sorter. The gating strategy: Gate out cellular debris and show FSC-SSC plot of live cells (P1); Gate three distinct populations: nucleated

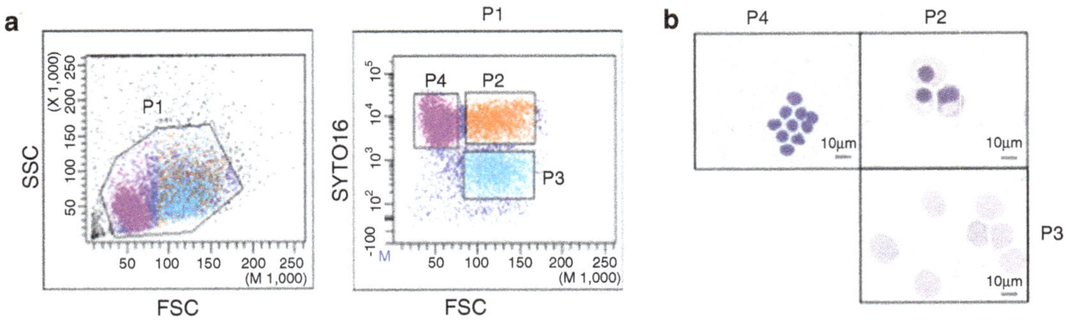

Fig. 3 Isolation of in vitro cultured human erythroblasts, reticulocytes, and extruded nuclei by FACS. (**a**) Nucleated erythroblast, reticulocytes, and extruded nuclei were separated based on SYTO16 staining and FSC. Nucleated erythroblasts, reticulocytes, and extruded nuclei are defined as FSC^{high} $SYTO16^{+}$ (P2), FSC^{high} $SYTO16^{-}$ (P3), $FSC^{low}SYTO16^{+}$ (P4), respectively. (**b**) Morphologies of the sorted nucleated erythroblasts (P2), reticulocytes (P3), and extruded nuclei (P4) are shown

erythroblasts, reticulocytes and extruded nuclei are defined as FSC^{high} $SYTO16^{+}$ (P2), $FSC^{high}SYTO16^{-}$ (P3), $FSC^{low}SYTO16^{+}$ (P4), respectively (Fig. 3a).

7. Collect the three populations in FBS.
8. Wash the cells with PBS and centrifuge at $300 \times g$ for 10 min at 4 °C.
9. Aspirate the supernatant and the purified cells are ready for downstream applications.
10. Take 5×10^4 cells from sorted populations to prepare cytospin, and perform with Giemsa staining to confirm the purity by cell morphology (Fig. 3b).

References

1. Ji P, Murata-Hori M, Lodish HF (2011) Formation of mammalian erythrocytes: chromatin condensation and enucleation. Trends Cell Biol 21(7):409–415
2. Keerthivasan G, Wickrema A, Crispino JD (2011) Erythroblast Enucleation. Stem Cells Int 2011:139851
3. Bell AJ, Satchwell TJ, Heesom KJ, Hawley BR, Kupzig S, Hazell M, Mushens R, Herman A, Toye AM (2013) Protein distribution during human erythroblast enucleation in vitro. PLoS One 8(4):e60300
4. Salomao M, Chen K, Villalobos J, Mohandas N, An X et al (2010) Hereditary spherocytosis and hereditary elliptocytosis: aberrant protein sorting during erythroblast enucleation. Blood 116:267–269
5. Patel VP, Lodish HF (1987) A fibronectin matrix is required for differentiation of murine erythroleukemia cells into reticulocytes. J Cell Biol 105:3105–3118
6. Lee JC, Gimm JA, Lo AJ, Koury MJ, Krauss SW et al (2004) Mechanism of protein sorting during erythroblast enucleation: role of cytoskeletal connectivity. Blood 103:1912–1919
7. May JM, Qu ZC, Qiao H, Koury MJ (2007) Maturational loss of the vitamin C transporter in erythrocytes. Biochem Biophys Res Commun 360:295–298
8. Soni S, Bala S, Gwynn B, Sahr KE, Peters LL et al (2006) Absence of erythroblast macrophage protein (Emp) leads to failure of erythroblast nuclear extrusion. J Biol Chem 281:20181–20189
9. Yoshida H, Kawane K, Koike M, Mori Y, Uchiyama Y, Nagata S (2005) Phosphatidylserine- dependent engulfment by macrophages of

nuclei from erythroid precursor cells. Nature 437(29):754–758

10. Ji P, Jayapal SR, Lodish HF (2008) Enucleation of cultured mouse fetal erythroblasts requires Rac GTPases and mDia2. Nat Cell Biol 10(3):314–321

11. Zhang J, Socolovsky M, Gross AW, Lodish HF (2003) Role of Ras signaling in erythroid differentiation of mouse fetal liver cells: functional analysis by a flow cytometry–based novel culture system. Blood 102:3938–3946

Chapter 12

High-Resolution Fluorescence Microscope Imaging of Erythroblast Structure

Alyson S. Smith, Roberta B. Nowak, and Velia M. Fowler

Abstract

During erythropoiesis, erythroblasts undergo dramatic morphological changes to produce mature erythrocytes. Many unanswered questions regarding the molecular mechanisms behind these changes can be addressed with high-resolution fluorescence imaging. Immunofluoresence staining enables localization of specific molecules, organelles, and membrane components in intact cells at different phases of erythropoiesis. Confocal laser scanning microscopy can provide high-resolution, three-dimensional images of stained structures, which can be used to dissect the molecular mechanisms driving erythropoiesis. The sample preparation, staining procedure, imaging parameters, and image analysis methods used directly affect the quality of the confocal images and the amount and accuracy of information that they can provide. Here, we describe methods to dissect erythropoietic tissues from mice, to perform immunofluorescence staining and confocal imaging of various molecules, organelles and structures of interest in erythroblasts, and to present and quantitatively analyze the data obtained in these fluorescence images.

Key words Erythroblast, Fetal liver, Bone marrow, Spleen, Immunofluorescence, Confocal microscopy

1 Introduction

Erythropoiesis entails a highly orchestrated process of regulated gene expression with a series of cell divisions leading to cell cycle exit, coupled to dramatic changes in cell and nuclear morphology, and culminating in nuclear expulsion (enucleation) [1–3]. After enucleation, the nascent reticulocyte undergoes further morphological remodeling of the plasma membrane and organelles to produce the mature biconcave erythrocyte lacking all internal organelles or transcellular cytoskeleton [4]. Modern genomic and proteomic studies [5, 6] have provided extensive information on the programs of gene regulation by transcription factors, epigenetic modifications, and miRNAs [7, 8] during mouse erythropoiesis

Alyson S. Smith and Roberta B. Nowak contributed equally to this work.

Joyce A. Lloyd (ed.), *Erythropoiesis: Methods and Protocols*, Methods in Molecular Biology, vol. 1698,
DOI 10.1007/978-1-4939-7428-3_12, © Springer Science+Business Media LLC 2018

in vivo, as well as during human erythropoiesis in vitro from hematopoietic stem cells. However, considerably less is understood regarding the molecular and structural basis for the dramatic morphological transformations of erythroblasts that create erythrocytes in the specialized erythropoietic tissues.

During embryonic development in mice, primitive erythropoiesis takes place in the yolk sac and the circulation from E7.5–16.5, followed by definitive erythropoiesis in the fetal liver at E12.5–17.5, subsequently moving to the bone marrow [1]. Stress erythropoiesis, a form of definitive erythropoiesis that produces large numbers of erythrocytes in response to crises such as heavy bleeding or hemolytic anemia, occurs in the spleen in mice, but the bone marrow in humans [9]. In all types of definitive erythropoiesis, erythroblasts are associated with erythroblastic islands, specialized macrophages surrounded by proliferating, differentiating, and enucleating erythroblasts [10, 11]. Starting with the earliest identifiable erythroid cell, the pro-erythroblast, erythropoiesis involves four or five mitotic cell divisions, accompanied by dramatic cytoplasmic and nuclear changes as cells differentiate from basophilic, to polychromatophilic and then orthochromatophilic erythroblasts [3]. Hemoglobin and erythroid-specific membrane and cytosolic proteins accumulate [8], nuclear size and transcriptional activity decrease, the nucleus moves to one side of the cell and is extruded [2], other organelles are lost, and the future erythrocyte membrane proteins are sorted to the membrane of the nascent reticulocyte [12]. All of these processes involve tightly regulated signaling cascades and directional movements of proteins, membranes, and organelles [13, 14]. Visualization of the locations of molecules, organelles, and cytoskeletal structures at submicron-scale resolution can provide unique insights into underlying mechanisms of terminal erythroid differentiation and establish how signaling, genetic, and epigenetic regulation programs control these morphological events. Detailed structural studies of the molecular and cellular features of erythroblasts isolated from diverse erythropoietic niches in vivo will contribute to understanding how various human conditions affect erythropoiesis, and provide rigorous standards to evaluate fidelity and efficiency of erythropoiesis protocols that aim to produce mature erythrocytes in vitro.

This chapter focuses on the dissection, preparation, imaging, and image analysis of erythroblasts from the mouse fetal liver, bone marrow, and spleen. We present an approach using confocal laser scanning microscopy, as this is a versatile and accessible method for high-resolution imaging of erythroblasts [15]. We outline factors to consider and pitfalls to avoid to obtain high-quality two or three-dimensional images with submicron-scale resolution and to extract accurate, quantitative information from them. Appropriate application of this technology will allow researchers to gain new molecular and cellular insights into the processes regulating erythropoiesis. While we focus on mouse erythroid cells here, similar approaches can be used to image human erythroid cells [16].

2 Materials

2.1 Dissection of Tissues

2.1.1 Animals

1. Pregnant mouse at embryonic day 13–15 (E13–15) gestation for dissection of fetal liver; E14.5 days gestation is ideal.
2. Adult mouse at least 6-weeks old for dissection of bone marrow or spleen.

2.1.2 Dissection Tools and Solutions

1. Standard dissecting Metzenbaum scissors.
2. Straight Dumont #5 microforceps.
3. Blunt-end microscissors.
4. Feather disposable scalpels (#11 or #15).
5. 1 mL syringe with 25-gauge 5/8in needle for bone marrow flushing.
6. Standard non-tissue culture treated (non-TC) dishes (10 cm).
7. 12-well non-TC plates.
8. 90 mm non-TC petri dishes.
9. 60 mm non-TC petri dishes.
10. Pipetman P1000 and standard tips.
11. Solution of 70% v/v histological grade ethanol.
12. Dulbecco's phosphate-buffered saline (DPBS).

2.2 Immunofluorescence Staining

2.2.1 Solutions and Supplies

1. Dulbecco's phosphate-buffered saline (DPBS).
2. 10× Dulbecco's phosphate-buffered saline (10×-DPBS).
3. Fixative—4% paraformaldehyde (PFA) in DPBS. Dilute 16% paraformaldehyde stock solution (EM grade 10 mL ampule, Electron Microscopy Sciences #15710) by the addition of 10 mL 16% PFA to 26 mL of dH_2O and 4 mL of 10× DPBS on the day of use, and store at 4 °C in the dark until needed. Discard diluted PFA at the end of 1 week.
4. Permeabilization Buffer: 0.3% Triton X-100 in DPBS. Dilute 7.5 mL of 20% (w/v) Triton X-100 stock solution into 450 mL of DPBS in a beaker at room temperature, stir until mixed and then bring to volume in a 500 mL graduated cylinder. Store at 4 °C until ready to use. Prepare 20% (w/v) Triton X-100 stock solution from a 100% solution.
5. Blocking Buffer (BB) Goat Serum: 3%BSA/1%Serum/DPBS. Weigh out 1.2 g of Bovine serum albumin (BSA fraction V, pH 7.0) into 30 mL of DBPS. Mix to dissolve BSA. Add 400 μL

goat serum and bring volume to 40 mL with DBPS. Filter with a 0.22 μm vacuum filter and store at 4 °C until ready to use.

6. Blocking Buffer (BB) Donkey Serum: 3%BSA/1%Serum/DPBS. Weigh out 1.2 g of BSA (fraction V, pH 7.0) into 30 mL of DBPS. Mix to dissolve BSA. Add 400 μL donkey serum and bring volume to 40 mL with DBPS. Filter with a 0.22 μm vacuum filter and store at 4 °C until ready to use.
7. Low retention 1.5 mL polycarbonate microfuge tubes.
8. Microscope slides with frosted ends and thickness 1.5 glass coverslips 22 × 22 mm (0.16–0.19 mm thickness). The slides should be electrostatically treated to prevent tissue or cell loss (such as Fisherbrand™ ColorFrost™ Plus Microscope Slides). Prepare thoroughly cleaned glass coverslips (*see* **Note 1**). Do not use coverslips with a different thickness, as most microscope objectives are designed for use with thickness 1.5 coverslips.
9. Antifade mounting media ProLong Gold (Molecular Probes).
10. Cytospin machine and cytofunnel.

2.2.2 Fluorescent Probes and Antibodies

Standard modern confocal laser scanning microscopes are equipped to capture images at four different wavelengths (blue, green, red, and far-red) [15]. The bone marrow, spleen, or fetal liver cell suspensions contain a heterogeneous cell population; therefore, erythroblast-specific markers are required to identify erythroblasts. The Ter119 rat monoclonal antibody binds to a 50 kDa antigen associated with glycophorin A (GPA) on mouse definitive erythroid cells from the pro-erythroblast stage through the mature erythrocyte. Since GPA levels increase during terminal differentiation, Ter119 binding increases and cells at later stages have increased fluorescent signal intensity. In addition, based on the gradual condensation of chromatin during terminal differentiation, followed by enucleation, a nuclear DNA stain can inform as to extent of terminal differentiation and distinguish between enucleating and non-enucleating erythroblasts. Using these two fluorescent probes (Ter119 and DNA) to identify erythroblasts leaves two possible channels for imaging other molecules in one staining experiment. In one staining set, multiple primary antibodies can be used as long as they are made in different species of host animals. Directly labeled primary antibodies as well as non-labeled primaries that require a labeled secondary antibody can be used together in one staining set. The type of blocking buffer (goat or donkey) is selected based on the host species of the fluorescent-conjugated secondary antibodies (*see* **Note 2**). In turn, this determines the range of possible host species for the primary antibodies (Table 1). In addition, small molecule dyes can be used to stain proteins of interest, such as rhodamine phalloidin to stain F-actin.

Table 1
Possible antibody and stain combinations for mouse erythroblasts

		Goat blocking buffer		Donkey blocking buffer	
Channel (emission wavelength)[a]	Stained molecule	Primary antibody	Fluorescent labeled secondary antibody or dye	Primary antibody	Fluorescent labeled secondary antibody or dye
Blue (400–500 nm)	DNA	NA	Hoechst 33342 (Life Technologies H3570)	NA	Hoechst 33342 (Life Technologies H3570)
Green (501–550 nm)	Erythroblast-specific marker[b]	FITC-conjugated rat anti-mouse Ter119 (BD Biosciences 557915)[c]	NA	FITC-conjugated rat anti-mouse Ter119 (BD Biosciences 557915)[c]	NA
Red (551–630 nm)	Protein-of-interest 1	Primary antibody against protein 1[d]	Secondary antibody raised in goat against primary antibody species conjugated to red fluorophore	Primary antibody against protein 1[e]	Secondary antibody raised in donkey against primary antibody species conjugated to red fluorophore
Far red (631–690 nm)	Protein-of-interest 2	Primary antibody against protein 2[d]	Secondary antibody raised in goat against primary antibody species conjugated to far red fluorophore	Primary antibody against protein 2[e]	Secondary antibody raised in donkey against primary antibody species conjugated to far red fluorophore

[a]The ranges of emission wavelengths of fluorophores typically used in each channel are given in parentheses. When choosing fluorophores, make sure the excitation and emission spectra are as distinct as possible to avoid fluorescence bleed-through between channels

[b]It is important to use an erythroblast-specific marker labeled in a visible channel (e.g., green or red) to aid in visual screening to locate the erythroblasts amidst the many other cell types in the dissociated cells from fetal liver, bone marrow, or spleen

[c]For human erythroblasts, the primary antibodies used as erythroblast-specific markers are generated in mouse (e.g., FITC-conjugated mouse anti-human CD235a, BD Pharmingen #559943). In this case, one cannot use mouse primary antibodies to stain other antigens. However, one can use primary antibodies from rat or other species

[d]Possible species for primary antibody: mouse, rabbit, chicken, hamster, swine, guinea pig

[e]Possible species for primary antibody: mouse, rabbit, goat, chicken, sheep, hamster, swine, guinea pig

Additional details as to selection and use of fluorescent antibodies are discussed in *Handbook of Biological Confocal Microscopy*, Third Edition, edited by James Pawley and published by Springer [15]. In this Chapter, for demonstration purposes, one staining set (set A) uses BB with goat serum and goat secondary antibody. The other staining set (set B) uses BB with donkey serum and donkey secondary antibody. The figures in this Chapter utilize images from these staining sets. Below are the specific antibodies and stains used (*see* **Note 3**).

1. Mouse-specific erythroid membrane marker: FITC-conjugated anti-mouse Ter119 (host species: rat).
2. DNA marker: Hoechst 33342 (Hoechst, 10 mg/mL) diluted to 1 mg/mL in water.
3. Primary antibodies specific for erythroblast proteins to be tested. The procedure is illustrated using a rabbit polyclonal anti-Ki67 (AbCam #ab15580) and a goat polyclonal anti-laminB antibody (Santa Cruz #sc-6217).
4. Secondary antibodies specific for primary antibodies selected. Alexa-594 conjugated goat anti-rabbit secondary and Alexa-594 conjugated donkey anti-goat secondary.
5. Staining Sets: Set A: FITC-Ter119, Rabbit anti-Ki67 + Alexa-594 goat anti-rabbit, Hoechst; Set B: FITC-Ter119, Goat anti-LaminB + Alexa-594 conjugated donkey anti-goat, Hoechst.

3 Methods

This section gives methods for dissecting, fixing, staining, and imaging of dissociated cells from the fetal liver, bone marrow, or spleen. Thin (12–14 μm) cryosections of an intact organ can also be prepared to study morphology and cellular organization (Fig. 1a) [17, 18]. In preparations made from dissociated cells (Fig. 1b), by contrast, it is easier to identify the membrane boundaries, shapes, and stage of maturation of each erythroblast. Figure 1 shows examples of images from each kind of preparation of a fetal liver, but the same considerations apply to the bone marrow and the spleen.

3.1 Dissection of Tissues

All the steps are performed at room temperature unless otherwise indicated.

3.1.1 Fetal Liver

1. Euthanize E13–15 pregnant mouse according to institutional protocols established by the veterinarians, and approved by the Institutional Animal Care and Use Committee (IACUC). Place mouse, abdomen side up, on a paper towel and spray abdomen with 70% (v/v) ethanol. Pinch and tent skin with forceps and cut through abdominal wall (being careful not to damage

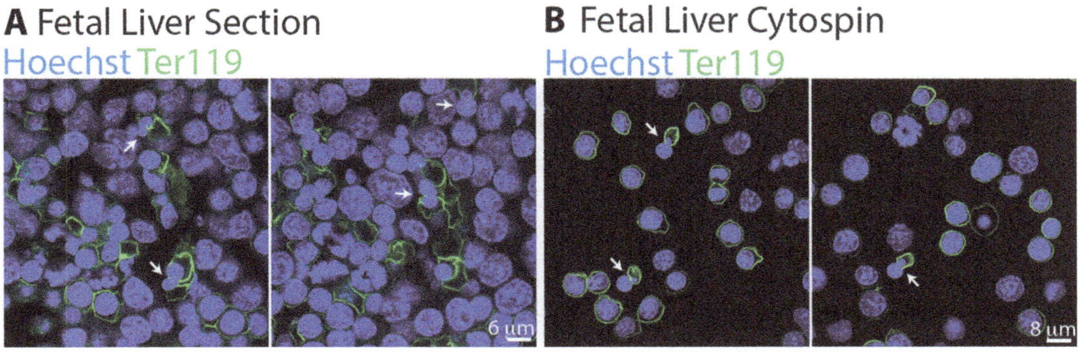

Fig. 1 Visualization of individual erythroblasts is improved in cells dissociated from tissues. Confocal images of (**a**) 12 μm cryosection of an intact fetal liver and (**b**) Cytospin of dissociated fetal liver cells fixed and stained in suspension, following the protocol in Subheading 3.1.1, acquired using a 100× oil objective, NA 1.4, zoom 2×. Both the samples were labeled with Hoechst and immunostained with Ter119. Enucleating erythroblasts (*white arrows*) can be identified in both preparations but are easier to distinguish in the Cytospin where the cells are more dispersed. Close proximity of cells to one another in the tissue in situ can make it difficult to determine the individual membrane boundaries and shapes of the erythroblasts

organs). With scissors, continue a midline incision in the abdominal wall to expose the abdominal cavity.

2. Locate and excise the uterus with conceptuses (which has the appearance of a string of pearls in a "u" shape) with scissors, and place uterus in a clean, 90 mm non-TC dish containing fresh DPBS.
3. With blunt-ended microdissection scissors, cut the uterus longitudinally and expose each conceptus. Carefully peel the placenta away from the uterus using the microforceps and move conceptuses into a new clean 90 mm non-TC dish containing DPBS. Place each conceptus into an individual clean 60 mm non-TC petri dish containing DPBS. Remove each fetus from the embryonic sac by carefully ripping the embryonic sac away from the embryo with sharp microforceps (*see* **Note 4**).
4. Remove the liver (dark red mass filling the abdominal cavity) by opening the abdominal cavity of the fetus with microforceps. Place microforceps behind the liver and pull it away from the body cavity.
5. Place each fetal liver in individual wells of a clean 12-well non-TC plate containing 2 mL of 4% PFA/DPBS at 4 °C.
6. With a P1000 set to 1000 μL, break up the fetal liver by drawing the fetal liver 15–20 times gently in and out through the tip until the cells are nearly all dispersed (*see* **Note 5**).
7. Fix cell suspension overnight at 4 °C (*see* **Note 6**).

3.1.2 Bone Marrow

1. Euthanize adult mouse according to institutional protocols approved by the IACUC.
2. Place mouse on its side and spray down the lower half of the body with 70% (v/v) ethanol. Pinch skin at ankle, and make a small incision in the skin of the hind leg above the ankle with scissors. Using forceps, pull up on the skin up to the hip to expose the musculature of the leg from the ankle to the hip. Feel for the hip joint, and cut through the pelvis with scissors to remove the leg and place in a clean 90 mm non-TC dish with DPBS to keep tissues hydrated.
3. Remove one of the legs from the DPBS, and place it on the dry lid of a 90 mm non-TC dish. Dissect muscle and connective tissues away from femur, making sure all the tissues are cleaned off the bone. Start at the knee joint, cutting around the knee cap and work up to the ball-and-socket joint of the hip. Use the scalpel to scrape off the muscle and connective tissues from the bone. Place cleaned femur in a fresh petri dish in DPBS while dissecting the opposite leg. Take one femur, cut each end off the long bone with scissors, so you are left with the straight shaft containing the marrow. Flush out marrow with 1 mL of cold 4% PFA in DPBS using a 25G5/8 needle into one well of a 12-well non-TC plate containing 1 mL of 4% PFA in DPBS (1 well/mouse) at 4 °C. The bone will turn white when all the marrow is flushed out. Draw up 1 mL of this suspension and use it to flush out the other femur from the same mouse. Draw bone marrow cell suspension gently up and down through the syringe a couple of times to disperse the bone marrow tissue. Final volume of cell suspension will be ~2 mL.
4. Fix cells overnight at 4 °C (*see* **Note 6**).

*3.1.3 Spleen (**See** Note 7)*

1. Euthanize adult mouse that has been treated to induce stress erythropoiesis, in protocols approved by the IACUC.
2. Place mouse, abdomen side-up, on a paper towel and spray down with 70% (v/v) ethanol. Pinch and tent skin with forceps and make an incision in the lower abdomen with scissors. Continue the midline incision in the abdominal wall to expose the abdominal cavity.
3. Dissect out the spleen, and place it in 2 mL of 4% PFA/DPBS at 4 °C in one well of a 12-well non-TC plate.
4. Cut up the spleen into 5–10 smaller pieces with scissors.
5. With a P1000 set to 1 mL, break up the spleen by drawing it in and out gently through the tip 15–20 times until cells are nearly all dispersed.
6. Fix cells overnight at 4 °C (*see* **Note 6**).

3.2 Immunofluorescence Staining

All the steps are performed at room temperature unless otherwise indicated. This immunofluorescence protocol is designed for cells in suspension. However, the same basic protocol can be applied to tissue cryosections [17, 18] or to cells adhered to poly-L-lysine-coated coverslips [19], with minor adjustments of times, permeabilization, and blocking steps (*see* **Note 8**).

3.2.1 Washing, Permeabilization, and Blocking

1. Label two 1.5 mL microfuge tubes "A" and "B." One tube will be blocked in BB with goat serum for staining set A. The other tube will be blocked in BB with donkey serum for staining set B (*see* Subheading 2.2.2, **item 5**).
2. Transfer one-half (1 mL) of fixed cell suspension from each well into tube A and 1 mL into tube B. Spin down the cells in a tabletop centrifuge at 2500 × *g* for 5 min. Pipette off supernatant containing fixative and discard into a hazardous waste container.
3. Gently flick the pellets to loosen cells before adding 1 mL of DPBS to resuspend the cells. Invert microfuge tube a few times after adding the DPBS to resuspend cells completely.
4. Centrifuge cells as above, discard the supernatant, and resuspend in DPBS as above, for a total of three washes to remove fixative.
5. Gently flick tubes to loosen cell pellets. Resuspend each cell pellet in 1 mL of permeabilization buffer by inverting the microfuge tube several times. Incubate cells for 15 min at room temperature.
6. Spin down cells at 2500 × *g* for 5 min as above and discard the supernatant.
7. Gently flick tubes to loosen cell pellets. Resuspend tube "A" of cells in 1 mL BB with goat serum and tube "B" in 1 mL BB with donkey serum by inverting the microfuge tube several times. Place at 4 °C overnight (*see* **Note 9**).

3.2.2 Fluorescent Probe and Antibody Staining

1. Pipette 200 μl of the cell suspension in Tube A (BB with goat serum) into a clean 1.5 mL microfuge tube marked "Set A." Pipette 200 μl of the cell suspension in Tube B (BB with donkey serum) into a clean 1.5 mL microfuge tube marked "Set B" (*see* **Note 9**).
2. To the "Set A" microfuge tube, add the unconjugated primary rabbit anti-Ki67 antibody (1 μL) and mix with gentle flicking. To the "Set B" microfuge tube, add the unconjugated primary goat anti-LaminB antibody (1 μL) and mix with gentle flicking. Incubate cells with primary antibody for 2 h.

3. Add 1 mL of DPBS to each cell-antibody mixture. Spin down cells in tabletop centrifuge at 2500 × *g* for 5 min. Discard the supernatant.
4. Gently flick the pellets to loosen cells before adding 1 mL of DPBS to wash away the unbound antibody.
5. Wash cells by resuspending and centrifuging two more times with 1 mL DPBS each for a total of three washes.
6. While washing cells, prepare the two mixtures of the FITC-conjugated Ter119 primary antibody, secondary antibodies and fluorescent probe all diluted into the appropriate BB at the desired final concentrations as follows:

   "Set A": 200 μL of BB with goat serum, 2 μL FITC-Ter119, 1 μL Alexa-594 goat anti-rabbit secondary, 1 μL of 1 mg/mL Hoechst.

   "Set B": 200 μL of BB with donkey serum, 2 μL FITC-Ter119, 1 μL Alexa-594 donkey anti-goat secondary, 1 μL of 1 mg/mL Hoechst.
7. Gently flick tubes to loosen cell pellets. Resuspend the washed cell pellets in the appropriate antibody/probe mixture from above.
8. Incubate cell-antibody mixture for 1–2 h.
9. Add 1 mL of DPBS to the cells, spin down as above, and wash cells two more times with 1 mL DPBS.
10. Resuspend cells in 1 mL DPBS (*see* **Note 10**).

3.2.3 Mounting Cells on Slides by Cytospin

1. Label the treated microscope slides to indicate the fluorescent antibodies and probes used for each sample.
2. Deposit the stained cells from suspension onto the slides using a Cytospin centrifuge. Assemble the cyto-funnel as directed. Load up to 200 μl of your cells into the funnel, and centrifuge at 113 x *g* (1000 rpm) for 3 min to deposit cells onto the slide.
3. Immediately remove slides and coverslip them with thoroughly cleaned coverslips [20] using a drop of ProLong Gold mounting media. The coverslip size is dependent on the cytofunnel hole size, but typically 22 mm × 22 mm coverslips are sufficient to cover the cells. Do not let the cells dry onto the slide (*see* **Note 1**).
4. After applying the coverslips, allow coverslipped slides to dry for 10–20 min. Carefully paint the edges of the coverslip with nail polish to seal each preparation. Once the nail polish has dried, the slides are ready to image (*see* **Note 11**).

3.3 Confocal Fluorescence Image Acquisition

Several fluorescence imaging modalities can be applied to erythroblasts. Conventional widefield microscopy is quick and relatively inexpensive, but out-of-focus light generated by structures above or below the focal plane may blur the images. New high-content imaging technologies such as ImageStream^X (Amnis Corporation) or InCell (GE Healthcare Life Sciences) allow for rapid and quantitative imaging of many thousands of cells, but do not capture images at sufficient resolution to detect submicron-scale changes in protein and organelle localization. Confocal laser scanning microscopy is one of the most common and useful methods for obtaining high-resolution images of various structures within cells, including erythroblasts. Resolution refers to the minimum distance between two points that can still be distinguished. The best resolution possible depends on the objective lens design and the wavelength of light used to illuminate the sample. In practice, it is not possible to achieve the optimum resolution, but addressing the following factors allows the actual resolution to approach this value, thus maximizing the amount of data an image can provide. For more extensive information on these concepts, refer to the *Handbook of Biological Confocal Microscopy* [15]. When acquiring images of two or more different samples to compare the amounts or distributions of proteins or organelles, identical microscope image acquisition parameters should be used for each sample. To control for non-specific secondary antibody staining, it is recommended to also image a sample stained with only the secondary antibodies.

3.3.1 Microscope Hardware Settings

1. Objective choice. Objective lens design involves optimizing image resolution, the amount of light captured, and the working distance at a given magnification while minimizing various optical aberrations. The most important feature of an objective is the numerical aperture (NA), which determines its ability to gather light and resolve image features. The higher the NA of the objective, the higher the optimal resolution of the images will be. The NA depends on the geometry of the lens and the refractive index of the objective immersion medium. Oil immersion objectives can achieve the highest NA values, water immersion objectives have intermediate NA values, and air objectives have the lowest NA values. Other important objective features include correction for optical aberrations and field curvature. For imaging cellular structures in erythroblasts, 100× plan apochromat oil immersion objectives with a numerical aperture of around 1.4 will produce the best images that provide the most information about cellular morphology.
2. Pinhole size. The size of the pinhole influences the amount of light detected and the resolution of the final image. Confocal microscopes have pinhole sizes measured in Airy units, a measurement that depends on the objective and the wavelength of

light. In most cases, setting the pinhole size to 1 Airy unit optimizes the signal-to-noise ratio and the resolution. Larger pinhole sizes let in more light and make the optical section thicker, which will enhance signal in dim samples but introduce more noise due to out-of-focus light. Smaller pinhole sizes decrease the amount of light detected but also eliminate more out-of-focus light, thus reducing noise.

3.3.2 Image Dynamic Range

Dynamic range refers to the distribution of pixels over the range of intensity values the detector can receive. To maximize the amount of quantitative information in an image, as much of the dynamic range as possible should be covered. Adjusting the following parameters can modulate the intensities collected. The dynamic range of an image can be tested in real time by the range indicator in the image acquisition software, which pseudocolors pixels based on intensity. The structure(s) to be imaged should contain pixels that are in the upper portion of the linear range but are not oversaturated.

1. Laser power. Higher laser powers allow collection of more light, but will photo-bleach fluorescent probes and antibodies faster, while lower laser powers protect the sample but limit the amount of information the final image contains. When acquiring many optical sections, bleaching can cause later sections to be dimmer than earlier ones, leading to incorrect interpretations of the data. To minimize photobleaching, use lower laser powers, use mounting media with antifade reagent(s), and use stable fluorophores (*see* **Note 12**).
2. Gain. Gain refers to the ratio of photons detected by the photomultiplier tube (PMT) detector to the output voltage. Increasing the gain will increase the sensitivity of the PMT detector, which will brighten the image and increase the dynamic range without increasing laser power. However, at very high gain values the signal-to-noise ratio decreases and the image begins to appear grainy.
3. Digital offset and digital gain. Digital offset determines the black level of the detector and is usually set to zero. Lowering the offset decreases the intensity of the background. Digital gain digitally enhances the signal after it leaves the detector and is usually set to 1. *For most applications, these values should not be changed, as this can lead to misrepresentation of the data.* Only the laser power (3.3.2.1) and the PMT gain (3.3.2.2) should be adjusted to optimize the image dynamic range.

3.3.3 Amount of Information Collected

The following parameters determine the amount of information in each pixel of the final image, which influences the resolution of the image and the amount of information it contains.

1. Laser scan speed, or pixel dwell time. The longer the laser dwells on each pixel, the more signal is detected and the less noise distorts the image. Faster laser scan speeds reduce photobleaching and reduce the amount of time to acquire an image, but also reduce the signal-to-noise ratio, producing dim, grainy images. The optimum laser scan speed depends on factors such as the type of detector, objective, digital zoom, and pixel size, and can be determined empirically for each configuration.
2. Digital zoom. Adjusting the zoom changes the area scanned without changing the number of pixels collected, thus changing the size of each pixel. Zooming out captures a larger field of view but can lead to smaller features in a sample being missed, while zooming in increases image resolution to a certain extent but will bleach the sample faster. In general, digital zooms between 1 and 3× work well. For imaging cellular structures in erythroblasts, a digital zoom of at least 2× with a 100× objective produces the best results.
3. Z-step size. Taking optical sections that are closer together than the optimal axial resolution, termed oversampling, will produce the best results, especially for three-dimensional measurements such as surface area and volume. In general, one should set the Z-step size between a third and a half of the optical slice thickness (determined by wavelength and objective used). For imaging erythroblasts at 100×, we recommend Z-step sizes of 0.25 or 0.3 μm. Taking optical sections further apart in the *Z* dimension can reduce imaging time and photobleaching but may result in important information being missed. Taking optical sections closer together will collect more information but could accelerate photobleaching, depending on the stability of the fluorophore (*see* **Note 13**).

3.4 Data Presentation

Addressing the factors described below will ensure accurate and clear communication of the information collected in each image. Optimal image views are shown in the figures, as are common mistakes and pitfalls. Working with 3D analysis software makes data presentation and data analysis more efficient. Most confocal microscope companies provide analysis software that can display and analyze images acquired on their microscopes. Third party analysis software packages, such as Volocity (Quorom Technologies) or Imaris (Bitplane), can open a wide variety of file types and offer more extensive analysis capabilities. Many free plugins are also available for ImageJ or Fiji (NIH) that can display and analyze images from a wide variety of file types. When presenting images from two or more samples, the same data presentation criteria should be used for each image to enable accurate comparison of the samples.

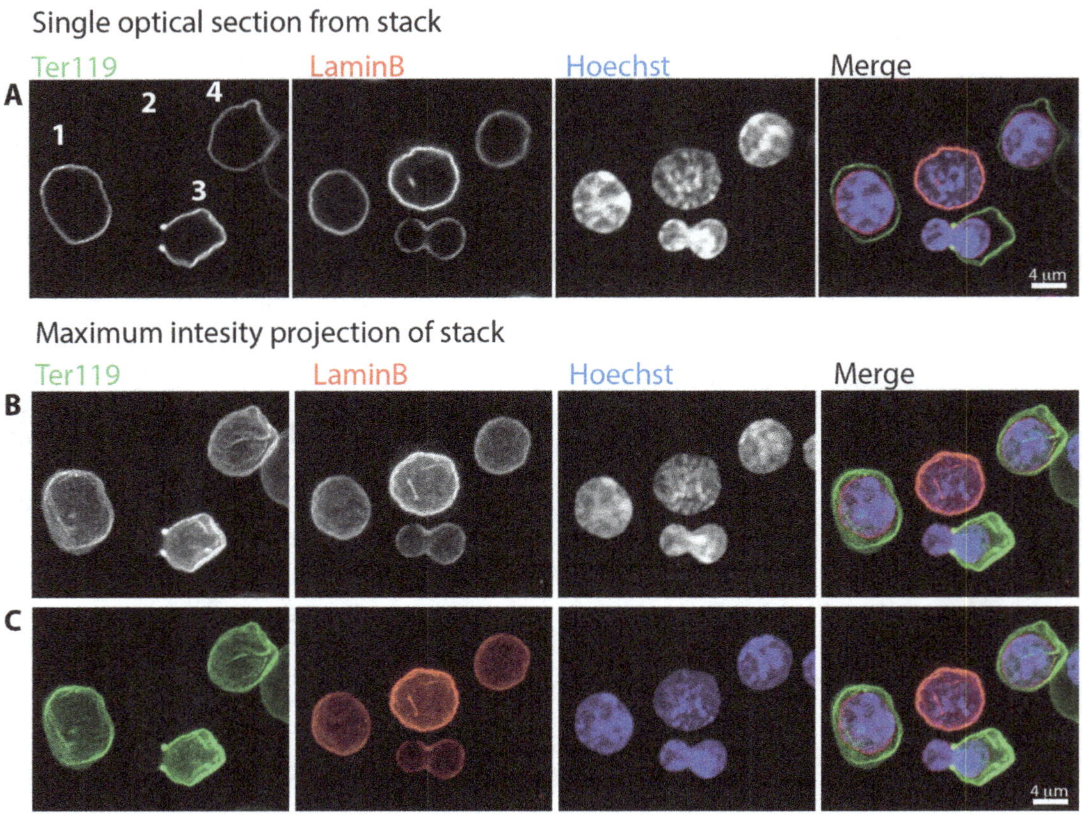

Fig. 2 Comparison of image presentation methods and how psuedocoloring affects visualization of image details. Z-stack of dissociated fetal liver cells immunostained for Ter119 in *green*, LaminB in *red*, and stained with Hoechst in *blue* was acquired as in Fig. 1. (**a**) A single optical section of an *XY* plane near the center of the cells shows the distribution of each molecule at that location, clearly revealing individual outlines of plasma membrane and nuclear envelope contours. (**b**, **c**) Maximum intensity projections display features from the entire Z-stack, including the tops and bottoms of the cells. (**b**) Each channel is presented separately in grayscale and as a merged pseudocolor image. (**c**) Each channel is presented separately in pseudocolor and as a merged psuedocolor image. Important features of each channel are more difficult to distinguish than in (**b**), particularly in the *red* and *blue* channels, which have low contrast with the *black* background. For example, the human eye can more readily distinguish the dim LaminB staining in cell 3 and the dim Hoechst staining in cell 2 from the black background

3.4.1 View

Information collected from 3D stacks can be presented in multiple ways.

1. One can show individual *XY*, *XZ*, and *YZ* planes from the stack. An example of a single *XY* plane is shown in Fig. 2a. This can provide 2D information about a particular area of the cell. Note that the *XZ* and *YZ* planes will not have as good resolution as the *XY* planes because the optimal axial resolution in *Z* (~0.6–0.8 µm) is not as good as the optimal lateral resolution in *X* and *Y* (~0.2–0.3 µm) [15]. One should be careful in

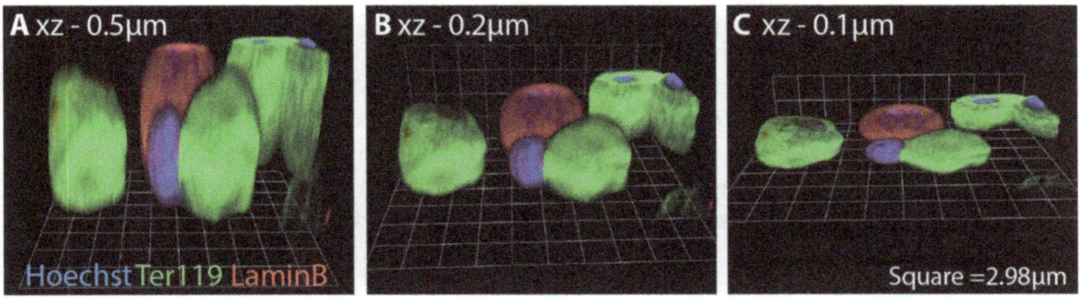

Fig. 3 Improper axial calibration distorts cellular morphology. 3D reconstructions of a Z-stack of the cells in Fig. 2, stained with Ter119 in *green*, LaminB in *red*, and Hoechst in *blue*, acquired with a 0.25 μm step size using the same objective and zoom used in previous figures. The step size was adjusted in the image analysis software to: 0.5 μm (**a**), 0.2 μm (**b**), or 0.1 μm (**c**). Based on the round shape of the condensed erythroblast nucleus in the center of the images, the calibration shown in (**b**) is likely the most accurate representation of the true cellular morphology. The tops of the cells appear flat in these images because the Z-stack does not encompass the entire cell in the axial (*Z*) dimension. Axial calibration can also be verified using fluorescent beads (not shown)

selecting which image planes to show because important features present in other planes may be missed.

2. To show information about the entire cell in one 2D image, one can show a 2D projection of image intensities from all the optical sections collected. There are multiple methods for doing these projections, but maximum intensity projections, such as those shown in Fig. 2b, c, work best for most confocal Z stacks. The resulting view shows all the features a wide-field image of the cell would show, but with out-of-focus light largely eliminated, revealing features that would otherwise be obscured.
3. Finally, it is also informative to present a 3D reconstruction or rendering of the object from the entire Z stack (Fig. 3), which provides more details about the 3D morphology of cells and subcellular structures (Fig. 3b) (*see* below, Subheading 3.4.2). When presenting 3D reconstructions, do not use a 2D scale bar, as 2D distances can be distorted due to the tilt of the reconstruction. Instead, use a scaled grid that tilts with the reconstruction to show perspective and size, as in Fig. 3.

3.4.2 Axial (Z) Distance Calibration

The optics of the microscope and refractive index mismatches between different objects in the light path can distort or stretch axial distances in 3D images. This will result in objects appearing stretched or longer than they really are in the axial dimension. An extreme example is shown in Fig. 3a, in which we have artificially introduced a Z-stretch. The condensed erythroblast nucleus in the center of the image, which should be round, now appears to be elliptical. Using stringently corrected objectives and matching the refractive indices of the objective immersion medium and the sample mounting medium can minimize but will not entirely eliminate this stretch. Software packages that include 3D viewing will allow for adjustment of the Z-step spacing, and this should be reduced until the structure(s) in the image has the expected shape. Fluorescent beads of a known diameter can be used to determine the correct spacing. Alternatively, one can use an object in the image that has a known shape, such as a round, condensed erythroblast nucleus. The correct Z-step spacing will depend on the microscope and objective used. In our experience, for images acquired with a 0.25 μm step size, adjustment of the step size in the 3D image viewer to 0.2 μm produces the best results (Fig. 3b). Overcorrection of the Z-step spacing will cause the cells to appear flattened (Fig. 3c). The chosen adjustment should be consistent across all images.

3.4.3 Image Pseudocoloring

The microscope collects each channel as a monochromatic, grayscale image. Many software packages will pseudocolor the image for each channel, making it easier to keep track of multiple channels. However, the human eye more readily detects differences in gray intensities than color intensities. For example, the human eye can readily see the Hoecsht-stained chromatin distribution in the grayscale image in Fig. 2b, but less easily in the pseudocolored blue image in Fig. 2c. When depicting the relationship between two or more channels, the most information can be conveyed by showing each channel in grayscale, along with a merged image of each channel in pseudocolor. A merged image in pseudocolor should not be presented alone, as this will obscure important features of each channel and make data interpretation difficult. For example, in the merged images in Fig. 2, the Hoechst staining obscures parts of the LaminB staining that can be seen in the individual grayscale images.

3.4.4 Intensity Levels

Image analysis software has various tools for adjustment of the range of intensity levels presented in the image. The software will allow for adjustment of both the minimum and maximum intensity values displayed in the image. Setting the maximum intensity to a lower value allows for the enhancement of the signal, enabling visualization of dim features that would otherwise not be visible by eye. However, it is critical to present the entire dynamic range of the image. Eliminating intensities at the top of the range can cause over-saturation, reducing the amount of information presented in the image. In extreme cases, this causes the image to be "blown out," making structures appear larger than they really are. For example, decreasing the maximum intensity value of Hoechst staining, as shown in Fig. 4c, obscures the distribution of chromatin within the nucleus and makes it appear as if the DNA protrudes past the LaminB staining of the nuclear envelope in certain locations. Cutting out intensities at the bottom of the range may eliminate dim image

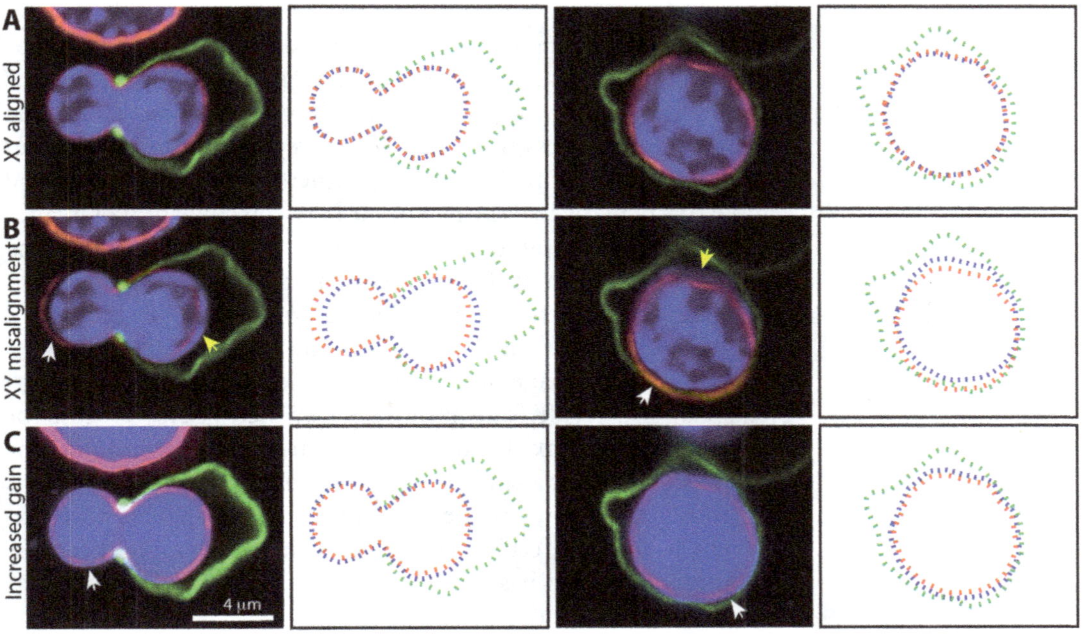

Fig. 4 Distortions in location and morphology of the nucleus introduced by improper image processing. (**a**) Correct *X–Y* registration of selected cells from the merged image in the single optical section presented in Fig. 2a, stained with Ter119 in *green*, LaminB in *red*, and Hoechst in *blue*. (**b**) *X–Y* registration misalignment makes it appear as though the DNA is not completely inside the nuclear envelope by causing gaps between Hoechst and LaminB staining (*white arrows*) and artifactual Hoechst staining outside the LaminB staining (*yellow arrows*). (**c**) Over-saturation of the Hoechst staining by increasing the digital gain obscures the variations in chromatin intensities apparent in (**a**) and (**b**), and causes the nuclei to appear larger in size. In places, this causes Hoechst staining to incorrectly appear outside LaminB staining (*white arrows*). *Dashed outlines* show the borders of fluorescence intensity in each channel to emphasize disparities between Hoechst and LaminB staining

features and falsely represent the data and the signal-to-noise ratio of the image. In general, one should only adjust the maximum intensity level, holding the minimum intensity level at zero.

3.4.5 Image Registration

Because multi-color confocal images are merged images of data collected separately in multiple channels, image registration problems can occur, in which the relative lateral and/or axial positions of structures in each channel are not accurately located in each optical section. This can lead to inaccurate data interpretation. A common source of inaccurate lateral (*X–Y*) image registration comes from image processing, especially when one splits, merges, and crops the individual image channels multiple times during figure preparation. For example, incorrect lateral image registration in Fig. 4b causes artificial gaps between the Hoechst and LaminB staining on one side of the nucleus, but artifactual DNA protrusion on the other side of the nucleus. To avoid this, one should be careful to present exactly the same portion of the image in each channel. Causes of axial (*Z*) image registration problems also include objective lens chromatic aberration, in which the microscope collects information for each channel at a slightly different *Z* position. Using an objective stringently corrected for chromatic aberration can address this issue. Image registration problems can also occur if vibrations in the environment or movements of the sample (particularly in live cells) occur between collection of images in each channel. Keeping the microscope on an air table can reduce vibrations coming from the environment, and collecting images in each channel at a particular *Z*-position before moving to the next *Z* position can improve image registration by minimizing the amount of time between image acquisitions in each channel. These problems can also result from improperly aligned lasers. To test for suspected improper axial registration, the user can image multicolor fluorescent beads, or a sample with one structure labeled with more than one fluorophore. If the bead or structure appears to be in different location in each channel, as is the case for the beads in Fig. 5, misalignment has occurred and one of the issues discussed above needs to be addressed.

3.4.6 Noise Removal

Although confocal microscopy removes the majority of out-of-focus light, other sources can introduce noise into images, which can result in a poor signal-to-noise ratio. Most software packages have noise removal capabilities, based on thresholding or deconvolution. This can enhance the signal-to-noise ratio of the image and make the structure of interest easier to see. However, overly stringent noise removal can introduce artifacts and misrepresent the data.

3.4.7 Image Compression

Most confocal microscopes produce images with large file sizes. Reducing file size with various image compression algorithms can make figure creation easier. Loss-less compression into .tif files is generally preferable to lossy compression into .jpg or .png files. Keep the original images for comparison with the compressed images to ensure that no visible information loss results.

3.5 Image Analysis

For most image analysis, avoid any changes in bit depth or image compression prior to analysis when possible. This will ensure the analysis takes into account all of the information collected in the image, making the analysis more accurate. Also important is keeping track of the pixel size and *Z*-axis spacing to ensure that geometric measurements reflect the actual distances between objects. Many image analysis software packages (examples listed in the previous section) will keep track of this for you. The majority of image analysis methods rely on the pixel intensities collected for each channel, which depend on the amount of fluorophore used to stain the sample and the imaging parameters discussed above. For the most accurate results, compare measurements from cells in the

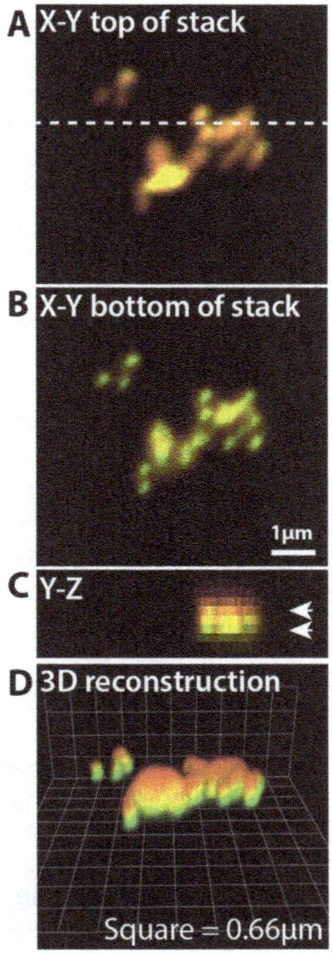

Fig. 5 Examples of *Z*-registration misalignment using fluorescent beads. Various views of a Z-stack of 0.2 μm-diameter TetraSpeck™ beads labeled with blue, green, red, and far red fluorophores (Fisher #T7280), with images acquired in the *red* and *green* channels, as in previous figures. In a properly aligned stack, all views would appear yellow to indicate complete colocalization of the red and green fluorophores, which label the same structures. In this case, axial (*Z*)-registration misalignment was caused by improperly aligned lasers. (**a**) Merged pseudocolor *X–Y* optical section near the top of the stack. Many areas appear *red.* (**b**) Merged pseudocolor *X–Y* optical section near the bottom of the stack. Many areas appear *green.* (**c**) *Y–Z* plane at the position indicated by the *dashed line* in (**a**). *Top arrow* indicates the position of the *X–Y* plane presented in (**a**), while *bottom arrow* indicates the position of the *X–Y* plane presented in (**b**). Red fluorescence is clearly displaced in the *Z* direction with respect to the green fluorescence. (**d**) 3D reconstruction of the Z-stack. As in (**c**), there is evident misalignment of the red and green fluorescence

same image or from different images of the immunofluorescence staining experiment collected with the identical imaging parameters. When this is not possible, great care must be taken to ensure that every sample is stained and imaged identically. Below are three common types of image analysis that can be applied to erythroblasts, and what to consider when applying these analyses.

3.5.1 Size and Shape Measurements

1. Useful 2D measurements include: area (of an object or an ellipse that approximates the size of the object), longest axis, shortest axis, aspect ratio (the ratio of the major and minor axis of the ellipse approximating the object shape) and circularity (which assesses shape by quantifying how close the object is to a perfect circle). 2D shape measurements performed on single optical sections may not accurately reflect the shape of the object as a whole because the size and shape of the object may vary between sections. Performing this analysis on maximum intensity projections of Z-stacks will produce more accurate results. However, if the cells do not lie flat, the dimensions of the projection will depend on the angle between the cells and slide or coverslip and inconsistent measurements can result. Adhering the cells to poly-L-lysine or using a Cytospin can minimize this problem.
2. Useful 3D measurements include: surface area, volume, and sphericity (which quantifies how close the object is to a perfect sphere). To get accurate 3D measurements, optical sections of the entire object need to be taken, and the *Z*-axis spacing should be adjusted correctly (*see* above, Subheading 3.4.2). The software measures the size and shape of the fluorescent signal and not of the labeled object, and there may be mismatch between the two. For example, when measuring the size of a cell stained with a membrane dye, the software will report the volume covered by the dye's fluorescent signal, not the volume of the cell contained by the membrane.

3.5.2 Colocalization

Colocalization can be defined in multiple ways, but it generally seeks to measure the degree of overlap between signals from two channels. This method, also called pixel intensity spatial correlation analysis, measures proximity of two molecules and cannot test for direct interactions. The Pearson's correlation coefficient, a common way of reporting colocalization, ranges from −1 for perfect anti-correlation to 0 for no correlation to 1 for perfect correlation. *Small values of this coefficient do not imply colocalization because two images with randomly distributed pixels will have values >zero* [21]. We do not consider correlation strong at coefficient values <0.5. The Pearson coefficient can also be skewed if there is only good colocalization in parts of the image. To minimize this effect, the coefficient should be measured only in the region of interest. Many

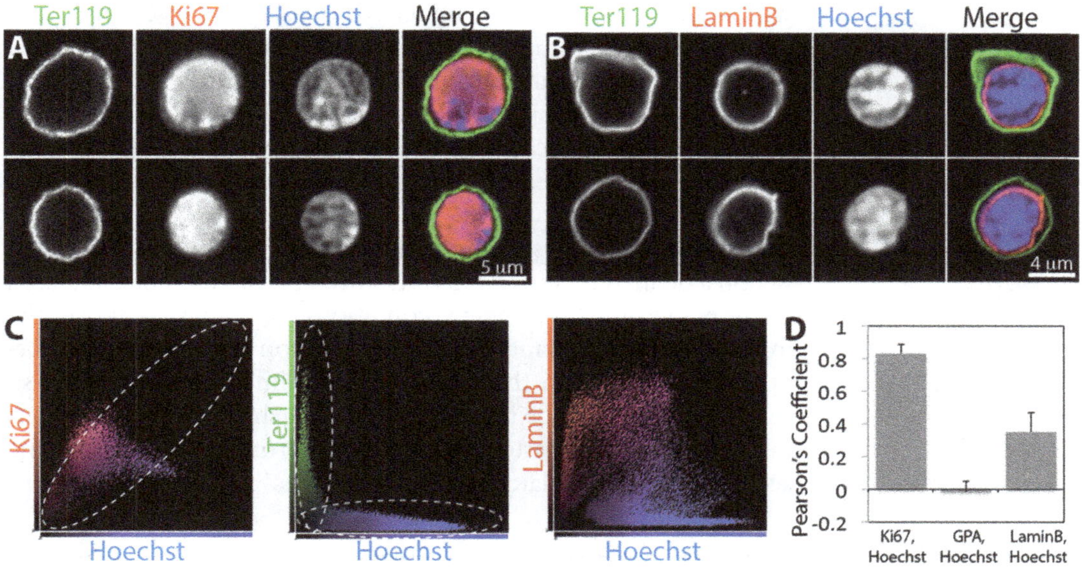

Fig. 6 Examples of pixel intensity correlation analysis using Pearson's coefficient. Single optical sections of fetal liver erythroblasts labeled with Hoechst and immunostained for (**a**) Ter119 in *green* and Ki67 in *red* or (**b**) Ter119 in *green* and LaminB in *red*. Images were acquired as in previous figures. (**c**) Scatterplots derived from pixel intensity correlation analysis displaying the intensities at each pixel in both channels of Ki67 vs. Hoechst, Ter119 vs. Hoechst, and LaminB vs. Hoechst. High colocalization is represented by points for each fluorophore close to a 45° diagonal from *lower left* to *upper right* across the figure (*dashed white oval* in *left panel*), while minimal colocalization by points distributed independently along the *X* and *Y* axes (*dashed ovals* in *middle panel*). (**d**) Pearson's coefficient for each analysis, showing strong colocalization with Ki67, no colocalization with Ter119, and weak colocalization with LaminB

software packages will generate a scatterplot along with colocalization analysis. This scatterplot allows for a more rigorous evaluation of the colocalization pattern (*see* **Note 14**). An example of colocalization analysis is shown in Fig. 6. As expected, Ki67, a nuclear marker of cell proliferation, and the Hoechst DNA stain colocalize (Fig. 6a). These channels have a high Pearson's coefficient, and the pixel intensities overlap at the center of the scatterplot (Fig. 6c, d). By contrast, the plasma membrane marker, Ter119, does not colocalize with Hoechst (6a, b), as these channels have a very low Pearson's coefficient and the majority of pixel intensities do not overlap (Fig. 6c, e). An intermediate situation is the nuclear envelope marker, LaminB, and Hoechst, which do not colocalize except at the interface of the nuclear envelope with the chromatin (Fig. 6b). Some of the pixel intensities overlap at the center of the scatterplot, but the majority do not (Fig. 6c). This results in a Pearson's coefficient well below 0.5 (Fig. 6d), and we would consider these molecules not colocalized. The precision of colocalization analysis depends heavily on the resolution of the image. Thus, apparent colocalization will be greater in images collected with lower-magnification, lower NA objectives than in images collected with higher-magnification, higher NA objectives. For example, the calculated colocalization of LaminB and Hoechst in Fig. 6 would have been larger if the image had been acquired with a 40×, 0.75 NA air immersion objective rather than a 100× 1.4 oil immersion objective. Thus, colocalization precision is limited by the achievable resolution of the objective and the microscope system.

3.5.3 Ratiometric Analysis

This analysis measures the ratio of intensities between two channels within a defined region of interest. This ratio can then be compared to other cells or experimental conditions. As discussed above, the intensities in each channel heavily depend on the amount of fluorophore used to stain the sample and the imaging parameters, so these factors should be held constant as much as possible. Correcting the intensity measurements for background will increase the accuracy of the calculated ratios.

4 Notes

1. Using thoroughly cleaned coverslips, termed "squeaky clean," can reduce imaging artifacts caused by any dirt or residues on the coverslips [20].
2. When using mouse and rat primary antibodies to stain two different antigens in the same sample, be sure to use secondary antibodies that have been tested by the manufacturer for minimal cross-reaction with antibodies from the other species.
3. Primary antibody and labeled small-molecule stain concentrations should be empirically determined for each system. To do this, stain identical samples with a range of concentrations and select the concentration that produces images with the best signal-to-noise ratio. Typical primary antibody concentrations range from 1:100 to 1:500 dilutions.
4. Each E13–15 pregnant wild-type mouse typically has between 3 and 12 embryos, depending on the strain of mouse.
5. You should end up with an evenly dispersed suspension of cells and a few larger whitish pieces of connective tissue, which can be removed with forceps.
6. This overnight 4% PFA fixation at 4 °C protocol to fix cells in suspension is optimal for the preservation of erythroblast morphology and antigenicity for most proteins, but may need to be modified for some epitopes whose signals may be reduced due to over-fixation. While lower PFA concentrations and/or shorter fixation times at room temperature (e.g., 4% PFA for 4 h at room temperature) may be tried, some erythroblasts may rupture and be lost during the subsequent labeling and

washing steps, which involve repeated centrifugation and resuspension steps.

7. Abundant erythroblasts can be obtained from the mouse spleens induced for stress erythropoiesis, e.g., either through phlebotomizing [22], phenylhydrazine drug treatment, or genetic manipulation [23].
8. Adhering cells to poly-L-lysine-coated coverslips before fixation, and other forms of handling (flow cytometry and sorting), can lead to subtle disruptions or changes in erythroblast morphology or protein localizations. These procedures may also introduce nonspecific background staining.
9. 2 mL of cells in BB are typically obtained from one fetal liver, bone marrow from two femurs, or one spleen. Use 200 μl aliquots for each staining condition from this stock solution. Cells can be stored in BB at 4 °C for up to 1 week, with relatively little loss of morphology and antigenicity for many proteins.
10. Some cells will be lost during the repeated washes and centrifugation steps required for permeabilization, blocking, and staining. There will still be plenty of cells left over to prepare cytospins for confocal imaging.
11. Cells located near the nail polish should not be imaged, as the nail polish will interfere with imaging. Properly centering the coverslip over the circular pattern of cells created by the Cytospin and only adding nail polish to the edge of the coverslip should minimize this issue.
12. Algorithms exist to computationally correct for photobleaching, but these should be used with caution and may need to be optimized for each fluorophore and staining condition [24].
13. To create a complete image of 3D cellular morphology, capture as much of the cells of interest as possible. For imaging erythroblasts at 100× and a step size of 0.25 μm, we typically capture around 20 optical sections. For imaging larger cells, such as megakaryocytes, or groups of cells, such as erythroblastic islands, more optical sections will be required.
14. We find that colocalization analysis is easiest to interpret when used to compare different conditions on similar samples (i.e., drug treatment, knockdowns). If the treatment changes the correlation coefficient, it can be said to affect the colocalization.

References

1. Palis J (2014) Primitive and definitive erythropoiesis in mammals. Front Physiol 5:3
2. Ji P, Murata-Hori M, Lodish H (2011) Formation of mammalian erythrocytes: chromatin condensation and enucleation. Trends Cell Biol 21:409–415
3. Migliaccio AR (2010) Erythroblast enucleation. Haematologica 95:1985–1988

4. Liu J, Guo X, Mohandas N et al (2010) Membrane remodeling during reticulocyte maturation. Blood 115:2021–2027
5. An X, Schulz VP, Li J et al (2014) Global transcriptome analyses of human and murine terminal erythroid differentiation. Blood 123:3466–3477
6. Bell AJ, Satchwell TJ, Heesom KJ et al (2013) Protein distribution during human erythroblast enucleation in vitro. PLoS One 8:e60300
7. Wong P, Hattangadi SM, Cheng AW et al (2011) Gene induction and repression during terminal erythropoiesis are mediated by distinct epigenetic changes. Blood 118: e128–e138
8. Hattangadi SM, Wong P, Zhang L, Flygare J, Lodish HF (2011) From stem cell to red cell: regulation of erythropoiesis at multiple levels. Blood 118:6258–6268
9. Paulson RF, Shi L, Wu DC (2011) Stress erythropoiesis: new signals and new stress progenitor cells. Curr Opin Hematol 18:139–145
10. Chasis JA, Mohandas N (2008) Erythroblastic islands: niches for erythropoiesis. Blood 112:470–478
11. Manawani D, Bieker JJ (2008) The erythroblastic island, Chapter 2. Curr Top Dev Biol 82:23
12. Lee JC-M, Gimm JA, Lo AJ et al (2004) Mechanism of protein sorting during erythroblast enucleation: role of cytoskeletal connectivity. Blood 103:1912–1919
13. Konstantinidis DG, Pushkaran S, Johnson JF et al (2012) Signaling and cytoskeletal requirements in erythroblast enucleation. Blood 119:6118–6127
14. Koury ST, Koury MJ, Bondurant MC (1989) Cytoskeletal distribution and function during the maturation and enucleation of mammalian erythroblasts. J Cell Biol 109:3005–3013
15. Pawley JB (ed) (2006) Handbook of biological confocal microscopy. Springer, New York, NY
16. Nowak RB, Papoin J, Gokhin DS, Casu C, Rivella S, Lipton JM, Blanc L, Fowler VM (2017) Tropomodulin 1 controls erythroblast enucleation via regulation of F-actin in the enucleosome. Blood 130(9):1144–1155
17. Sui Z, Nowak RB, Bacconi A, Kim NE, Liu H, Li J, Wickrema A, An X, Fowler VM (2014) Tropomodulin3-null mice are embryonic lethal with anemia due to impaired erythroid terminal differentiation in the fetal liver. Blood 123:758–767
18. Gokhin DS, Fowler VM (2011) Cytoplasmic γ-actin and tropomodulin isoforms link to the sarcoplasmic reticulum in skeletal muscle fibers. J Cell Biol 194:105–120
19. Fischer AH, Jacobson KA, Rose J, Zeller R (2008) Preparation of slides and coverslips for microscopy. Cold Spring Harb Protoc 2008: pdb.prot4988–pdb.prot4988
20. Waterman-Storer CM (2001) Microtubule/organelle motility assays. Curr Protoc Cell Biol. Chapter 13, Unit 13.1
21. Costes SV, Daelemans D, Cho EH, Dobbin Z, Pavlakis G, Lockett S (2004) Automatic and quantitative measurement of protein-protein colocalization in live cells. Biophys J 86:3993–4003
22. Pantel K, Loeffler M, Bungart B, Wichmann HE (1990) A mathematical model of erythropoiesis in mice and rats. Part 4: Differences between bone marrow and spleen. Cell Tissue Kinet 23:283–297
23. Latunde-Dada GO, McKie AT, Simpson RJ (2006) Animal models with enhanced erythropoiesis and iron absorption. Biochim Biophys Acta 1762:414–423
24. Bernas T, ZarÉBski M, Cook RR, Dobrucki JW (2004) Minimizing photobleaching during confocal microscopy of fluorescent probes bound to chromatin: role of anoxia and photon flux. J Microsc 215:281–296

Chapter 13

Chromatin Immunoprecipitation (ChIP) with Erythroid Samples

Ivan Krivega and Ann Dean

Abstract

Chromatin immunoprecipitation (ChIP) allows determination of the locations to which a select protein is bound in chromatin. Chemical crosslinking of DNA and protein with bi-functional reagents such as formaldehyde and precipitation of the protein with a specific antibody permit PCR amplification (ChIP) or sequencing (ChIP-seq) to identify the bound sites. Here, we present methodology for these approaches that are widely applicable to erythroid cell lines, progenitor cells, and tissues.

Key words Chromatin, DNA-binding protein, ChIP, ChIP-seq, Crosslinking, Antibody, Immunoprecipitation

1 Introduction

Chromatin immunoprecipitation (ChIP) is a powerful approach to study the localization of regulatory proteins in the genome. Interaction between these proteins and DNA is critical to direct multiple nuclear processes including DNA replication, maintaining genomic stability, and regulation of gene expression. Transcription is regulated by general and lineage-specific factors, DNA modifications, histones, and histone/DNA modifying enzymes that together prepare the chromatin template for this process. Transcription factors dynamically interact with regulatory DNA elements stimulating or repressing gene expression. Histones compact DNA and shape chromatin affecting the accessibility of regulatory regions to transcription factors. Chromatin-modifying enzymes affect chromatin accessibility by modifying histone tails. These important transcription factors and co-factors can be captured in residence on chromatin using the ChIP assay.

ChIP assays require fixation of cells with formaldehyde to stabilize DNA/protein interactions [1]. However, a native ChIP technique without formaldehyde fixation has been described. It is suitable for detecting DNA/histone modifications only and cannot

Joyce A. Lloyd (ed.), *Erythropoiesis: Methods and Protocols*, Methods in Molecular Biology, vol. 1698, DOI 10.1007/978-1-4939-7428-3_13,

be used for transcription factor ChIP [2]. Despite the necessity of a fixation step in ChIP, the fixation time should be minimized to avoid technical artifacts [3]. After sequential cell and nuclei lysis, chromatin is subject to sonication into small fragments. The protein or complex of interest associated with DNA is pulled down by a specific antibody. Immunoprecipitated complexes are washed under stringent conditions, DNA/protein complexes are eluted and de-crosslinked, proteins are digested, and DNA is purified. Enrichment of a DNA region of interest can be assayed quantitatively by real-time quantitative PCR (ChIP-qPCR). The genome-wide binding pattern of a protein of interest can be detected by next-generation sequencing (ChIP-seq). We successfully used the protocol presented below for transcription factors or histone ChIP-qPCR and ChIP-seq using immortalized and primary mouse [4] and human [5] erythroid cells (Fig. 1).

2 Materials

1. 37% formaldehyde.
2. 1.25 M glycine.
3. 1× phosphate-buffered saline (PBS): 137 mM NaCl, 2.7 mM KCl, 10 mM Na_2HPO_4, 2 mM KH_2PO_4, pH 7.4.
4. Protease inhibitors. These can be purchased as a ready-to use mix or prepared separately as described below:
   (a) 50 mM 4-(2-aminoethyl) benzenesulfonyfluoride hydro chloride (AEBSF) (100× stock solution, final concentration 500 μM).
   (b) 2 mM pepstatin A (1000 × stock solution, final concentration 2 μM). 2 mg/mL aprotinin (1000 × stock solution, final concentration 2 μg/μL).
   (c) 1 mM leupeptin (1000 × stock solution, final concentration 1 μM).
5. Cell lysis buffer: 10 mM Tris–HCl (pH 8.0), 10 mM NaCl, 0.2% NP-40. Protease inhibitors should be added before use. It should be kept at 4 °C.
6. Nuclei lysis buffer: 50 mM Tris–HCl (pH 8.0), 10 mM EDTA, 1% SDS, Protease inhibitors should be added before use. It should be kept at room temperature.
7. IP dilution buffer: 20 mM Tris–HCl (pH 8.0), 150 mM NaCl, 2 mM EDTA, 0.01% SDS, 1% Triton X-100. Protease inhibitors should be added before use. It should be kept at 4 °C.
8. Low Salt wash buffer: 20 mM Tris–HCl (pH 8.0), 150 mM NaCl, 2 mM EDTA, 0.1% SDS, 1% Triton X-100. Protease

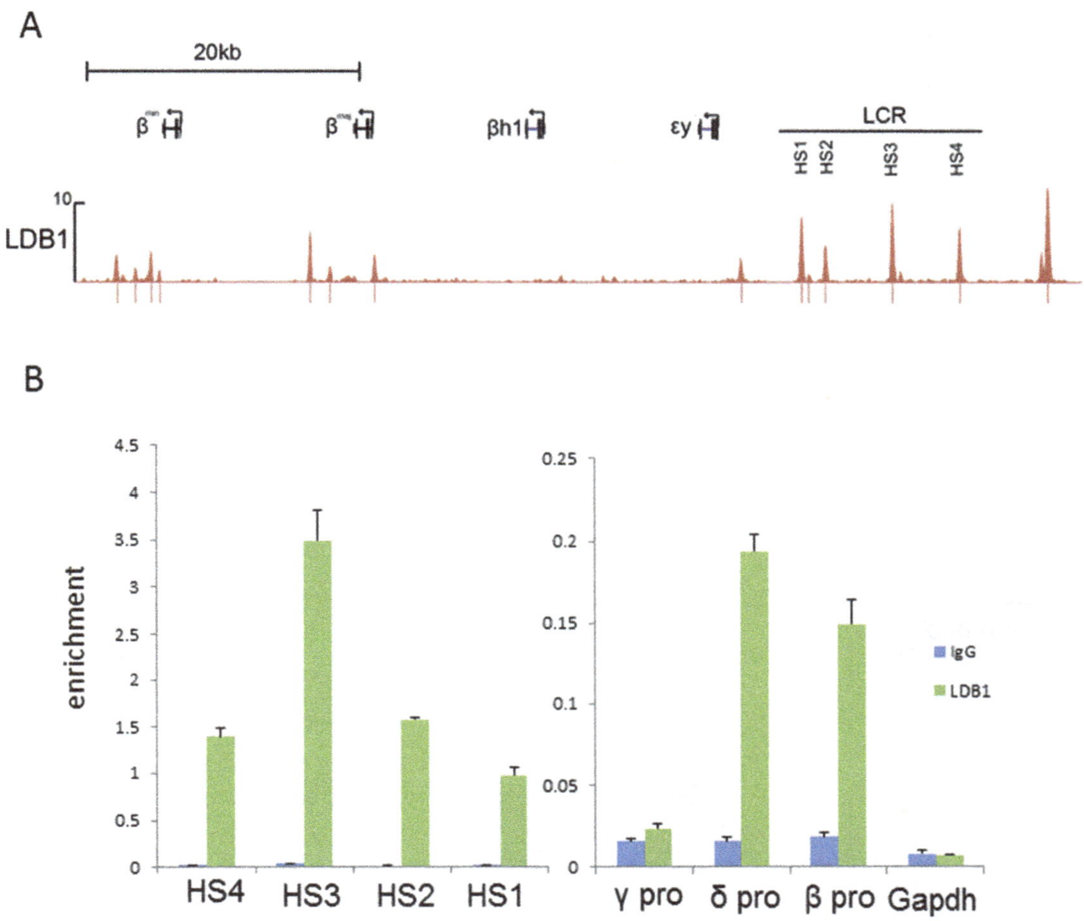

Fig. 1 LDB1 protein occupancy in the β-globin gene locus. (**a**) Representative LDB1 ChIP-seq track from mouse fetal liver E14.5 cells. Above the horizontal line is the signal and below the line are the called peaks. (**b**) Representative LDB1 ChIP-qPCR data from erythroblasts differentiated from primary human $CD34^+$ hematopoietic progenitor cells. Error bars indicate SEM of 3 independent experiments. In (**a**) and (**b**), ChIP was performed as described using antibody against LDB1

inhibitors should be added before use. It should be kept at 4 °C.

9. High Salt wash buffer: 20 mM Tris–HCl (pH 8.0), 500 mM NaCl, 2 mM EDTA, 0.1% SDS, 1% Triton X-100. Protease inhibitors should be added before use. It should be kept at 4 °C.
10. LiCl wash buffer: 10 mM Tris–HCl (pH 8.0), 0.25 M LiCl, 1 mM EDTA, 1% NP-40, 1% desoxycholate. Protease inhibitors should be added before use. It should be kept at 4 °C and protected from light.
11. Protein A/G PLUS-Agarose, Santa-Cruz Biotechnology.
12. Elution buffer: 0.1 M $NaHCO_3$, 1% SDS. It should be made fresh before use.

13. PCR Clean-Up and Gel Extraction Kit. Clontech/Takara.
14. NTB buffer, Clontech/Takara.
15. QIAquick PCR Purification Kit, Qiagen.
16. Ethylene glycol bis(succinimidyl succinate) (EGS).
17. ChIP-grade antibody and normal IgG.
18. Sonicator.
19. Rotating shaker.
20. Micrococcal nuclease (MNase).
21. 10 mg/μL Proteinase K (PK).
22. PK buffer: 50 mM Tris-Cl (pH 7.5), 100 mM NaCl, 1 mM EDTA, 0.5% SDS.
23. 0.5 M EDTA.
24. 0.1 M $CaCl_2$.

3 Methods

3.1 Day 1

1. Harvest 3×10^7 of cultured cells or primary cells from two embryonic day 14.5 (E14.5) fetal livers in 40 mL of culture medium or 1× PBS. Tissue samples such as E14.5 fetal livers can be easily disrupted by pipetting in 1× PBS or using a cell strainer. Add 1.09 mL of 37% formaldehyde (*see* **Note 1**). Incubate the cells with gentle agitation for 10 min at room temperature.
2. Add 4.6 mL of 1.25 M glycine and incubate cells with gentle agitation for 10 min at room temperature.
3. Centrifuge the cells at 4 °C for 5 min at 600 × *g*. Remove the supernatant and resuspend cell pellet in 40 mL of cold 1× PBS.
4. Centrifuge the cells at 4 °C for 5 min at 600 × *g*. Remove the supernatant, resuspend the cells in 1 mL of cold 1× PBS, and transfer the cells in a 1.5 mL tube.
5. Centrifuge the cells at 4 °C for 5 min at 600 × *g* and aspirate the supernatant (*see* **Note 2**).
6. Resuspend the cells in 1 mL of ice-cold cell lysis buffer and incubate on ice for 10 min.
7. Centrifuge the nuclei at 4 °C for 5 min at 600 × *g* (*see* **Note 3**).
8. Resuspend the nuclei in 1 mL of nuclei lysis buffer and incubate on ice for 10 min.
9. Depending on sonication conditions and sonicator manufacturer's recommendation, add 0.5–1 mL of IP dilution buffer.
10. Sonicate the lysate. Exact sonication parameters depend on sonication instrument and cell type used and should be

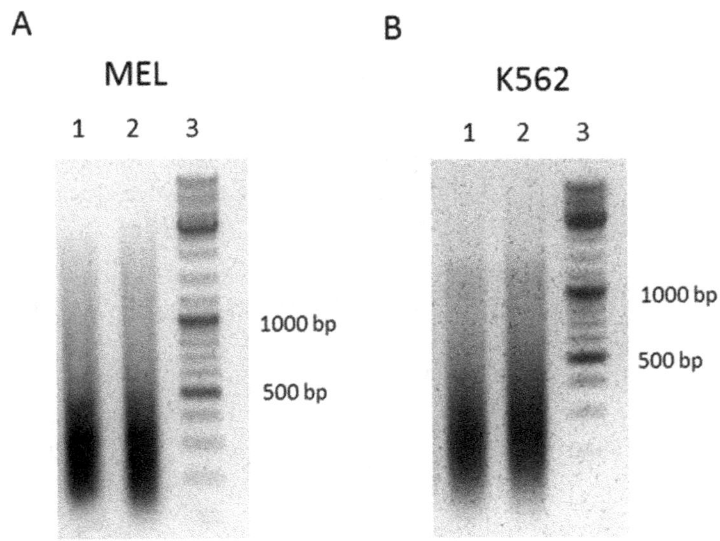

Fig. 2 Sonication of chromatin for ChIP analysis. Chromatin was prepared from (**a**) MEL cells induced with 2% DMSO for 4 days and (**b**) K562 cells. Representative preparations are shown in *lanes 1* and *2*. GeneRuler DNA Ladder (Thermo Fisher Scientific) in *lane 3*

determined experimentally. Optimal sonication conditions would result in obtaining chromatin fragments up to 500 bp for ChIP-seq or up to 1 kb for ChIP-qPCR (Fig. 2). *See* **Note 4** for instructions on checking sonication efficiency.

11. Centrifuge the sonicated chromatin at 4 °C for 10 min at 13,000 × *g* to pellet cellular debris. Transfer the supernatant to a clean 15 mL conical tube, and dilute the soluble chromatin with IP dilution buffer up to 5 mL. Add 100 μL of washed A/G agarose to preclear the chromatin (*see* **Note 5**). Incubate on a rotating shaker at 4 °C for 2 h.
12. Centrifuge the precleared chromatin at 4 °C for 5 min at 900 × *g*. Transfer the supernatant to a clean 15 mL conical tube. Transfer 160 μL of precleared chromatin to a 1.5 mL tube to serve as the INPUT control sample. Store INPUT at 4 °C until Day 2.
13. Distribute remaining precleared chromatin in 6 × 800 μL aliquots. Add specific antibody against protein of interest (2.5–5 μg depending on antibody quality) to chromatin aliquot and an equal amount of normal IgG (raised in the same animal species as specific antibody) to another chromatin aliquot for an isotype control. Incubate samples at 4 °C overnight on a rotating shaker. If not all chromatin aliquots have been used in the experiment, the remaining chromatin can be stored for future use at −80 °C for up to 3 months.

3.2 Day 2

1. Add 100 μL of washed A/G agarose to each sample (*see* **Note 5**). Incubate samples at 4 °C during 4 h with rotation.
2. Wash beads twice with 500 μL of Low Salt Buffer, once with 500 μL of High Salt buffer, once with 500 μL of LiCl buffer and thrice with TE (*see* **Note 6**).
3. After last TE wash resuspend agarose in 300 μL of elution buffer. Incubate at room temperature during 30 min with rotation (*see* **Note 7**).
4. Add 140 μL of elution buffer to the INPUT sample.
5. Add 15 μL of 5 M NaCl (final concentration 0.25 M) and 10 μL of 10 mg/mL PK to each sample including INPUT and incubate samples at 65 °C overnight to reverse cross-links.

3.3 Day 3

1. Clean DNA by PCR Clean-Up and Gel Extraction Kit using NTB buffer following the manufacturer's instructions (*see* **Note 8**).
2. Elute DNA with 50 μL of 5 mM Tris–HCl pH 8.0 and increase the volume to 300 μL with TE.
3. Dilute INPUT DNA 1:25 times with TE.
4. Use 2 μL of immunoprecipitated DNA or diluted INPUT for 10 μL qPCR reaction. SYBR Green or TaqMan chemistry can be used. For TaqMan probe and primer design, we have successfully used Primer Express, Primer3, and Primer Blast.
5. The comparative Ct method can be used to calculate relative enrichment of the protein of interest [6], or the standard curve method can be used for absolute quantification of immunoprecipitated DNA.

4 Notes

1. If protein of interest does not bind DNA directly, cells can be fixed with 150 mM EGS in 40 mL of 1× PBS during 20 min at room temperature before formaldehyde fixation [7].
2. Fixed cells can be used for 3C assays as well (*see* Chapter 14). If necessary, the cell pellet can be kept at −80 °C for several months.
3. To obtain single-nucleosome resolution, cells can be treated with MNase before nuclear lysis. Resuspend the cells in 500 μL cell lysis buffer containing 1 mM $CaCl_2$. Add 40–200 units of MNase and incubate the cells at 37°C for 5–10 min. Add EDTA to a final concentration of 10 mM and centrifuge the nuclei at 4 °C for 5 min at 600 × *g*. Exact MNase treatment conditions

should be determined experimentally for each cell type used for ChIP assay. We do not recommend MNase treatment if immunoprecipitated DNA will be used for NGS library construction.

4. To check sonication efficiency, remove 50 μL of chromatin and add 150 μL of PK buffer. Add NaCl to 0.25 M and 100 μg of PK and incubate at 65 °C for 4 h to reverse cross-links. Clean up DNA using QIAquick PCR Purification Kit (Qiagen) or any other appropriate DNA purification kit. Run the sample on a 1.5% agarose gel. As an alternative to using a DNA purification kit, phenol/chloroform purification and ethanol precipitation of DNA can be used. In this case, RNAse treatment should be performed before agarose gel electrophoresis.
5. To wash A/G agarose, briefly centrifuge required amount of A/G agarose at 13,000 × *g* 4 °C, aspirate supernatant, resuspend agarose in 500 μL of IP dilution buffer. Repeat wash three times. After last wash resuspended agarose in an equal volume of IP dilution buffer before use.
6. To wash A/G agarose, briefly centrifuge the sample at 13,000 × *g* 4 °C, aspirate supernatant, resuspend agarose sequentially in buffers as indicated in Subheading 3.2, **step 2**, keep 5 min on ice with occasional rotation. Briefly centrifuge, aspirate the supernatant, and resuspend in next washing buffer.
7. To increase the efficiency of precipitated DNA elution from A/G agarose, DNA can be eluted from the A/G agarose by two sequential incubations in 150 μL of elution buffer at 65 °C with shaking during 15 min. This elution method is essential if eluted DNA will be used for ChIP-seq library construction.
8. As an alternative to the DNA purification kit, phenol/chloroform purification and ethanol precipitation of DNA can be used. After precipitation, the DNA pellet should be resuspended in 300 μL of TE.

Acknowledgments

This work was supported by the Intramural Program of the National Institute of Diabetes and Digestive and Kidney Diseases, National Institutes of Health (DK015508 to AD).

References

1. Hoffman EA, Frey BL, Smith LM, Auble DT (2015) Formaldehyde crosslinking: a tool for the study of chromatin complexes. J Biol Chem 290:26404–26411
2. O'Neill LP, Turner BM (2003) Immunoprecipitation of native chromatin: NChIP. Methods 31:76–82
3. Baranello L, Kouzine F, Sanford S, Levens D (2016) ChIP bias as a function of cross-linking time. Chromosom Res 24:175–181

4. Krivega I, Dale RK, Dean A (2014) Role of LDB1 in the transition from chromatin looping to transcription activation. Genes Dev 28:1278–1290
5. Krivega I, Byrnes C, de Vasconcellos JF, Lee YT, Kaushal M, Dean A, Miller JL (2015) Inhibition of G9a methyltransferase stimulates fetal hemoglobin production by facilitating LCR/gamma globin looping. Blood 126:665–672
6. Schmittgen TD, Livak KJ (2008) Analyzing real-time PCR data by the comparative CT method. Nat Protoc 3:1101–1108
7. Zeng PY, Vakoc CR, Chen ZC, Blobel GA, Berger SL (2006) In vivo dual cross-linking for identification of indirect DNA-associated proteins by chromatin immunoprecipitation. Biotechniques 41:694, 696, 698

Chapter 14

Chromosome Conformation Capture (3C and Higher) with Erythroid Samples

Ivan Krivega and Ann Dean

Abstract

Chromosome conformation capture (3C) allows for the determination of the proximity in nuclei of DNA sequences that are linearly distant from one another in the genome. Proximity that is above that expected from random interaction provides evidence for potential long-range functional interactions such as between enhancers and their target genes. Many controls are required to convincingly demonstrate increased frequency of interaction between sequences and stringent functional tests must also be applied. Here, we present methodology suitable for 3C experiments that can also be applied as the basis for related 4C, 5C, and Hi-C approaches. These procedures are widely applicable to erythroid cell lines, progenitor cells, and tissues.

Key words Chromosome conformation capture, 3C, Chromatin, Enhancer, Long-range interactions, Gene regulation

1 Introduction

Gene expression is regulated by distantly located DNA elements. A growing body of evidence suggests that regulatory elements make direct contact with targeted genes. Based on this idea, it was proposed that the genome is organized through specific interactions between genomic elements. The vast majority of recent descriptions of 3D genome organization have been made using assays based on chromatin capture (3C) [1]. 3C allows the detection of interactions between linearly separated genomic regions of interest. In recent years, the 3C technique was adjusted to detect multiple interactions of one element with many elements (4C) [2], many elements with many (5C) [3], and interactions genome-wide (HiC) [4]. Here, we present a protocol suitable for performing 3C assays using erythroid mouse MEL and fetal liver E14.5 cells or human K562 and $CD34^+$ derived erythroblast cells. We have successfully applied this protocol to detect long-range interactions in multiple genomic loci including the β-globin locus [5, 6] (Fig. 1).

Joyce A. Lloyd (ed.), *Erythropoiesis: Methods and Protocols*, Methods in Molecular Biology, vol. 1698,
DOI 10.1007/978-1-4939-7428-3_14, © Springer Science+Business Media LLC 2018

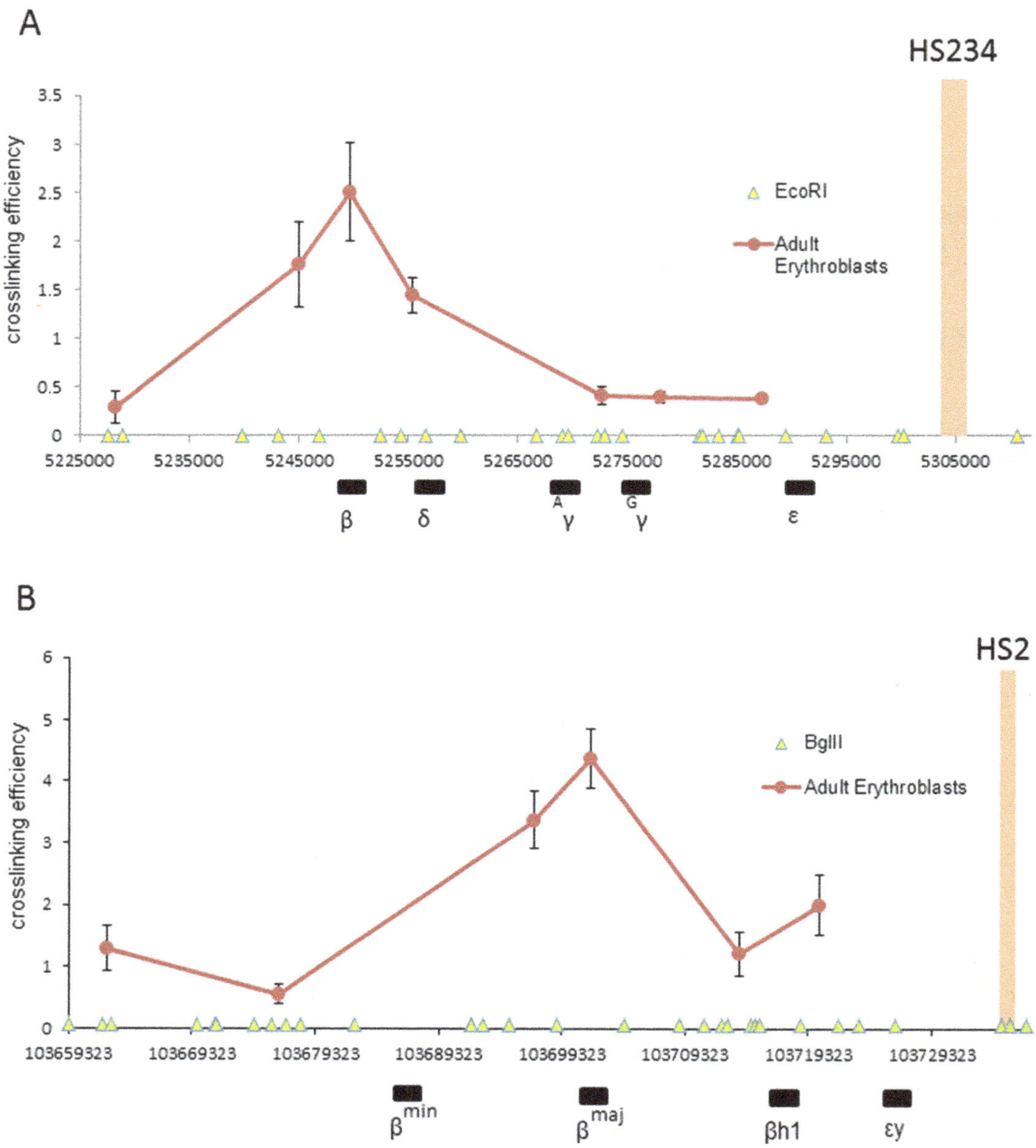

Fig. 1 Locus-wide interactions in the β-globin locus. (**a**) Interaction frequency determined by 3C between locations across (**a**) human β-globin locus using HS2-4 enhancer as the anchor observed for erythroblasts differentiated from primary human CD34$^+$ hematopoietic progenitor cells and (**b**) mouse β-globin locus using HS2 enhancer as the anchor (*orange rectangle*) observed for MEL cells induced with 2% DMSO for 4 days. Anchor fragments are marked by orange rectangle. Restriction sites are depicted by yellow triangles. Error bars indicate SEM of three independent experiments

Briefly, cells are fixed with formaldehyde to preserve 3D genome structure. Routinely, we use 10 million cells for each assay; however, we were able to scale down to 2 million cells without a negative effect on the assay sensitivity [7]. Then chromatin is digested by a restriction enzyme and ligated under dilute conditions allowing the detection of specific interactions between genomic elements in close spatial proximity. After ligation, chromatin is de-crosslinked, proteins are digested, and DNA is purified. Specific interactions between elements of interest can be detected by qPCR. Alternatively, 3C DNA can be used as a source material for 4C, 5C, and HiC library construction.

2 Materials

1. 37% formaldehyde.
2. 1.25 M glycine.
3. 1× phosphate-buffered saline (PBS): 137 mM NaCl, 2.7 mM KCl, 10 mM Na_2HPO_4, 2 mM KH_2PO_4, pH 7.4. Cell lysis buffer: 10 mM Tris–HCl (pH 8.0), 10 mM NaCl, 0.2% NP-40. Protease inhibitors should be added before use. Store at 4 °C.
4. Protease inhibitors (cOmplete, EDTA-free, Roche). These may also be prepared individually (*see* Chapter 13, Subheading 2, **item 4**).
5. Restriction enzyme (*see* **Note 1**).
6. 10× restriction enzyme buffer.
7. 20% SDS.
8. 20% Triton X-100.
9. Highly concentrated T4 DNA ligase.
10. 10× T4 DNA ligase buffer.
11. 100 mM ATP.
12. 10 mg/μL Proteinase K (PK).
13. PK buffer: 50 mM Tris-Cl (pH 7.5), 100 mM NaCl, 1 mM EDTA, 0.5% SDS.
14. UltraPure™ Phenol:Chloroform:Isoamyl Alcohol (25:24:1, v/v) pH 8.0 (Thermo Fisher Scientific).
15. Chloroform.
16. 96% Ethanol.
17. 3 M $NaHCO_3$.
18. TE (10 mM Tris-Cl, pH 7.5, 1 mM EDTA).
19. Thermomixer-R (Eppendorf).
20. Cell lysis buffer: 10 mM Tris-HCl (pH 8.0), 10 mM NaCl, 0.2% NP-40. Keep at 4°C and add protease inhibitor before use.

3 Methods

3.1 Day 1

1. Harvest 0.2–10 million cells in 40 mL of culture medium or 1× PBS. Add 1.09 mL of 37% formaldehyde. Incubate cells with gentle agitation for 10 min at room temperature.
2. Add 4.6 mL of 1.25 M glycine and incubate cells with gentle agitation for 10 min at room temperature.
3. Centrifuge the cells at 4 °C for 5 min at 600 × *g*. Remove the supernatant and resuspend the cell pellet in 40 mL of cold 1× PBS.
4. Centrifuge the cells at 4 °C for 5 min at 600 × *g*. Remove the supernatant, resuspend cells in 1 mL of cold 1× PBS and transfer cells to a 1.5 mL tube.
5. Centrifuge the cells at 4 °C for 5 min at 600 × *g* and aspirate the supernatant (*see* **Note 2**).
6. Resuspend fixed cells in 2 mL of ice-cold lysis buffer. Incubate on ice for 30 min with occasional mixing.
7. Collect the nuclei by centrifugation at 4 °C for 5 min at 600 × *g*. Discard the supernatant.
8. Resuspend the collected nuclei in 2 mL of cold 1.2× restriction enzyme buffer.
9. Centrifuge the sample at 4 °C for 5 min at 600 × *g*. Discard the supernatant.
10. Resuspend the collected nuclei in 500 μL of cold 1.2× restriction enzyme buffer.
11. Add 7.5 μL of 20% SDS (to a final concentration of 0.3%). Incubate with shaking at 1400 rpm on a Thermomixer (or similar) for 1 h at 37 °C.
12. Add 50 μL of freshly prepared 20% Triton X-100 (to a final concentration of 2%). Incubate with shaking at 1400 rpm for 1 h at 37 °C.
13. Take a 10 μL aliquot of the sample as an undigested genomic DNA control and store at −20 °C.
14. Add 300–1000 U of Restriction Enzyme of choice. Digest with shaking at 1400 rpm overnight at 37 °C.

3.2 Day 2

1. Take a 10 μL aliquot of the sample as a digested genomic DNA control. Check cutting efficiency (*see* **Note 3**).
2. Add 40 μL of 20% SDS (to a final concentration of 1.6%). Incubate for 25 min at 65 °C with shaking at 1400 rpm to inactivate the restriction enzyme.

3. Transfer the sample to a 15 mL tube containing 750 μL of 10× ligation buffer, 375 μL of 20% Triton X-100 (to a final concentration of 1%), 5700 μL of water. Incubate for 1 h at 37 °C with shaking.
4. Add 75 μL of 100 mM ATP.
5. Add 3 μL of highly concentrated T4 DNA ligase. Incubate for 4 h at 16 °C with occasional mixing.
6. Incubate for 30 min at room temperature.
7. Add proteinase K to a final concentration of 100 μg/mL. Incubate overnight at 65 °C with mixing.

3.3 Day 3

1. Cool the samples to room temperature.
2. Add 7.5 mL of phenol:chloroform:isoamyl alcohol and vortex. Centrifuge for 10 min at 16,000 × *g* at room temperature.
3. Transfer the aqueous phase to a clean tube. Add 7.5 mL of chloroform to the aqueous phase and vortex. Centrifuge for 10 min at 16,000 × *g* at room temperature.
4. Collect the aqueous phase. Transfer to 50 mL tube and add 7 mL of water.
5. Add 1.5 mL of 3 M sodium acetate and 30 mL of 96% ethanol. Incubate 30 min at −80 °C.
6. Centrifuge the samples at 14,000 × *g* for 30 min at 4 °C. Discard the supernatant.
7. Wash the pelleted DNA with 10 mL of cold 70% ethanol.
8. Centrifuge the sample at 14,000 × *g* for 5 min at 4 °C. Discard the supernatant.
9. Air-dry the pellet. Resuspend the pellet in 100 μL of TE (*see* **Note 4**).
10. Perform qPCR using TaqMan chemistry with primers to anchor fragment and restriction fragments of interest, and Taqman probes. Use internal control interaction and digested/ligated BAC to normalize interaction frequency (*see* **Note 5**). For primers and Taq-Man probe design suggestions, *see* Chapter 13, Subheading 3.1.

4 Notes

1. Different restriction enzymes can be used for 3C assay. We successfully utilized BglII, EcoRI, HindIII, ApoI, and BstYI in our lab. One should pick suitable restriction enzyme based on restriction pattern of locus of interest. The restriction enzyme should separate anchor fragment from restriction fragment containing element of interest by at least four restriction

fragments. To increase the ratio of restriction enzyme to glycerin in the restriction enzyme reaction, one should consider using the highly concentrated or fast-digest version of restriction enzymes.

2. Fixed cells can be used for ChIP assay as well (*see* Chapter 13). If necessary cell pellet can be kept at −80 °C for several months.

3. Checking cutting efficiency. Increase digested and undigested sample volume to 200 μL by PK buffer. Add 20 μg of PK to each sample and incubate for 30 min at 65 °C. Clean DNA by phenol/chloroform and precipitate with ethanol. Dilute DNA pellet in 200 μL of TE. Run qPCR using primer pairs amplifying fragments containing restriction sites of interest along with primer pair without restriction site. Use the following formula to calculate cutting efficiency (should be more than 85%):

$$100 - \left(\frac{2^{\Delta Ct \text{ digested}}}{2^{\Delta Ct \text{ undigested}}}\right) \times 100$$

4. Sometimes, one round of phenol/chloroform purification is not enough to get DNA of sufficient purity for qPCR. In this case, a second round of phenol/chloroform purification is required.

5. Interaction frequency between fragments of interest should be normalized by the interaction frequency of an internal control. Primers for the internal control should be designed to two restriction fragments separated by two internal restriction fragments. We have successfully used the ERCC3 and Tubulin gene loci for internal controls for mouse and human samples, respectively. To normalize primer efficiency, DNA from a BAC clone containing the locus of interest (such as β-globin) and from a BAC clone with the internal control locus (ERCC3 or Tubulin) should be digested with restriction enzymes and ligated. This ligated DNA from the BAC controls should be diluted to provide Ct values comparable to the Ct values from sample DNA. Interaction frequency should be calculated using the following formulas:

   For the 3C sample:
   ΔCt (sample) = Ct (primers for the anchor/restriction fragment of interest) – Ct (primers for internal control locus)

   For the BAC sample:

   ΔCt (primer efficiency) = Ct (primers for the anchor/restriction fragment of interest) – Ct (primers for internal control locus)
   Interaction frequency $= 1/2^{\Delta \text{Ct(sample)} - \Delta \text{Ct(primer efficiency)}}$.

Acknowledgments

This work was supported by the Intramural Program of the National Institute of Diabetes and Digestive and Kidney Diseases, National Institutes of Health (DK015508 to AD).

References

1. Dekker J, Rippe K, Dekker M, Kleckner N (2002) Capturing chromosome conformation. Science 295:1306–1311
2. Simonis M, Klous P, Splinter E, Moshkin Y, Willemsen R, de Wit E, van Steensel B, de Laat W (2006) Nuclear organization of active and inactive chromatin domains uncovered by chromosome conformation capture-on-chip (4C). Nat Genet 38:1348–1354
3. Dostie J, Richmond TA, Arnaout RA, Selzer RR, Lee WL, Honan TA, Rubio ED, Krumm A, Lamb J, Nusbaum C, Green RD, Dekker J (2006) Chromosome conformation capture carbon copy (5C): a massively parallel solution for mapping interactions between genomic elements. Genome Res 16:1299–1309
4. Lieberman-Aiden E, van Berkum NL, Williams L, Imakaev M, Ragoczy T, Telling A, Amit I, Lajoie BR, Sabo PJ, Dorschner MO, Sandstrom R, Bernstein B, Bender MA, Groudine M, Gnirke A, Stamatoyannopoulos J, Mirny LA, Lander ES, Dekker J (2009) Comprehensive mapping of long-range interactions reveals folding principles of the human genome. Science 326:289–293
5. Krivega I, Dale RK, Dean A (2014) Role of LDB1 in the transition from chromatin looping to transcription activation. Genes Dev 28:1278–1290
6. Krivega I, Byrnes C, de Vasconcellos JF, Lee YT, Kaushal M, Dean A, Miller JL (2015) Inhibition of G9a methyltransferase stimulates fetal hemoglobin production by facilitating LCR/gamma globin looping. Blood 126:665–672
7. Deng W, Rupon JW, Krivega I, Breda L, Motta I, Jahn KS, Reik A, Gregory PD, Rivella S, Dean A, Blobel GA (2014) Reactivation of developmentally silenced globin genes by forced chromatin looping. Cell 158:849–860

Chapter 15

Genome Editing of Erythroid Cell Culture Model Systems

Jinfen J. Yik, Merlin Crossley, and Kate G.R. Quinlan

Abstract

Genome editing to introduce specific mutations or to knock out genes in model cell systems has become an efficient platform for research in the fields of molecular biology, genetics, and cell biology. With recent rapid improvements in genome editing techniques, bench-top manipulation of the genome in cell culture has become progressively easier. The application of this knowledge to erythroid cell culture systems now allows the rapid analysis of the downstream effects of virtually any engineered gene disruption or modification in cell systems. Here, we describe a CRISPR/Cas9-based approach to making genomic modifications in erythroid lineage cells which we have successfully used in both murine (MEL) and human (K562) erythroleukaemia immortalized cell lines.

Key words Genome editing, Genome engineering, CRISPR/Cas9, Erythrocyte, MEL, K562

1 Introduction

According to the World Health Organization, approximately 5% of the world's population carry some form of hemoglobin disorder. β-thalassemia and Sickle Cell Disease (SCD), resulting from mutations in genes and regulatory regions of the *β-globin* locus, make up a significant subset of these hemoglobinopathies [1]. Individuals with β-thalassemia have mutations that reduce or obliterate the synthesis of normal β-globin. Individuals with SCD have a single base pair substitution in the *β-globin* gene that results in the substitution of valine for glutamate at the sixth position of the β-chain. This substitution produces HbS, abnormal haemoglobin that polymerizes in the deoxygenated state and forces the red blood cell (RBC) into a deformed sickle shape. The sickling of RBCs can lead to blood vessel occlusion and anemia due to hemolysis of the dysfunctional RBCs.

Interestingly, not all *β-globin* mutations produce a detrimental effect. During development, hemoglobin switching occurs, and foetal haemoglobin (HbF) is transiently produced in the human fetus. In some individuals, however, HbF continues to be highly

Joyce A. Lloyd (ed.), *Erythropoiesis: Methods and Protocols*, Methods in Molecular Biology, vol. 1698,
DOI 10.1007/978-1-4939-7428-3_15, © Springer Science+Business Media LLC 2018

expressed throughout adulthood. Investigations revealed the presence of mutations in the promoter or elsewhere in the locus that allow HbF to continue to be expressed into adulthood. This Hereditary Persistence of Fetal Hemoglobin (HPFH) is asymptomatic, but in individuals with a β-hemoglobinopathy HPFH mutations ameliorate the effects of SCD and β-thalassemia [2–6]. The compensatory nature of the HbF protein is also seen in the fact that sufferers of *β-globin* diseases do not manifest symptoms until infancy when HbF to adult haemoglobin (HbA) switching takes place and HbF levels rapidly decline. With the help of genome engineering technology, research into specific mutations in the erythroid cell lineage genome has helped to provide new insights into hemoglobin switching and potential avenues for using gene therapy in treating β-hemoglobinopathies [7].

The CRISPR/Cas9 system is an efficient gene editing system involving use of a Cas9 nuclease and a single guide RNA (sgRNA): made up of a fusion of the CRISPR RNA (crRNA) and trans-activating crRNA (tracrRNA) [8–10]. Traditionally, crRNA directs Cas9 to complementary target sequences (~20 bp) while the tracrRNA allows Cas9 to cleave the targeted site. Each target cut site needs to contain a protospacer adjacent motif (PAM) at the 3′ end, *5′-NGG 3′*, which the Cas9 recognises and cuts. Once recruited to a target site with the aid of the sgRNA, the Cas9 nuclease cuts the DNA creating a double-stranded break (DSB). Innate cell mechanisms can repair the break by non-homologous end joining (NHEJ), an imperfect process that generally introduces small insertions or deletions (indels) that can cause a frameshift, and, if directed to occur in an early codon exon of a gene, effectively knockout the gene. Alternatively, if a repair template is provided, in the form of a double-stranded or single-stranded DNA oligonucleotide (ssODN) [11], the cell toolbox is able to repair the break via homology-directed repair (HDR), consequently introducing the specific mutation directed by the template. To date, much research utilizing CRISPR/Cas9 and other gene editing tools has shown successful introduction of mutations that have either helped elucidate molecular and disease mechanisms [12–15] or shown potential for clinical use in gene therapy [16–21].

This chapter describes the CRISPR/Cas9 genome editing method that we have optimized for creating gene knockouts and site-directed mutations in the erythroid lineage MEL and K562 cells. An illustration of the steps involved in this protocol is shown in Fig. 1.

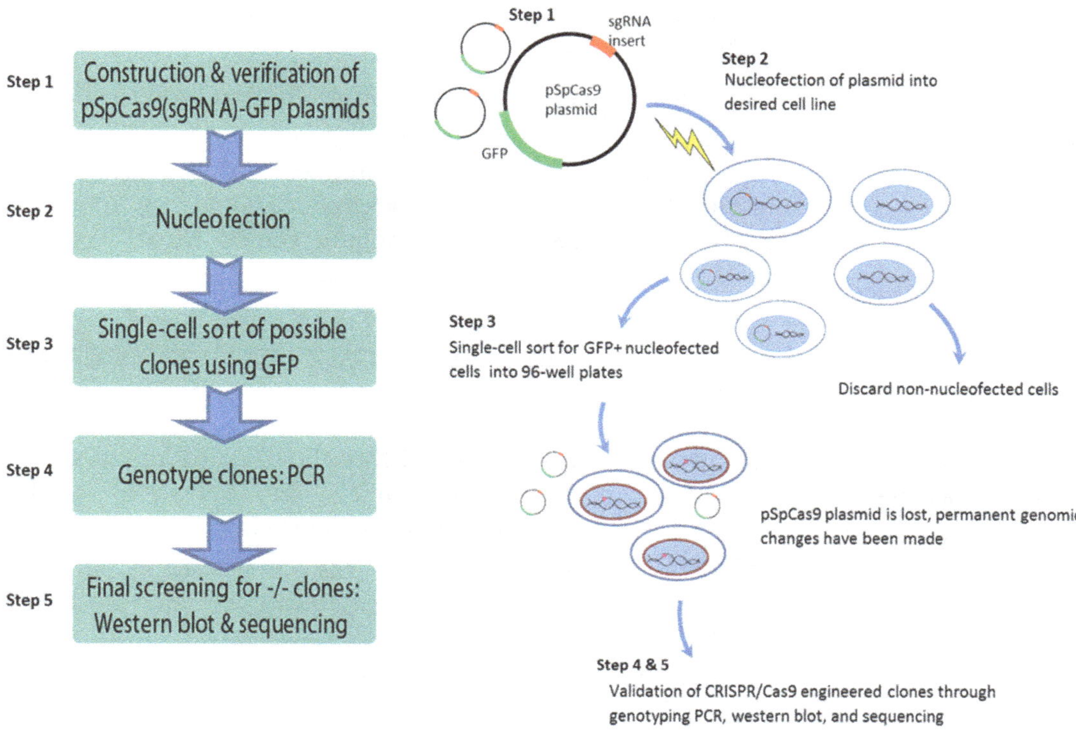

Fig. 1 General diagram of the CRISPR/Cas9 genome editing protocol on the erythroid cell lines, MEL and K562, that will be discussed in this chapter. This includes: *Step 1*. The design, construction and verification of the pSpCas9(sgRNA)-GFP plasmid to be used to; *Step 2*. Nucleofection delivering the pSpCas9 plasmid to the cell; *Step 3*. Sorting for single colonies of GFP positive clones indicating successful uptake of pSpCas9 plasmid; *Step 3*. Preliminary screening of clones through PCR genotyping; and *Step 4*. Screening of the right clones through western blot and sequencing

2 Materials

2.1 Design of sgRNA Oligonucleotides (Oligos)

1. CRISPR/Cas9 sgRNA design program (http://crispr.mit.edu/) [22].
2. sgRNA oligos (Sigma-Aldrich).
3. ssODN for HDR (Integrated DNA Technologies, IDT).
4. Cloning and sequence alignment software (SnapGene®).

2.2 Construction of pSpCas9(sgRNA)-GFP Plasmid

1. PX458 (pSpCas9(BB)-2A-GFP): Cas9 from S. pyogenes with 2A-EGFP, and a cloning backbone for sgRNA using BbsI restriction sites (*see* **Note 1**). pSpCas9(BB)-2A-GFP (PX458) was a gift from Feng Zhang (Addgene plasmid # 48138) [22].
2. T4 Polynucleotide Kinase.
3. BbsI restriction enzyme.
4. T4 DNA Ligase.
5. Antarctic phosphatase (New England Biolabs).

6. Gel and PCR clean-up kit (Promega Wizard SV Gel and PCR Clean-up system).
7. Competent *Escherichia coli* (*E. coli*) cells for transformation.
8. Luria broth agar plates: Add 5 g NaCl, 5 g Tryptone, 2.5 g Yeast Extract and 7.5 g Agar, into 500 mL deionised water. Autoclave, cool to 55 °C, and add 500 μL of Ampicillin (100 μg/mL). Swirl to mix and pour out 30 mL per plate.
9. PureLink® HiPure Plasmid Midiprep Kit (ThermoFisher Scientific).
10. U6-Fwd sequencing primer: GAGGGCCTATTTCCCATGA TTCC.

2.3 Nucleofection of pSpCas9(sgRNA)-GFP into MEL and K562 Cells

1. Mouse erythroleukaemia (MEL) cell and K562 cell medium: RPMI1640 supplemented with 10% fetal calf serum and 1% penicillin, streptomycin and L-glutamine (Life Technologies).
2. Nucleofection apparatus: Neon Transfection System (Life Technologies) or any similar system capable of delivering a pulse of 1400 V for 30 ms.
3. Nucleofection buffer R and T (Neon Transfection Kit, Life Technologies).

2.4 Single-Sort of Nucleofected Cells for Single Clones

1. BD FACSAria™ IIIu cell sorter or any cell sorter equipped with filters for 7AAD live-dead stain (561 nm) and GFP fluorescence (488 nm).
2. 1× Phosphate-buffered saline (PBS).
3. Fluorescence-activated cell sorting (FACS) Buffer: To 465 mL PBS, add 10 mL EDTA (0.5 M), and 25 mL FCS. Filter sterilize and store at 4 °C.
4. 7AAD live-dead stain (561 nm).
5. 96-well tissue culture plates.

2.5 Screening for Mutations: Genotyping, Sequencing, and Western Blot

1. QuickExtract™ DNA Extraction Solution (or phenol-chloroform for extraction of gDNA).
2. Q5® High-Fidelity DNA Polymerase.
3. PCR primers to genotype and sequence CRISPR/Cas9 modified region.
4. NanoDrop spectrophotometer.
5. Agarose gel electrophoresis equipment.
6. Apparatus for protein electrophoresis and western blot.

A.

Orientate chosen 20 nucleotide sgRNA guide in the 5'-3' direction:

Forward strand 5' – NNNNNNNNNNNNNNNNNNNN – 3'

B.

Generate the reverse strand and add BbsI sites to the forward and reverse sgRNA oligo:

Forward strand 5' – **CACCG**NNNNNNNNNNNNNNNNNNNN – 3'
Reverse strand 3' – CNNNNNNNNNNNNNNNNNNNN**CAAA** – 5'

Fig. 2 Generation of sgRNA for annealing and ligation into the PX458 plasmid. (**a**) Regardless of sgRNA guide location on the top or bottom strand, orientate chosen sgRNA in the 5′ to 3′ direction, do not include the PAM site at the 3′ end. This 20 nucleotide guide will be known as the forward strand. (**b**) Add a G to the 5′ end of the sgRNA and generate the corresponding sgRNA reverse strand. Construct BbsI restriction sites as shown, such that the overhangs are on the 5′ ends

3 Methods

Carry out the following procedures at room temperature unless stated otherwise.

3.1 Design of sgRNA Guide Oligos

The following CRISPR/Cas9 design protocol is adapted from Ran et al., 2013 [22]

1. To engineer gene knockouts via non-homologous end joining (NHEJ), select an exon sequence that is closest to the start codon as possible (but shared in any splice variant isoforms that you intend to disrupt). Introduced insertions or deletions are able to cause a frameshift mutation early in the coding sequence, causing premature stop codons to be generated and thus effectively creating a knockout. To engineer specific mutations via homology-directed repair (HDR), choose the sequence comprising the area to be edited.
2. Input selected section of genome sequence to be modified into a CRISPR/Cas9 sgRNA guide design platform such as http://crispr.mit.edu/, or any other suitable program. From the program, choose guides that have high scores (*see* **Note 2**).
3. Orient chosen 20 nucleotide sgRNA guides in the 5′ to 3′ direction, this will be called the forward sequence. Generate the complementary reverse sequence using any suitable program. Add restriction enzyme BbsI sites at the 5′ and 3′ end such that the guides can be later ligated into the PX458 plasmid; to the 5′ end, insert the sequence **CACC**. If the 5′ end of the guide is not a G, design the guide to have an extra G. Likewise, to the reverse complementary strand, insert **AAAC** into the 5′ end (Fig. 2). Order designed guides from preferred oligo provider (*see* **Note 3**).

4. If choosing the option to engineer specific mutations via homology-directed repair (HDR), design an extra single-stranded donor oligonucleotide (ssODN) that will act as the repair template. The ssODN should overlap the CRISPR/Cas9 cut site, with ~50–80 bp homology arms on each side of the CRISPR/Cas9-targeted cut site. As the ssODN also contains the sgRNA PAM site, design the template with a silent mutation in the PAM site so that the Cas9 will not re-cut the repaired sequence (*see* **Note 4**). Use a sequence alignment software like SnapGene® or similar to design this oligo.

3.2 Construction of pSpCas9(sgRNA)-GFP Plasmid

1. To prepare the sgRNA guide for cloning into the pSpCas9 vector PX458, perform a kinase treatment to add a phosphate group to the 5′ end of each forward and reverse sgRNA guide oligo. In a 1.5 mL micro-centrifuge tube, set up the following reaction: mix 1 μL sgRNA oligo (10 μm), 1 μL 10× PNK buffer, 1 μL T4 PNK (10 U/μL), 1 μL ATP (10 mM), and 6 μL MilliQ water for a total volume of 10 μL. Lightly flick to mix and incubate the kinase reaction mix at 37 °C for 45 min.
2. Briefly centrifuge tubes to collect liquid at the bottom of the tube. Combine the forward and reverse oligo reactions into one tube and incubate at 100 °C for 5 min. Allow the mixture to cool slowly to room temperature (*see* **Note 5**). Briefly centrifuge tubes again to collect liquid at the bottom of tube. The sgRNA oligos will have annealed to form the sgRNA insert.
3. Digest the backbone plasmid, PX458 with BbsI. In a 1.5 mL micro-centrifuge tube, combine 2.5 μL BbsI (10 U/μL), 4 μL NEB Buffer 2.1, 5 μL PX458 plasmid (1 μg/μL), and 28.5 μL MilliQ water to a total volume of 40 μL. Gently mix and incubate digestion reaction at 37 °C for 1 h.
4. Phosphatase treat the digested plasmid to prevent plasmid re-circularization. Add 1 μL Antarctic phosphatase (5 U/μL) and 4.5 μL 10× phosphatase buffer to the digested plasmid and incubate at 37 °C for 1 h. Heat inactivate the phosphatase at 65 °C for 20 min.
5. Purify the treated plasmid using Promega Wizard® SV Gel and PCR Clean-Up System kit to a final volume of 40 μL. Other similar DNA clean-up kits can also be used.
6. Ligate the annealed sgRNA insert from **step 2** with the purified digested PX458 plasmid from **step 5**. In a 1.5 mL micro-centrifuge tube, add 1 μL of annealed sgRNA insert (**step 2**), 1 μL digested plasmid (**step 5**), 1 μL T4 DNA ligase (400 U/μL), 1 μL 10× Ligase buffer, and 6 μL MilliQ water to a total volume of 10 μL. Incubate the ligation mix at 16 °C for 2 h or 4 °C overnight. The product from this reaction will be the

sgRNA insert ligated into the PX458 plasmid, pSpCas9 (sgRNA)-GFP (*see* **Note 6**).

7. Using standard transformation methods, transform the ligation product containing pSpCas9(sgRNA)-GFP from **step 6** into a competent *E. coli* strain. Spread transformed *E. coli* onto an LB plate with 100 μg/mL ampicillin. Incubate at 37 °C, overnight.
8. Next day, inspect plate for colonies and pick 5–10 clones to miniprep. The recommended negative control plate should have none to a few colonies. Miniprep the pSpCas9(sgRNA)-GFP plasmid from the bacterial culture (*see* **Note 7**).
9. Sequence the prepared plasmids to check for correct insertion of guide into the vector. The following primer that starts from the U6 promoter in the PX458 plasmid can be used: U6-FWD, GAGGGCCTATTTCCCATGATTCC.
10. Check sequence for correct insertion of guide into plasmid. Perform a midiprep for the desired pSpCas9(sgRNA)-GFP plasmid to obtain transfection-grade plasmid DNA.

3.3 Nucleofection of pSpCas9(sgRNA)-GFP into MEL and K562 Cells

1. Check that cell viability is at least 90% before nucleofection. MEL and K562 cells are best maintained between 1×10^5 and 1×10^6 cells/mL.
2. Prepare 6-well plates with 3 mL growth media without PSG.
3. For each nucleofection, aliquot out 5×10^5 cells into 1.5 mL microcentrifuge tubes. Pellet cells at 500 $\times g$, 2 min and wash with PBS. Resuspend pellet in 9 μL of nucleofection buffer (Buffer R using the Neon Transfection Kit).
4. To engineer knockout clones via NHEJ, add 1 μL of appropriate guide plasmid (2.5 μg/μL) to the cell and Buffer R mix.
5. To engineer direct mutations via HDR, add 0.8 μL guide plasmid (2.5 μg/μL) and 0.2 μL donor oligo (50 μM) to the cell mixture. If replicate nucleofections are to be performed, simply multiply amounts by number of replicates (*see* **Note 8**).
6. Nucleofect cells according to nucleofection apparatus instructions with settings of: one pulse of 1400 V for 30 ms.
7. Place nucleofected cells in the pre-prepared 6-well plate of PSG-free media and incubate at 37 °C, 5% CO_2 for 72 h. For a negative control, use a mix of non-nucleofected cells and guide plasmid (*see* **Note 9**).

3.4 Single-Sort of Nucleofected Cells for Single Clones

1. Set up 96-well plates with 180–200 μL of cell media in each well.
2. Spin down and wash nucleofected cells in PBS and resuspend nucleofected cells in 500 μL cold sterile FACS buffer. Keep on ice.

3. Using the BD FACSAria™ IIIu cell sorter or any suitable cell sorter, sort aseptically for single-live cells with 7AAD live-dead stain (561 nm) and GFP fluorescence (488 nm) into each well of the prepped 96-well plates from **step 1**. Cells that have been successfully nucleofected with pSpCas9(sgRNA)-GFP will transiently produce GFP.
4. Incubate the 96-well plates of single-sorted cells at 37 °C, 5% CO_2. The cells will take at least 2–3 weeks to recover and multiply to be of sufficient quantity for screening (*see* **Note 10**).

3.5 Screening for Mutations

Wherever appropriate, the following steps are useful for identifying clones with the desired mutations or knockouts.

1. Harvest cells for gDNA extraction (*see* **Note 11**). Resuspend each 96-well of cells and aliquot out a portion, leaving enough cells in the well for future expansion. Quickly centrifuge to pellet cells, and discard the supernatant (*see* **Note 12**). Add 50 μL (or appropriate volume) of QuickExtract™ DNA Extraction Solution to each pellet, quick vortex to mix samples, spin and run on a thermocycler with the following program: step 1 (65 °C for 5 min); step 2 (98 °C for 2 min), step 3 (hold at 4 °C) (*see* **Note 13**).
2. Measure the amount of gDNA from the extraction with the NanoDrop spectrophotometer.

   Genotyping of each clone can be carried out to get an approximate estimation of any deletion, insertion or mutation occurring in each isolated clone.
3. To genotype for clones with insertions or deletions, design forward and reverse primers to amplify a short ~200–300 bp sequence that includes the site of CRISPR/Cas9 modification.
4. Perform a PCR run with the primers from **step 3** on the extracted gDNA. For controls, use a wild-type sample (from unmodified cells). In each PCR tube setup, add 5 μL of gDNA (50–200 ng, from **step 2**), 0.25 μL Q5® High-Fidelity DNA Polymerase (2 U/μL), 5 μL 5× Q5 Buffer, 1.25 μL forward primer (10 μM), 1.25 μL reverse primer (10 μM), 2 μL dNTP (2.5 mM), and 10.25 μL MilliQ water to a final volume of 25 μL.
5. Run PCR with the following program: step 1 (98 °C for 30 s); step 2 (98 °C for 10 s); step 3 (50–72 °C for 30 s); step 4 (72 °C for 20 s); cycle through steps 2–4 (34×); step 5 (72 °C for 5 min); step 6 (4 °C hold).
6. Run PCR amplicons from **step 5** on a 3% agarose gel at 120 V for 50–60 min. Clones with a deletion or insertion will show a faster or slower migration compared to the wild type (Fig. 3).

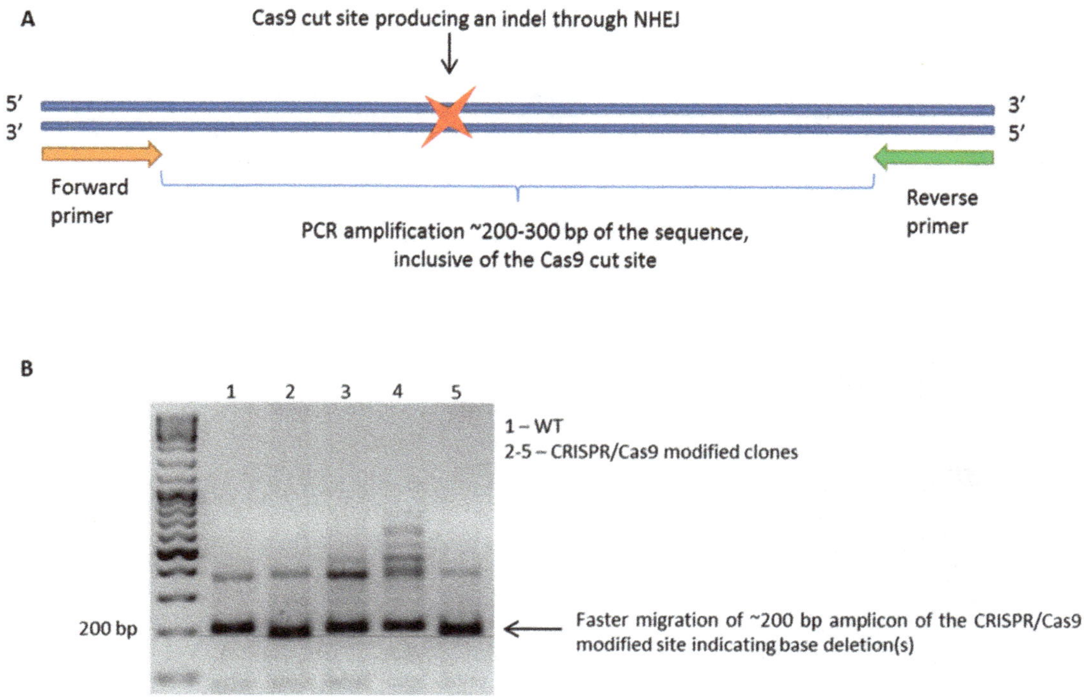

Fig. 3 PCR genotyping of clones for indel mutations. (**a**) A ~200 bp amplicon including the CRISPR/Cas9 modified site is made. (**b**) Running the amplicons on an electrophoresis gel (3%, 120 V, 50 min) shows clones which have maintained the WT sequence (clone 3, 5) have a ~200 bp band similar to the WT, whereas clones with successful indels show a band that runs smaller (clone 2) or larger (clone 4) than the WT

7. To genotype for clones with directed mutations, design forward and reverse primers to amplify a short ~200–300 bp sequence that includes the site of CRISPR/Cas9 modification. Use PCR program as in **step 5**. DNA polymerases extend only from the 3′ end of the primer (*see* **Note 14**), design the 3′ end of the forward or reverse primer to match either only the wild-type sequence (wild-type primer) or the modified sequence (modified primer) [23]. Amplicons will only be produced when the primer is able to bind to the template strand (Fig. 4).
8. Run PCR amplicons on a 1% agarose gel at 100 V for 45–60 min. Clones with a modification will produce a band with the modified primer, which can be confirmed with an absence of band with the wild-type primer.
9. Sequencing individual clones. Larger PCR amplicons (~700–800 bp) including sites of CRISPR/Cas9 modification in selected clones can be sent for sequencing using preferred laboratory protocol or service to double-check gene editing results (*see* **Note 15**).

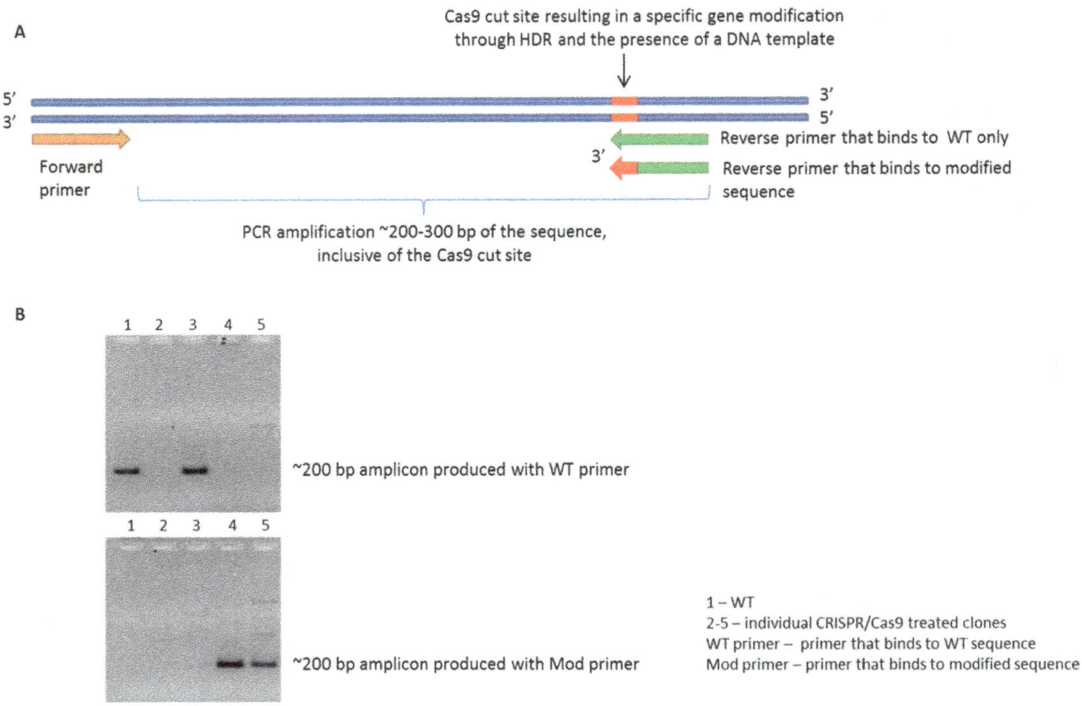

Fig. 4 PCR genotyping of clones for specified mutations. (**a**) A ~200 bp amplicon of the CRISPR/Cas9 modified site is made using either the WT primer that bind to the WT sequence only or the modified primer that binds to the modified sequence only due to the difference in the bp at the 3′ end of the primers. (**b**) Running the amplicons on an electrophoresis gel (1%, 100 V, 45 min) show clones which have maintained the WT sequence have a ~200 bp band produced with the WT primer only (clone 1 & 3), compared to clones with successful modification showing a band with the modified primer only (clone 4 & 5)

10. Knockout clones should also be confirmed with standard western blot protocols using the appropriate antibody and positive controls to check for protein absence in the knockout clone.

4 Notes

1. There are alternative pSpCas9 plasmids available with reporter genes/selectable markers other than GFP, which also use the BbsI restriction site for the insertion of sgRNA, for example plasmids encoding puromycin resistance [22], BFP, and Cherry [24].
2. The sgRNA directs the Cas9 to cut at a targeted site complementary to the sgRNA. It is beneficial to choose guides not only with high scores but with a PAM sequence that directs the CRISPR/Cas9 to cut as close to the desired mutation as possible. We recommend making a few guides for each experiment to find the most efficient guide.

3. When designing the sgRNA oligos, the PAM site is not included as part of the oligo.
4. To increase efficiency of CRISPR/Cas9 modifications using HDR, it is best to design the homology junction to be as close as possible to the CRISPR/Cas9 cut site.
5. Turn off the heat-block and leave the reaction tubes to cool with the heat-block. This will take a few hours; alternatively, it can be left to cool overnight. A thermocycler can also be programmed for this step.
6. Set up a concurrent negative control with the same reagents excluding the annealed sgRNA insert. The negative control should contain cut plasmid only and transformation using this negative control should yield fewer colonies than the plates with the ligated whole plasmid plus insert.
7. With the same tool used to pick the colonies for miniprep, patch a separate LB-Amp plate to make a master-plate. This should be grown at 37 °C overnight, and then stored at 4 °C and can be used later for midiprep when the plasmid sequence has been confirmed from the miniprep.
8. Performing replicate nucleofections allows for a bigger pool of cells to be used in downstream cell-sorting of positively nucleofected individual clones. A master-mix of plasmid and cells can be used here.
9. Nucleofection protocol can also be used for electroporation of TALENS [16].
10. During the 2–3 week recovery incubation, the cell media does not need to be changed. The initial setup of 180 μL cell media per well is enough to support the expansion and growth of a single cell into a visible colony in each well of the 96-well plate. Once a visible colony is formed, standard cell expansion or maintenance can be done as needed.
11. Before taking an aliquot of cells for gDNA extraction, ensure that there are enough cells left to maintain for further expansion of cells and other downstream analyses like western blots.
12. For convenience, the gDNA extraction can be done on a 96-well PCR plate as a replica of the 96-well tissue culture plate, centrifuge the aliquoted plate with a plate centrifuge, and flick the supernatant. The pelleted cells should still remain in the well, after which gDNA extraction can be performed.
13. In place of using QuickExtract, standard DNA extraction protocols like phenol/chloroform extraction can also be used.
14. DNA polymerases only extend from the matched 3′ end of a primer complementary to the reading strand. Mismatches

inhibit primer extension that will result in decreased PCR efficiency.

15. When aligning sequencing results of potential knockout clones, it is best to select clones with frameshift mutations, which give increased chances of total obliteration of the protein. Examination of the sequencing chromatogram will provide information on the cleanliness of the sequencing. Chromatogram trace peaks at or after the CRISPR/Cas9 modification site may help give indication on the zygosity of the clone; homozygous clones have single peaks, while heterozygous clones will show double peaks at a single peak position.

Acknowledgment

This work has been supported by funding from the Australian Research Council and the National Health and Medical Research Council to M.C.

References

1. WHO|Sickle-cell disease and other haemoglobin disorders. http://www.who.int/mediacentre/factsheets/fs308/en/. Accessed 12 Dec 2015
2. Thein SL, Menzel S, Lathrop M, Garner C (2009) Control of fetal hemoglobin: new insights emerging from genomics and clinical implications. Hum Mol Genet 18:R216–R223
3. Hemminger J (2012) Identification and reporting of common hemoglobin disorders: a review. Ibnosina J Med Biomed Sci 4:8–10
4. Bauer DE, Orkin SH (2011) Update on fetal hemoglobin gene regulation in hemoglobinopathies. Curr Opin Pediatr 23:1–8
5. Bauer D, Kamran S, Orkin Š (2015) Reawakening fetal hemoglobin: prospects for new therapies for the B-globin disorders. Blood 125:1061–1072
6. Bank A (2006) Regulation of human fetal hemoglobin: new players, new complexities. Blood 107:435–443
7. Wilber A, Nienhuis AW, Persons DA, Dc W (2013) Transcriptional regulation of fetal to adult hemoglobin switching: new therapeutic opportunities. Blood 117:3945–3953. doi:10.1182/blood-2010-11-316893
8. Jinek M, Chylinski K, Fonfara I et al (2012) A programmable dual-RNA–guided DNA endonuclease in adaptive bacterial immunity. Science 337:816–822
9. Sander JD, Joung JK (2014) CRISPR-Cas systems for editing, regulating and targeting genomes. Nat Biotechnol 32:347–355
10. Hsu PD, Lander ES, Zhang F (2014) Development and applications of CRISPR-Cas9 for genome engineering. Cell 157:1262–1278
11. Richardson CD, Ray GJ, DeWitt MA et al (2016) Enhancing homology-directed genome editing by catalytically active and inactive CRISPR-Cas9 using asymmetric donor DNA. Nat Biotechnol 34:339–344
12. Masuda T, Wang X, Maeda M et al (2016) Transcription factors LRF and BCL11A independently repress expression of fetal hemoglobin. Science 351:285–289
13. Domcke S, Bardet AF, Adrian Ginno P et al (2015) Competition between DNA methylation and transcription factors determines binding of NRF1. Nature 528:575–579
14. Paquet D, Kwart D, Chen A et al (2016) Efficient introduction of specific homozygous and heterozygous mutations using CRISPR/Cas9. Nature 533:125–129
15. Komor AC, Kim YB, Packer MS et al (2016) Programmable editing of a target base in genomic DNA without double-stranded DNA cleavage. Nature 61:5985–5991
16. Wienert B, Funnell APW, Norton LJ et al (2015) Editing the genome to introduce a beneficial naturally occurring mutation

associated with increased fetal globin. Nat Commun 6:7085

17. Kleinstiver BP, Pattanayak V, Prew MS et al (2016) High-fidelity CRISPR–Cas9 nucleases with no detectable genome-wide off-target effects. Nature 529:490–495
18. Finotti A, Breda L, Lederer CW et al (2015) Recent trends in the gene therapy of β-thalassemia. J Blood Med 6:69–85
19. Puthenveetil G, Scholes J, Carbonell D et al (2004) Successful correction of the human β-thalassemia major phenotype using a lentiviral vector. Blood 104:3445–3453
20. DeWitt M, Magis W, Bray NL, et al (2016) Efficient correction of the sickle mutation in human hematopoietic stem cells using a Cas9 ribonucleoprotein complex. bioRxiv 036236
21. Chang JC, Ye L, Kan YW (2006) Correction of the sickle cell mutation in embryonic stem cells. Proc Natl Acad Sci U S A 103:1036–1040
22. Ran F, Hsu P, Wright J, Agarwala V (2013) Genome engineering using the CRISPR-Cas9 system. Nat Protoc 8:2281–2308
23. Ayyadevara S, Thaden JJ, Shmookler Reis RJ (2000) Discrimination of primer 3′-nucleotide mismatch by taq DNA polymerase during polymerase chain reaction. Anal Biochem 284:11–18
24. Chu VT, Weber T, Wefers B et al (2015) Increasing the efficiency of homology-directed repair for CRISPR-Cas9-induced precise gene editing in mammalian cells. Nat Biotechnol 33:543–548

Chapter 16

In Vitro Erythroid Differentiation and Lentiviral Knockdown in Human CD34+ Cells from Umbilical Cord Blood

Anna Kovilakath, Safa Mohamad, Farrah Hermes, Shou Zhen Wang, Gordon D. Ginder, and Joyce A. Lloyd

Abstract

Human umbilical cord blood is a rich source of hematopoietic stem and progenitor cells. CD34+ cells in umbilical cord blood are more primitive than those in peripheral blood or bone marrow, and can proliferate at a high rate and differentiate into multiple cell types. In this protocol, a dependable method is described for the isolation of fetal CD34+ cells from umbilical cord blood and expanding these cells in culture. The cells can then be in vitro differentiated along an erythroid pathway, while simultaneously performing knockdown of a gene of choice. The use of lentiviral vectors that express small hairpin RNA (shRNA) is an efficient method to downregulate genes. Flow cytometric analyses are used to enrich for erythroid cells. Using these methods, one can generate in vitro differentiated cells to use for quantitative reverse transcriptase PCR and other purposes.

Key words Umbilical cord blood, CD34+ cells, Lentiviral infection, shRNA knockdown, Erythroid differentiation

1 Introduction

Hematopoietic stem and progenitor cells are defined as having the presence of cell surface marker CD34 [1]. The need to study hematopoietic stem and progenitor cells arose over the course of the late twentieth century in order to understand and treat metabolic as well as benign and malignant hematological disorders. Human umbilical cord blood is a rich source of hematopoietic stem and progenitor cells [2, 3]. CD34+ cells can also be isolated from bone marrow or peripheral blood or umbilical cord blood [3].

There are major technical advantages to isolating CD34+ cells from umbilical cord blood as compared to bone marrow. These include a higher proportion of more primitive hematopoietic cells [4], which respond to cytokines by proliferating at a higher rate [5, 6]. Additionally, there are ~8000 Erythroid burst-forming units

Joyce A. Lloyd (ed.), *Erythropoiesis: Methods and Protocols*, Methods in Molecular Biology, vol. 1698,
DOI 10.1007/978-1-4939-7428-3_16,

(BFU-E) in 1 ml of umbilical cord blood, which is ~3× more than in an equal volume of bone marrow [2].

Recently, immortalized erythroid progenitor cell lines such as HUDEP and HiDEP have been developed. These originate from umbilical cord blood hematopoietic stem/progenitor cells and induced pluripotent stem cells, respectively [7]. Although these cell lines are experimentally convenient, they are immortalized and hence not biologically representative of all erythropoietic processes [8].

Bone marrow cells were first isolated by Ficoll-Hypaque gradient separation in 1987 [9]. Human CD34+ hematopoietic stem cells were isolated from bone marrow and umbilical cord blood in 1988 using an enrichment method called "Panning" with Anti-My10 (Anti-CD34) antibody [10, 11]. Panning fractionates specific cell types by binding them to a specific antibody coated on plastic dishes, and results in 70–90% enrichment [12]. Erythropoietin and hIL-3 were added to semisolid medium with the panned cells, and the CD34+ cells proliferated for a short time and then began to differentiate and form erythroid burst-forming units [10, 11]. These first experimental methods for growing CD34+ cells from umbilical cord blood failed to yield large CD34+ populations, failed to isolate pure and synchronized (with respect to differentiation) CD34+ cells, and failed to allow for the quantitative analysis of the CD34+ cells after collection [8].

These shortcomings were later overcome by growing early erythroid progenitor cells in liquid culture. In 1991, a two-step liquid culture protocol was developed using peripheral blood mononuclear cells isolated using Percoll. These were placed in the first phase medium for growth and differentiation of the progenitors (erythropoietin-independent). The second phase was erythropoietin-dependent, and the cells matured into normoblasts and erythrocytes. Approximately 7.5×10^6 cells/ml and a total of 7×10^8 cells could be harvested per blood unit. Identification of hemoglobin-containing cells was performed by benzidine staining, and hemoglobin content was quantitated using HPLC or flow-cytometry [13, 14].

In 2002, the two-step liquid culture method most widely followed currently was developed. Doxycycline, dexamethasone, and the growth factors SCF, IL-3, and EPO are used to culture $CD34^+$ cells ex vivo. In the first (proliferative) phase medium, the cultures can generate $1–2 \times 10^7$ erythroblasts from 1×10^5 CD34+ cells. At day 7, the cells are transferred from the first phase to the second (differentiation) phase medium, which contains EPO and insulin. By day 11 of culture, the cells become mature erythroblasts. This method enables better control over proliferation and differentiation of the cells, as well as increased production of erythroid cells [15].

A method to use short-hairpin RNA (shRNA) constructs in lentiviral vectors and transfected into HEK 293T cells using calcium phosphate transfection was also described [16]. A method to isolate, expand, differentiate, and manipulate primary CD34+ cells from umbilical cord blood integrated the above methods [17]. The Easysep CD34 positive selection is used to isolate CD34+ cells from cord blood. The two-phase liquid medium is modified to include StemSpan SFEM medium with 1× CC100 cytokine mix [18]. shRNAs can be cloned and packaged in a lentivirus with GFP via calcium phosphate transfections in 293T cells [19]. CD34+ cells can be sorted using flow cytometry to enrich for GFP positive cells, and then differentiated for 10 days and sorted again by flow cytometry for cells with erythroid markers [17].

This chapter describes in detail the methods to isolate human CD34+ cells from umbilical cord blood, and expand and differentiate them in culture to obtain erythroid cells. The hematopoietic stem and progenitor cells can be manipulated by knockdown of any gene of choice. This is accomplished by transfecting HEK 293T cells with lentiviral packaging plasmids, to produce lentiviral particles expressing negative control (scramble) and gene-specific shRNAs [20–22]. The lentiviruses are then used to infect the expanded CD34+ cells, and CD34+ cells that express GFP are selected by Fluorescence-Activated Cell Sorting (FACS) [23]. After 8 days of erythroid differentiation culture, the majority of cells express transferrin receptor and Glycophorin A. FACS using CD71 and CD235a antihuman antibodies [24], for these two cell surface markers, respectively, is performed to enrich for differentiated erythroid cells. The enriched erythroid cells treated with the test and control shRNAs can then be compared using a variety of methods, including quantitative reverse transcriptase PCR, other quantitative methods to measure RNA or protein, and cytospin analyses and other morphological tests.

2 Materials

2.1 Media

All media components are stored long term at −20 °C, and for short term (a few weeks) at 4 °C, unless otherwise noted. The temperature for short-term storage is referred to as working temperature below. If only a single-storage temperature is noted, the reagent should be kept at that temperature at all times.

1. UCB Medium I: 40 ml Serum-Free Expansion Medium II (SFEM II) with 800 μl of 10 kU/ml Penicillin-Streptomycin and 400 μl Cytokine Cocktail (CC100). Always store CC100 at −20 °C, even for brief periods. Required on Days 0, 2, 4, 6, 7 and 8.
2. UCB Medium II: 40 ml SFEM II with 800 μl of 10 kU/ml Penicillin-Streptomycin and 600 μl CC100. Required on Day 8.

3. UCB Growth Medium: 32 ml Iscove's Modified Dulbecco's Medium (IMDM), 8 ml fetal bovine serum (FBS), 40 μl of 10 ng/μl Stem Cell Factor (SCF, Store at −80 °C, working at −20 °C), 4 μl of 10 ng/μl Interleukin-3 (store at −80 °C, working at −20 °C), 20 μl of 2 kU/ml Erythropoietin (store at 4 °C), 15.7 μl of 1 mg/ml Dexamethasone (store at −80 °C), 10.9 μl of 1 mg/ml Estradiol (store at −80 °C), and 800 μl of 10 kU/ml Penicillin-Streptomycin. Required on Days 10 and 12. Store at 4 °C. Make fresh for each new batch of UCB erythroblasts.
4. UCB Differentiation Medium: 32 ml IMDM, 8 ml FBS, 20 μl of 2 kU/ml Erythropoietin, 20 μl of 10 ng/ml Insulin, and 800 μl of 10 kU/ml Penicillin-Streptomycin. This is required on Days 13, 15, and 17.
5. 293T Medium I: 40 ml Dulbecco's Modified Eagle Medium (DMEM) with 4 ml FBS and 400 μl of 10 kU/ml Penicillin-Streptomycin. This is required for starting cultures of 293T cells from frozen stock, and on Days 0, 2, and 4.
6. 293T Medium II: 40 ml DMEM with 800 μl FBS and 400 μl of 10 kU/ml Penicillin-Streptomycin. This is required on Days 6 and 7.

2.2 293T Cell Culture

1. HEK 293T (293T) Cells: A 1 ml aliquot of frozen stock harvested from one T75 flask at 80–100% confluence, stored in liquid nitrogen in FBS with 10% DMSO.
2. T75 flasks.
3. 15 ml Conical tubes.
4. Hemocytometer.
5. Round Tissue Culture Dish [100 mm diameter].

2.3 Isolation of CD34+ Cells from Umbilical Cord Blood

1. One unit of umbilical cord blood (approximately 90–150 ml).
2. 70% ethanol.
3. Dulbecco's Phosphate-Buffered Saline (DPBS).
4. Ficoll-Paque (store at room temperature in the dark).
5. Human CD34 Positive Selection Cocktail [StemCell™ Technologies].
6. Magnetic Nanoparticles for CD34 Positive selection [StemCell™ Technologies].
7. 5 ml Polystyrene Round-Bottom Tubes, such as Falcon.
8. Separation Solution: PBS with 2% FBS and 1 mM EDTA.
9. Microcentrifuge.
10. 12-Well Culture Plates.
11. EasySep™-Magnet [StemCell™ Technologies].
12. 0.25% Trypsin.

2.4 Calcium Phosphate Transfection of 293T Cells

1. Four Plasmids (store at −20 °C). These can be purchased from a plasmid repository such as Addgene. The packaging plasmid pCMV-dR8.74 [25], the envelope plasmid pMD2.G, the scramble control shRNA expression plasmid, and the knockdown shRNA expression plasmid. The final two plasmids are in the pRRL backbone. The knockdown shRNA expression plasmid is designed specifically for your gene of interest.
2. 2 M $CaCl_2$ (store at −20 °C).
3. Cell Culture Grade Water.
4. For 25 ml of 2× Hepes-Buffered Saline (HBS) Solution: 20 ml of Tissue Culture Grade Water, 2.1 ml of 0.5 M Hepes, 2 ml of 3.42 M NaCl, 0.18 ml of 0.25 M Na_2HPO_4, 0.25 ml of 1 M KCl, adjust pH to 7.0 by adding 5 M NaOH. Prepare HBS Solution fresh for every new transfection.
5. 5×10^6 293T cells per transfection.
6. 60 ml Syringe with 0.20 μm syringe filter to sterilize and filter HBS Solution before use.

2.5 Calculating Transfection Efficiency of 293T

1. Fluorescent Microscope, blue filter with 400× magnification.
2. 0.45 μm filter to remove cell debris.
3. 4 mg/ml Polybrene (Hexadimethylene bromide) (store at −80 °C).

2.6 Flow Cytometry and Differentiation of UCB Cells

1. Iscove's Modified Dulbecco's Medium (IMDM).
2. UCB Growth Medium. One milliliter is required for every $1–1.25 \times 10^6$ CD34+ cells.
3. BD FACS Aria™ II High-Speed Cell Sorter or similar.

2.7 Isolation of In Vitro UCB Erythroblasts

1. Antihuman CD71 (APC, Transferrin receptor) [Clone: OKT9, BioLegend, Conc: 100 μg/ml].
2. PE/Cy7 antihuman 235a (Glycophorin A) [Clone: HI264, BioLegend, Conc: 0.06 μg/ml].
3. PBS with 10% FBS.
4. BD FACS Aria™ II High-Speed Cell Sorter or similar.

3 Methods

3.1 Starting Cultures of 293T Cells from Frozen Stock

1. Begin this step 2 days before receiving your unit(s) of cord blood.
2. Remove one vial of 293T cells from liquid nitrogen and warm at 37 °C until thawed. The vial should have $0.6–1 \times 10^6$ 293T cells. Add contents of the vial to a conical 15 ml tube and add 14 ml of 293T Medium I. Pellet the tube at 300 rcf for 3 min.

Remove the supernatant and dissolve the pellet in 1 ml of 293T Medium I and place in one T75 flask.

3. Add 14 ml of 293T Medium I to the T75 flask to make 15 ml total volume.
4. Place flask in an incubator at 37 °C with 5% CO_2.

3.2 Day 0: Isolation of CD34+ Cells from Umbilical Cord Blood

1. Number the cord blood bags received if more than one bag. Usually, one bag contains 90–150 ml of cord blood. Sterilize the hood, scissors, and blood bag with 70% ethanol.
2. Lay down paper towels in the hood (in case of a blood spill) and holding the blood bag upright cut the tubing at the neck of the bag. Decant approximately 30 ml of blood into each of approximately five 50 ml tubes by pointing the neck into the tube and lifting the bag to reduce the chance of blood spillage. Measure and record the total volume of blood from each sample.
3. Dilute the blood with DPBS (1:0.5 ratio), i.e., 30 ml blood + 15 ml DPBS.
4. Pipette 27 ml of the blood/DPBS over 20 ml Ficoll in 50 ml tubes. Tilt the tube at a 45° angle for a steady slow stream while adding the blood to the Ficoll. Release blood very slowly at first, and be careful to avoid disturbing the layers. You can speed up a little after several milliliters of blood are layered. Watch the boundary between the Ficoll and blood phases and immediately stop if it begins wavering. If a limited number of 50 ml tubes can be centrifuged at one time, stagger the layering and centrifugation (for example, 12 tubes at a time).
5. Centrifuge at 400 rcf for 30 min at room temperature with 0 brake. Acceleration can be left at 9 (*see* **Note 1**).
6. Remove tubes carefully from the centrifuge without disturbing the layers.
7. Remove the white buffy layer using a plastic Pasteur pipette (second layer from top) into 50 ml tubes, 10 ml per tube. Avoid drawing Ficoll into the pipette. Plasma is fine. Get as much as you can but do not waste time trying to get every last bit.
8. Add 30 ml DPBS into each tube containing buffy coat. Invert to mix.
9. Centrifuge at 300 rcf for 10 min to wash. Brake should be on again for this spin.
10. Decant the supernatant and resuspend the pellet in 10 ml DPBS. Combine up to 50 ml of cells of the same UCB unit to reduce number of tubes.
11. Spin down at 300 rcf for 10 min.

12. Repeat combining and washing until there is only one tube per sample. Using a 10 ml pipette, remove the supernatant completely and resuspend in 1 ml of the Separation Solution and transfer to a 5 ml Falcon tube.
13. Briefly vortex CD34 Positive Selection Cocktail to mix, and add 100 μl to each 5 ml Falcon tube. Mix well and incubate at room temperature for 15 min.
14. Mix magnetic nanoparticles well by pipetting. Do not vortex. Add 55 μl of suspended magnetic nanoparticles to each 5 ml Falcon tube. Mix well and incubate at room temperature for 10 min.
15. Add 1.35 ml of separation solution, giving 2.5 ml total volume in each 5 ml Falcon tube.
16. Mix by pipetting up and down several times. Remove the cap and place the 5 ml Falcon tube into Easysep magnet for 5 min.
17. Pick up the magnet and invert the 5 ml Falcon tube over a beaker without shaking for 3 s. The supernatant will flow out. Return the 5 ml Falcon tube to upright position. Remove the 5 ml Falcon tube from the magnet and add 2.5 ml of separation solution. Mix well and place the tube in magnet for 5 min. Repeat this **step 4** more times for a total of five times.
18. Add 2.5 ml of separation solution to the 5 ml Falcon tube and transfer this to a 15 ml conical tube. Centrifuge the sample at 300 rcf for 10 min.
19. Remove the supernatant and resuspend cells in 1 ml of UCB Medium I (*see* **Note 2**).
20. Load 10 μl of cells on a hemocytometer. Count total cells in one of the corner 4 × 4 grids; each cell represents 10,000 cells/ml. Dilute to 1×10^6 cells/ml and transfer to a 12 well plate (1 ml/well). The number of cells isolated varies but somewhere between 0.6 and 5×10^6 cells is normal. Figure 1 depicts the expected appearance of the undifferentiated CD34+ cells.
21. Incubate CD34+ cells at 37 °C with 5% CO_2.

3.3 Day 0: Expansion of 293T Cells

1. 293T cells should be 70% confluent after 2 days of growth. Remove the media from 293T cells. Add 3 ml of 0.25% trypsin.
2. Keep in the incubator at 37 °C for 5 min (*see* **Note 3**).
3. Decant the medium and cells into a 15 ml conical tube and take 10 μl and load on a hemocytometer. Count total cells in one of the 4 × 4 corner grids; each cell represents 10,000 cells/ml. You should have 4–5 × 10^6 293T cells in total.
4. Centrifuge the cells at 300 rcf.
5. Wash T75 flask with 5 ml of PBS to remove residual trypsin. Discard the PBS.

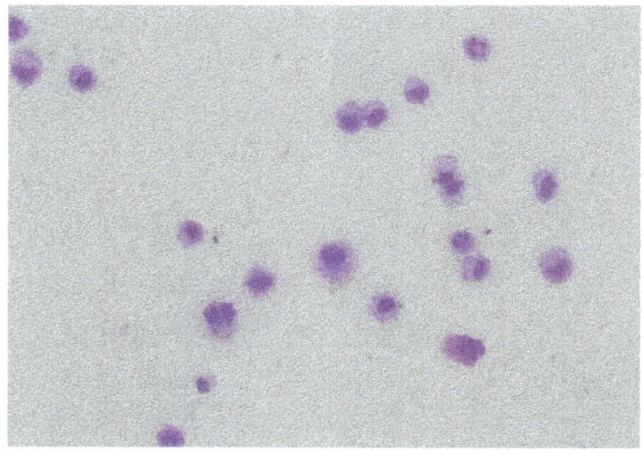

Fig. 1 UCB Cells at Differentiation day 0. The progenitors are nucleated and roughly all the same shape and size

6. Remove the supernatant from the 15 ml conical tube and resuspend the pellet in 1 ml of 293T Medium I.
7. Place 14 ml of 293T Medium I in the T75 flask and add the 1 ml of cell suspension.
8. Incubate 293T cells at 37 °C with 5% CO_2.

3.4 Day 2: UCB Cell Expansion

1. Gently mix the cells with a P1000 pipetteman and pool wells from the same UCB unit into one 15 ml conical tube.
2. Load 10 μl of cells on a hemocytometer. Count the number of cells per ml of sample (the count should be nearly double the count on day 0).
3. Add UCB Medium I to dilute the sample to 1×10^6 cells/ml (*see* **Note 4**).
4. Transfer the cells to a 12-well plate with 1 ml per well.
5. Incubate UCB cells at 37 °C with 5% CO_2.

3.5 Day 2: 293T Cell Expansion

1. Trypsinize the 293T cells as in Subheading 3.3, **steps 1–5**, pool into a single 15 ml conical tube.
2. Count on a hemocytometer. You should have 8–10 $\times$ 10^6 293T cells in total.
3. Split cells into two T75 flasks if ~50–60% confluent or three flasks if >70% confluent. Add the medium so there is 15 ml 293 Medium I in each T75 flask (*see* **Note 5**).
4. Incubate 293T cells at 37 °C with 5% CO_2.

3.6 Day 4: UCB Cell Expansion

1. Mix, pool, count, spin down, and resuspend the cells in UCB Medium I as in Subheading 3.4, **steps 1** and **2**.
2. Resuspend the cells in UCB Medium I at 2×10^6 cells/ml. Cells can be maintained at this higher concentration to conserve CC100.
3. Incubate UCB cells at 37 °C with 5% CO_2.

3.7 Day 4: 293T Cell Expansion

1. Trypsinize the 293T cells as in Subheading 3.3, **steps 1–5**, pool into a single 15 ml conical tube.
2. Count on a hemocytometer. You should have 15–20 × 10^6 293T cells in total. (15 × 10^6 cells is the minimum number of cells you will need for one replicate for three categories).
3. Spin down and resuspend the cells in 293T Medium I at 5 × 10^6 cells/ml.
4. In each round tissue culture dish, add 10 ml of 293T Medium I and 1 ml of cell suspension evenly around the plate. You need at least one round tissue culture dish of 5 × 10^6 293T cells per category (*see* **Note 6**).
5. Shake the plate laterally and vigorously to evenly disperse cells. Do not rotate or swirl the plate; this will concentrate cells around the edge.
6. Incubate 293T cells at 37° with 5% CO_2.

3.8 Day 5

1. The UCB Cells must be transferred to an incubator with *5% CO_2* and 37 °C.
2. The 293T cells are divided evenly into three categories for the calcium phosphate transfection: mock, scramble, and knockdown. If doing knockdowns with multiple shRNAs, then add one additional category for each additional shRNA.
3. The transfection solution is prepared separately in a 5 ml Falcon tube for each of the categories: mock, scramble, and knockdown. Each tube is used to transfect one round tissue culture dish (*see* **Note** 7). For all categories, you will need 5.6 μg of pCMVRd8.74 and 2.3 μg of pMD2.G. This is all that is needed for Mock. For scramble, you will additionally need 7.5 μg of pRRL encoding scramble shRNA. For each knockdown, you will need 7.5 μg of pRRL encoding-specific shRNA for knockdown. Combine appropriate plasmids with sterile water to a volume of 437 μl in Falcon tube, and mix gently. Tilt the tube and add 63 μl of 2 M $CaCl_2$ down the side of the tube. Add the DNA/$CaCl_2$ solution dropwise to 500 μl HBS. Mix by pipetting with a P1000 pipetteman.
4. Using the round tissue culture plates from Day 4, still containing the 293T Phase I media, add the 1 ml of transfection solution dropwise to one plate for each respective category. Start from the center of the plate and continue in a spiral pattern of approximately six concentric circles toward the edge of the plate, distributing the drops evenly. DO NOT SHAKE.
5. Incubate each plate at room temperature for 5 min.
6. Grow the plates for 16–18 h in a 37 °C incubator at *3% CO_2*. Do not grow longer than 16–18 h. $CaPO_4$ disrupts cell membranes to allow DNA uptake and is thus toxic. Cells will begin to die after this point.

3.9 Day 6: Expansion of UCB Cells

1. Mix, pool, count, spin down, and resuspend the cells in UCB Medium I as in Subheading 3.4, **steps 1** and **2**.
2. Resuspend the cells at 2 × 10^6 cells/ml.
3. Place cells in the 37 °C, *5%* CO_2 incubator.

3.10 Day 6: Removal of Viral Medium in 293T Plates

1. Caution: May contain live virus. Do not open plates outside of hood.
2. After 16–18 h, tilt each plate 1–2 times and then remove medium from each plate.
3. Wash cells with ~5 ml PBS to remove leftover particles.
4. Warm 293T Medium II in 37 °C water bath prior to use.
5. Gently add 10 ml of warm 293T Medium II to each round tissue culture dish. It is very easy to dislodge cells from the plate and this should be avoided. Tilt plate and slowly add to the bottom.
6. Incubate 293T plates at 37°, *10%* CO_2.

3.11 Day 7: Calculating Transfection Efficiency of 293T Cells

1. It is best not to infect UCB cells until late afternoon, to allow 293T cells to produce the maximum amount of virus (*see* **Note 8**).
2. Check under the fluorescent microscope to observe 293T cells expressing GFP (*see* **Note 9**). To do this, tape plates shut. Transport in a plastic autoclave tub with several layers of paper towels under the plates. Count 100 cells; good efficiency is about 50% of cells showing fluorescence (Fig. 2).
3. The bulb must be left on for at least 30 min before shutting it off to preserve its life. Set a timer and remember to turn it off.

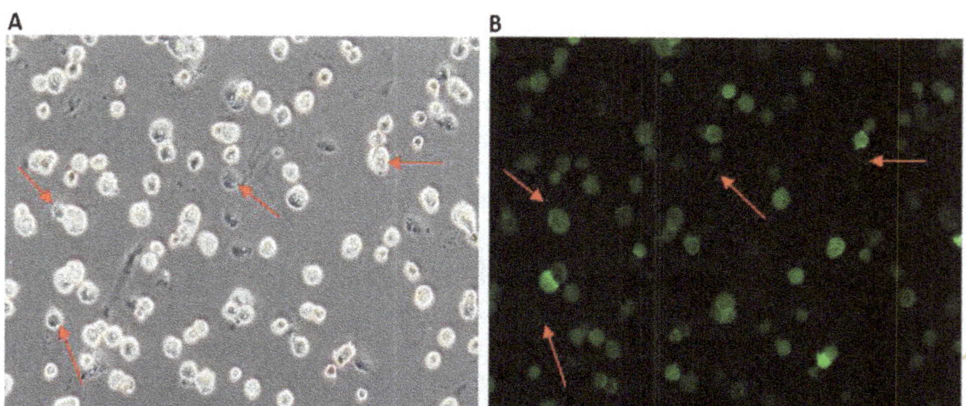

Fig. 2 Transfected 293T cells on the fluorescent microscope under (**a**) normal light to see all (non-transfected + transfected) 293T cells and (**b**) blue filter to see transfected 293T cells. The *red arrows* indicate a cell which is visible in (**a**) but not (**b**) due to failure of transfection. Transfection efficiency is calculated using the first and second images (*see* **Note 9**)

4. Tilt and collect the media from three round tissue culture flasks into one 50 ml tube per category. You will have 30 ml of medium collected from 15×10^6 293T cells in each 50 ml tube. This should be enough virus for simultaneous infection of multiple cord blood samples, if desired.
5. Filter the virus through a 0.45 μm filter using a 25–50 ml syringe into a fresh 50 ml tube to remove cell debris (*see* **Note 10**). Add 32 μl polybrene to each 50 ml tube containing 30 ml of medium with virus. Virus is ready for infection.
6. Tape 293T tissue culture plates shut before placing in an autoclave bag. For safety, autoclave tissue culture plates on the same day.

3.12 Day 7: Infecting UCB Cells

1. Pool, count, spin down, and resuspend in UCB Medium I at 6×10^6 cells/ml.
2. Divide the cells evenly between scramble and knockdown categories. Use one 12-well plate for each category minimum, unless transfection efficiency is less than 50%, then 24–36 wells is required. Three to four wells for the Mock category are sufficient. Draw your plating plan on the 12-well plates.
3. For infection, add 50 μl (3×10^5) of UCB cell suspension as a drop in the middle of each well of a 12-well plate. Add 600 μl of 293T supernatant containing virus above the drop.
4. For the mock cells, which remain uninfected, add 50 μl of UCB cell suspension in the middle of each well and add 600 μl of 293T Medium II. They will grow faster than infected cells.
5. Grow plates in an incubator at 37 °C with 5% CO_2.

3.13 Day 8: Growth of Infected UCB Cells

1. Wells still contain virus. Do not open outside hood.
2. Add 600 μl of UCB Medium II in each knockdown and scramble category well.
3. Add 600 μl of UCB Medium I in the plates containing mock category (*see* **Note 11**).

3.14 Day 10: Flow Cytometry to Isolate Transduced UCB Cells (Differentiation Day 0/DD0)

1. Wells still contain virus. Do not open outside hood.
2. Pool, count, and spin down scramble (control) and knockdown (test) cells. Take one well of mock as a negative control to calibrate the sorter.
3. Remove the media and resuspend in 20 ml IMDM.
4. Spin down the cells at 300 rcf for 5 min.
5. Remove the media, add 400 μl of SFEM-II (No additives). Keep on ice.
6. Deliver the first two tubes for fluorescent cell sorting [9, 10], along with two 50 ml tubes of IMDM with 20% FBS (for making the final cell suspension).

7. Obtain the final number of cells sorted to calculate infection efficiency which should be ≥50%. It should have greater than 2×10^6 cells expressing GFP in the knockdown and scramble categories. The Mock (unsorted) should have more than 4×10^6 cells (that is, twice as many as the lentivirus-infected cells).
8. Spin down the sorted GFP-positive cells and resuspend in UCB Growth Medium at $1–1.25 \times 10^6$ cells/ml. Plate in a 12-well plate at 1 ml/well.
9. Dispose of plates containing virus by taping shut and autoclaving as before. Do not reuse any plates that had virus in them.

3.15 Day 12: Differentiation Day 2 (DD2)

1. Pool, count, and spin down the cells. The number of cells should have doubled in each category.
2. Resuspend in UCB Growth Medium at $1–1.25 \times 10^6$ cells/ml.
3. Add 1 ml in each well of a 12-well plate.

3.16 Day 13: Differentiation Day 3 (DD3)

1. Pool, count, and spin down the cells. You should have a minimum of $6–8 \times 10^6$ cells in the knockdown and scramble categories and a minimum of 14×10^6 in the mock category. The cells will not multiply any more after adding the UCB Differentiation Medium.
2. Resuspend in UCB Differentiation Medium at $1–1.25 \times 10^6$ cells/ml.
3. Add 1 ml in each well of a 12-well plate.

3.17 Day 15: Differentiation Day 5 (DD5)

1. Pool count and spin down the steps as in Subheading 3.16, **step 1**.
2. Plate the cells at 2×10^6 cells/ml.
3. Add 1 ml in each well of a 12-well plate.

3.18 Day 17: Differentiation Day 7 (DD7)

1. Pool, count, and spin down the steps as in Subheading 3.16, **step 1**.
2. Plate cells at 3×10^6 cells/ml here because flow sorting is the next day (*see* **Note 12**).
3. Add 1 ml in each well of a 12-well plate.

3.19 Day 18: Isolation of UCB Erythroblasts at Differentiation Day 8 (DD8)

1. Pool and count the cells. The cell pellet will be red, and the expected appearance for erythroblasts at various stages of differentiation is shown in Fig. 3.
2. Spin down the cells at 300 rcf for 5 min, remove the media, and add 14 ml of PBS with 10% FBS. Wash cells.
3. Resuspend in 100 μl of PBS with 10% FBS.

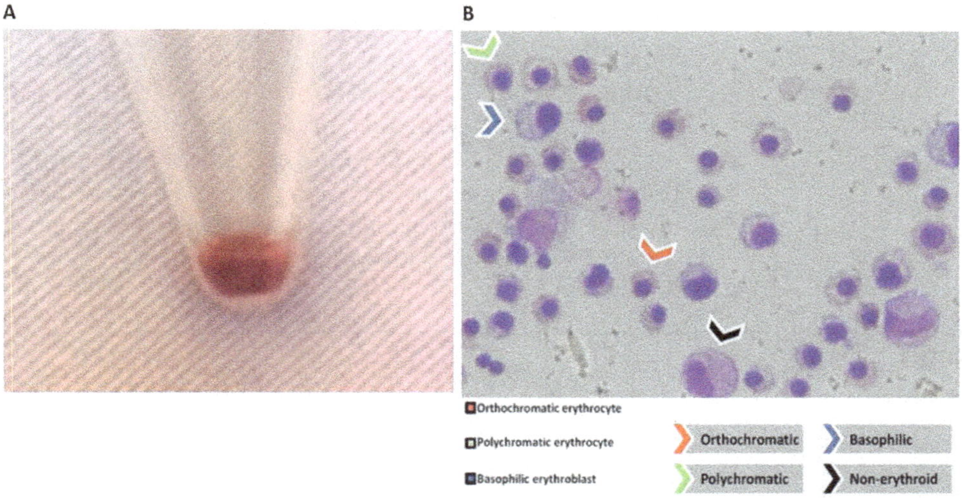

Fig. 3 CD34+ cells from fresh umbilical cord blood units were expanded for 8 days then in vitro differentiated to erythroid cells for 8 days. At the end of the expansion period cells were treated with lentiviral particles carrying either scramble (Scr) or gene-specific shRNAs. UCB Erythroblasts at day 8 of Differentiation (Day 18 of the protocol) are *red* (**a**) and cytospin slides prepared have erythroid cells at various different stages of differentiation (**b**). This is prior to selection with CD235a and CD71 markers so ~10% of all cells are non-erythroid. Cells are stained with Giemsa

4. For every 0.5×10^6 cells, add 1.25 μl of Anti-Human CD71 (APC, transferrin receptor) Antibody, Clone OKT9.
5. For every 0.5×10^6 cells, add 1.8 μl of Anti-Human CD235a PE/Cy7a (Glycophorin A) Antibody, Clone HIR2 (*see* **Note 13**).
6. Incubate tubes on ice in darkness for 20 min.
7. Along with the cells on ice, prepare two 50 ml tubes of PBS with 10% FBS (for making the final cell suspension). Take note of the final number of cells sorted.
8. After flow sorting, the double positive cells need to be spun down and washed once more using only PBS before using for RNA extraction and/or Western blots.

3.20 Analyses of Sorted Erythroid Cells

1. Giemsa Staining: Cytospins from the collected cells can be stained with Giemsa and observed under a microscope to visually identify the stages of differentiation. To carry out staining, dip the slides first in methanol for approximately 10 s to fix them. Stain the slides with Giemsa for 2 min. Wash the slides with distilled water. Air-dry them and observe on a light microscope at a magnification of 100×. Take pictures.
2. RNA extraction: Add 300 μl Trizol reagent for greater than 5×10^6 cells in each category. Let it sit for 5 min. Store at

−80 °C or continue with RNA extraction per manufacturer's instructions.

3. Storing cells for Western blotting: Wash with 1 ml PBS per 10×10^6 cells. Resuspend and pipette into Eppendorf tubes, between 10 and 20×10^6 cells per tube. Spin down, decant PBS, and pipette out all remaining liquid. Hold the bottom of the tube in liquid nitrogen for a few seconds to freeze. Store at −80 °C or continue with western blot protocol.

4 Notes

1. Do not forget to remove the brake. The sudden deceleration will ruin the layers if the brake is on.
2. At the end of Day 0 all plastic-ware and culture dishes that came in contact with the blood and no longer in use need to be autoclaved the same day.
3. You do not want to incubate the 293T cells with Trypsin for longer than 5 min or the cells may no longer remain viable due to over-digestion of proteins.
4. You will need about 15×10^6 UCB cells per knockdown, scramble and mock category for infection by the lentivirus on Day 7.
5. The 293T cells should be at ~80% confluence because at least 5×10^6 293T cells per category are needed by day 4.
6. If more than 5×10^6 cells per category you should make multiple round tissue culture dishes with 5×10^6 293T cells per category. But ensure that number of round tissue culture dishes with these cells is equal for each category.
7. To calculate the volume of each plasmid to be added to the transfection solution, divide the given μg quantities of plasmid by the respective plasmid concentration. The μg quantities of plasmid needed per transfection are as follows: pCMVdR 8.74–5.6 μg; pMD2.G—2.3 μg, pRRL knockdown shRNA—7.5 μg; pRRL scramble shRNA—7.5 μg.
8. Infection is best to do in the late afternoon. This is to allow the 293T cells transfected on Day 5 (Subheading 3.8) to produce more lentiviral product to obtain transfection efficiency of ≥50% to infect the UCB cells on Day 7 (Subheading 3.12).
9. To calculate the transfection efficiency: Mark four areas of the plate to count under the fluorescent microscope. Use a cell counter and manually count the number of transfected (green or whichever FP marker you have used) cells then switch to the normal light microscopy and calculate the total number of cells. Do this in four different areas and then calculate the average.

Transfection Efficiency = [(Sum total number of transfected cells/Sum total number of cells) × 100]. Good Transfection efficiency ~50% or greater and one plate per category is adequate. If Transfection Efficiency is less than ~50% 2–3 plates per category are required.

10. Do not use a 0.22 μm filter because it could shear the viral capsid.
11. Since the Mock category is uninfected the infected cells require less CC100 to grow than the affected scramble and knockdown category cells, i.e., 1% versus 1.5%. Any remaining UCB Medium II can be converted to UCB Medium I by diluting 1.5% CC100 to 1% CC100 by adding SFEM II and Penicillin/Streptomycin.
12. Because we were doing KLF1 knockdown, the erythroid cells were collected at DD8. The knockdown cells showed a lag in differentiation compared to scramble at DD12, but not at DD8. At DD8 cells were transferrin receptor and glycophorin A positive, but at DD12, they were further differentiated [19]. Depending on the gene being knocked down, later stages of differentiation beyond DD8 may be more appropriate.
13. Here you need to note which fluorescent protein you used and make sure you add the corresponding antibody. Example CD235 Pe is for GFP and CD235 Pe/Cy7 is for RFP. Also check concentrations on antibodies.

Acknowledgments

We would like to thank the St. Louis Cord Blood Bank (SLCBB, St. Louis, MO) for providing us with fresh umbilical cord blood.

References

1. Simmons DL, Satterthwaite AB, Tenen DG, Seed B (1992) Molecular cloning of a cDNA encoding CD34, a sialomucin of human hematopoietic stem cells. J Immunol 148 (1):267–271
2. Sidney LE, Branch MJ, Dunphy SE, Dua HS, Hopkinson A (2014) Concise review: evidence for CD34 as a common marker for diverse progenitors. Stem Cells 32(6):1380–1389
3. Huang S, Law P, Young D, Ho AD (1998) Candidate hematopoietic stem cells from fetal tissues, umbilical cord blood vs. adult bone marrow and mobilized peripheral blood. Exp Hematol 26(12):1162–1171
4. Smogorzewska EM, Barsky LW, Crooks GM, Wienberg KI (1997) Purification of hematopoietic stem cells from human bone marrow and umbilical cord blood. Cent Eur J Immunol 22:232–239
5. Hordyjewska A, Popiołek Ł, Horecka A (2015) Characteristics of hematopoietic stem cells of umbilical cord blood. Cytotechnology 67 (3):387–396
6. Zhou G, Chen J, Lee S, Clark T, Rowley JD, Wang SM (2001) The pattern of gene expression in human CD34+ stem/progenitor cells. Proc Natl Acad Sci 98(24):13966–13971
7. Kurita R, Suda N, Sudo K, Miharada K, Hiroyama T, Miyoshi H, Tani K, Nakamura Y (2013) Establishment of immortalized human erythroid progenitor cell lines able to produce enucleated red blood cells. PLoS One 8(3):e59890

8. Fibach E, Prus E (2005) Differentiation of human erythroid cells in culture. Curr Protoc Immunol Chapter 22:Unit 22F.7
9. Loken MR, Shah VO, Dattilio KL, Civin CI (1987) Flow cytometric analysis of human bone marrow. II normal B lymphocyte development. *Blood* 70(5):1316–1324
10. Saeland S, Caux C, Favre C, Aubry JP, Mannoni P, Pebusque MJ, Gentilhomme O, Otsuka T, Yokota T, Arai N (1988) Effects of recombinant human interleukin-3 on CD34-enriched normal hematopoietic progenitors and on myeloblastic leukemia cells. Blood 72 (5):1580–1588
11. Caux C, Favre C, Saeland S, Duvert V, Mannoni P, Durand I, Aubry JP, de Vries JE (1989) Sequential loss of CD34 and class II MHC antigens on purified cord blood hematopoietic progenitors cultured with IL-3: characterization of CD34-, HLA-DR+ cells. Blood 74 (4):1287–1294
12. Wysocki LJ, Sato VL (1978) "Panning" for lymphocytes: a method for cell selection. Proc Natl Acad Sci 75(6):2844–2848
13. Fibach E, Manor D, Oppenheim A, Rachmilewitz EA (1989) Proliferation and maturation of human erythroid progenitors in liquid culture. Blood 73:100–103
14. Fibach E, Rachmilewitz EA (1993) The two-step liquid culture: a novel procedure for studying maturation of human normal and pathological erythroid precursors. Stem Cells 11 (S1):36–41
15. Migliaccio G, Di Pietro R, Di Giacomo V, Di Baldassarre A, Migliaccio AR, Maccioni L, Galanello R, Papayannopoulou T (2002) In vitro mass production of human erythroid cells from the blood of normal donors and of thalassemic patients. Blood Cell Mol Dis 28 (2):169–180
16. Li Y, Tsun A, Gao Z, Han Z, Gao Y, Li Z, Lin F, Wang Y, Wei G, Yao Z, Li B (2013) 60-kDa Tat-interactive protein (TIP60) positively regulates Th-inducing POK (ThPOK)-mediated repression of eomesodermin in human CD4+ T cells. J Biol Chem 288(22):15537–15546
17. Amaya M, Desai M, Gnanapragasam MN, Wang SZ, Zu Zhu S, Williams DC, Ginder GD (2013) Mi2β-mediated silencing of the fetal -globin gene in adult erythroid cells. Blood 121(17):3493–3501
18. Gnanapragasam MN, Scarsdale JN, Amaya ML, Webb HD, Desai MA, Walavalkar NM, Wang SZ, Zu Zhu S, Ginder GD, Williams DC (2011) p66α–MBD2 coiled-coil interaction and recruitment of Mi-2 are critical for globin gene silencing by the MBD2–NuRD complex. Proc Natl Acad Sci 108 (18):7487–7492
19. Kingston RE, Chen CA, Rose JK (2003) Calcium phosphate transfection. Curr Protoc Mol Biol Chapter 9:Unit 9F.1
20. Singer O, Verma IM (2008) Applications of lentiviral vectors for shRNA delivery and transgenesis. Curr Gene Ther 8(6):483–488
21. Vinjamur DS, Alhashem YN, Mohamad SF, Amin P, Williams DC Jr, Lloyd JA (2016) Krüppel-like transcription factor KLF1 is required for optimal γ-and β-globin expression in human fetal erythroblasts. PLoS One 11(2): e0146802
22. Naldini L, Blömer U, Gage FH, Trono D, Verma IM (1996) Efficient transfer, integration, and sustained long-term expression of the transgene in adult rat brains injected with a lentiviral vector. Proc Natl Acad Sci 93 (21):11382–11388
23. Dijkers PF, Medema RH, Lammers JWJ, Koenderman L, Coffer PJ (2000) Expression of the pro-apoptotic Bcl-2 family member Bim is regulated by the forkhead transcription factor FKHR-L1. Curr Biol 10(19):1201–1204
24. Olivier EN, Qiu C, Velho M, Hirsch RE, Bouhassira EE (2006) Large-scale production of embryonic red blood cells from human embryonic stem cells. Exp Hematol 34 (12):1635–1642
25. Wiznerowicz M, Trono D (2003) Conditional suppression of cellular genes: lentivirus vector-mediated drug-inducible RNA interference. J Virol 77(16):8957–8951

Chapter 17

Growing and Genetically Manipulating Human Umbilical Cord Blood-Derived Erythroid Progenitor (HUDEP) Cell Lines

Divya S. Vinjamur and Daniel E. Bauer

Abstract

The recently established human umbilical cord blood-derived erythroid progenitor (HUDEP) cell lines have equipped red blood cell researchers with valuable in vitro models of erythroid development. Of the three established HUDEP cell lines, HUDEP-2 cells express predominantly adult β-globin and most closely resemble adult erythroid cells. This chapter describes culture protocols for the maintenance and erythroid differentiation of HUDEP-2 cells. Methods to genetically manipulate HUDEP-2 cells using a CRISPR/Cas9 nuclease-based approach are also discussed.

Key words HUDEP, Immortalized erythroid cell line, Erythroid differentiation, Hemoglobin switching, Fetal hemoglobin, CRISPR/Cas9

1 Introduction

The use of in vitro cultured cell lines as models to study various in vivo phenomena is an integral part of scientific research. These cellular systems allow researchers to manipulate specific molecular targets or cellular signaling pathways to better understand their role in various biological processes. Until recently, the cellular model systems available for the study of erythropoiesis consisted primarily of erythroleukemic cell lines such as mouse erythroleukemic cells (MEL cells) and K562 cells, a human erythroleukemic cell line [1]. A major limitation of erythroleukemic cell lines such as K562 cells as models of erythroid development lies in the fact that upon induction of differentiation into erythroid cells, these cells exhibit an embryonic/fetal pattern of globin gene expression, rather than expressing adult hemoglobin. Likewise, protocols to differentiate embryonic stem cells or induced pluripotent stem cells toward erythroid fates recapitulate embryonic/fetal but not adult stages of globin gene expression [2, 3]. Another shortcoming of embryonic stem cells and cell lines is that terminal maturation to produce reticulocytes or mature RBCs is relatively inefficient. Primary

Joyce A. Lloyd (ed.), *Erythropoiesis: Methods and Protocols*, Methods in Molecular Biology, vol. 1698,
DOI 10.1007/978-1-4939-7428-3_17,

hematopoietic CD34+ progenitor cells derived from human bone marrow, mobilized peripheral blood, or umbilical cord blood may be differentiated in vitro into reticulocytes and are considered the best available cellular model of erythroid development [4, 5]. However, the scope of their application is limited by access to human donors, their expense to procure, and suitability only for short-term culture.

The human umbilical cord blood-derived erythroid progenitor (HUDEP) cell line was established by immortalization of umbilical cord blood CD34+ cells using lentiviral delivery of a tetracycline-inducible HPV16-E6/E7 transgene [6]. The human papillomavirus 16 (HPV16) E6 and E7 proteins induce immortalization of cells by inhibiting the function of the p53 and retinoblastoma (pRb) tumor suppressor proteins, respectively [7]. Using this method, Kurita et al. established 3 HUDEP cell lines from three different cord blood samples: HUDEP-1, HUDEP-2, and HUDEP-3. There are differences in the β-like globin gene expression profiles of these three HUDEP cell lines. The HUDEP-2 cells most closely resemble adult erythroid cells since they express predominantly β-globin upon culture in erythroid differentiation media, whereas the HUDEP-1 and HUDEP-3 lines express predominantly γ-globin [6]. Additionally, the HUDEP-2 (but not HUDEP-1 or HUDEP-3) line expresses BCL11A, a known repressor of γ-globin. Thus for the first time researchers have access to an immortalized erythroid cell line that more accurately portrays adult erythropoiesis than previously available cellular models. A frequently observed limitation of immortalized or transformed cell lines is that they deviate from the normal diploid state. It is extremely valuable to ascertain the genetic make-up of a cellular model system, especially for the purpose of genetic manipulation. To analyze the genetic complement of the HUDEP-2 cells, we karyotyped the HUDEP-2 cells by GTG-banding. At this resolution, the karyotype of the HUDEP-2 cells is 50,XY,+6,+psu dic (8;2)(q24;q21),+19,+21 (Fig. 1). This system has opened up a plethora of avenues for interrogation that were challenging with previously available cellular systems [8–10].

We follow a 3-phase liquid culture protocol for the maintenance and differentiation of HUDEP cells. This protocol has been adapted with modifications from the protocols for HUDEP cell culture described by Kurita et al. [6] and erythroid differentiation of CD34+ hematopoietic progenitor cells described by Giarratana, et al. [5]. For the purpose of maintenance of the cell line in a self-renewing state, the HUDEP cells are cultured in a serum-free expansion medium consisting of an SFEM base supplemented with stem cell factor (SCF), EPO, doxycycline, dexamethasone, and penicillin-streptomycin. To further differentiate the HUDEP cells to polychromatophilic and orthochromatic erythroblasts, the cells are moved from expansion medium to an erythroid

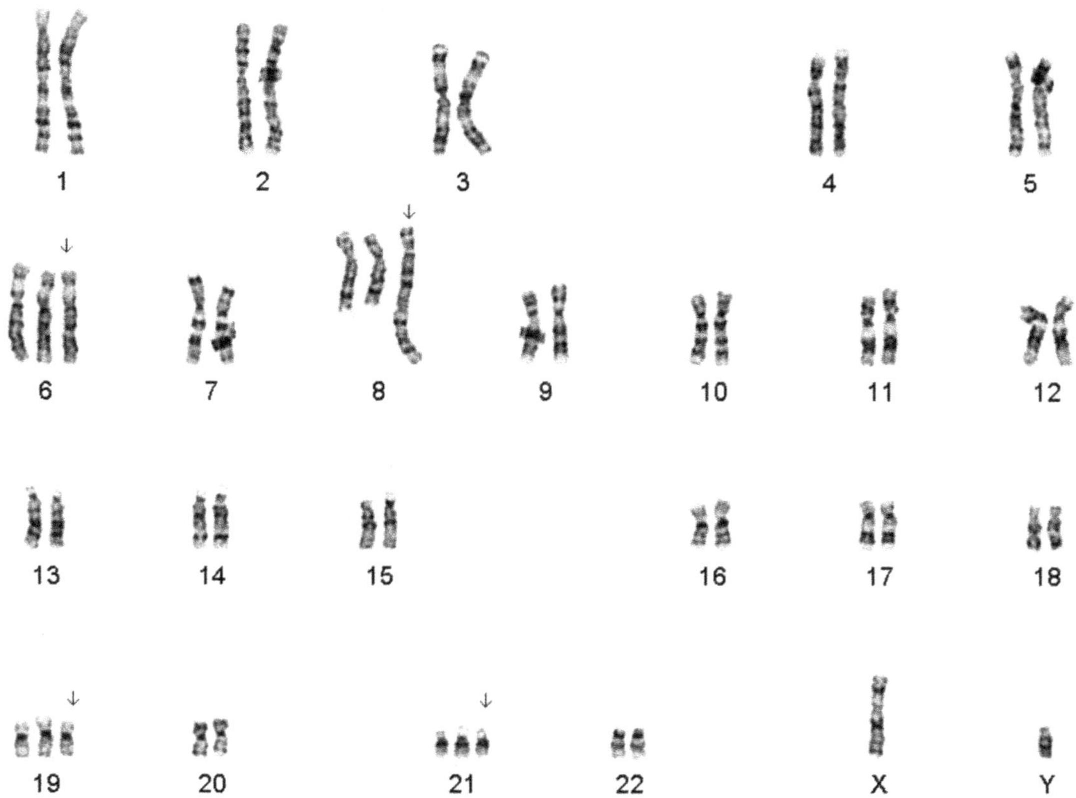

Fig. 1 Karyotype of HUDEP-2 cells. HUDEP-2 cells were karyotyped by GTG-banding. 18 out of 20 metaphase cells analyzed had the karyotype 50,XY,+6,+psu dic(8;2)(q24;q21),+19,+21

differentiation medium. Differentiation medium consists of two phases: a "growth/proliferation" phase, where the basal erythroid differentiation medium (EDM) is supplemented with stem cell factor and doxycycline (EDM-2) and a "maturation" phase medium in which the stem cell factor is withdrawn (EDM-3 (+dox)).

The HUDEP cells are easy to transduce which makes it possible to manipulate gene expression either by RNA interference using short hairpin RNAs (shRNAs) or by gene disruption using the CRISPR/Cas9 system. The combination of using the HUDEP cells and the CRISPR/Cas9 system has opened up the very exciting avenue of obtaining a clonal cell population that has biallelic disruption of a gene of interest. An approach to genetically manipulate HUDEP cells using an integrating lentivirus and the CRISPR/Cas9 system is discussed here. Non-integrating and non-lentiviral methods of delivery are also possible, though the HUDEP cells are relatively sensitive to DNA plasmid toxicity following electroporation.

A two-component system consisting of Cas9 nuclease and target site-specific chimeric single-guide RNA (sgRNA) is required to produce disruption of a gene of interest. These two components can be introduced either separately into cells (two-vector system) or together (one-vector system) [11]. We typically prefer the two-vector system described here, because the viral titer tends to be low for the single-vector system.

The protocols described below for the culture and manipulation of HUDEP cells have been developed using the HUDEP-2 cell line, but may be adapted for the other HUDEP lines. Requests for the HUDEP cells should be directed to Yukio Nakamura, RIKEN BioResource Center, Tsukuba, Ibaraki, Japan [6].

2 Materials

2.1 Cell Culture

1. Phosphate-buffered saline (PBS).
2. HUDEP expansion medium: StemSpan Serum-Free Expansion Medium (SFEM, Stemcell Technologies), 2% Penicillin-Streptomycin solution (10,000 U/mL stock), 50 ng/mL recombinant human stem cell factor (SCF), 3 IU/mL Epoetin alfa (Epogen, Amgen), 0.4 μg/mL dexamethasone, 1 μg/mL doxycycline.
3. Erythroid differentiation medium (EDM): Iscove's modified Dulbecco's medium (IMDM), 1% L-glutamine (this is in addition to the L-glutamine present in IMDM), 2% Penicillin-Streptomycin solution (10,000 U/mL stock concentration), 330 μg/mL human holo-transferrin, 10 μg/mL recombinant human insulin solution, 2 IU/mL heparin, 5% inactivated human plasma, 3 IU/mL Epoetin alfa (Epogen, Amgen).
4. Plasma from an individual with the AB blood type is recommended, to avoid interactions between antibodies in the donor serum and antigens on the HUDEP-2 cells (which are from an individual with blood type A).
5. "Growth" phase media (EDM-2): EDM with 100 ng/mL SCF and 1 μg/mL doxycycline.
6. "Maturation" phase media (EDM-3): EDM with 1 μg/mL doxycycline.

2.2 Cytospins

1. PBS.
2. May-Grünwald stain.
3. Giemsa stain.
4. Cytospin centrifuge.

2.3 Flow Cytometry

1. Cell staining buffer: PBS with 0.1% bovine serum albumin (BSA).
2. Anti-human CD71 (Transferrin Receptor) PE-Cyanine7.

3. Anti-human CD235a (Glycophorin A) APC.
4. Hoeschst 33342, Trihydrochloride Trihydrate, 10 mg/mL solution in water.

3 Methods

3.1 Culturing HUDEP Cells for Expansion Phase

1. Begin culture of HUDEP cells from frozen stock by plating in HUDEP expansion media at 100,000 cells/mL.
2. The HUDEP-2 cells typically double every 24–36 h. Perform media changes every 3–4 days, while ensuring that the cell density remains below 800,000 cells/mL. Although the cells may be recovered from single cell culture, we typically do not split below 50,000 cells/mL to maintain optimal growth rates.
3. Perform cytospin preparation and cell surface marker assessment by flow cytometry, as described in Subheadings 3.3 and 3.4. HUDEP-2 cells maintained in expansion medium consist primarily of proerythroblasts and basophilic erythroblasts (Fig. 2a) and should be double positive for cell surface expression of the transferrin receptor (CD71) and Glycophorin A (CD235a) (Fig. 2b).

3.2 Culturing HUDEP Cells for Erythroid Differentiation Phase

1. Transfer the cells from expansion medium to EDM-2 and culture for 4 days (*see* **Note 1**).
2. Transfer the cells to EDM-3 and culture for 3 days.
3. Transfer cells to EDM (no doxycycline). The length of culture for this phase is variable and dependent on the stage of

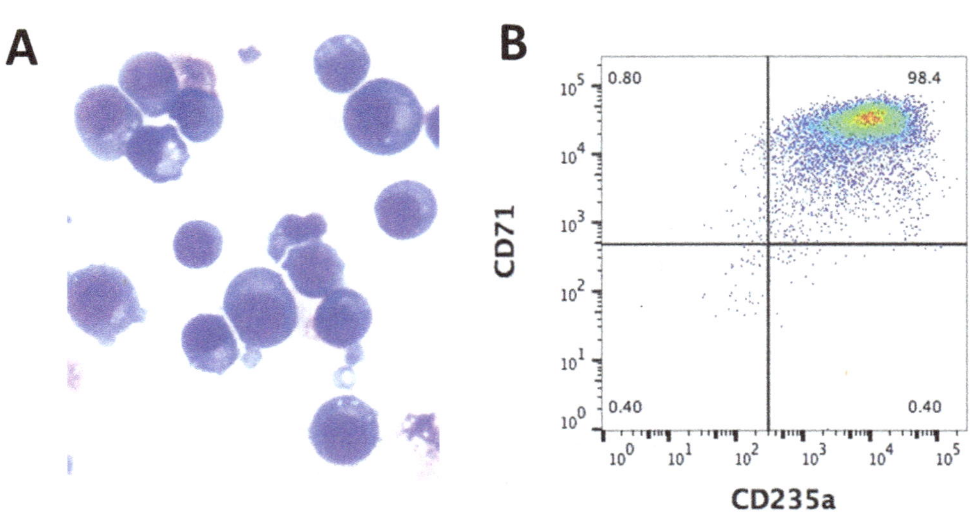

Fig. 2 HUDEP-2 cells in expansion medium. (**a**) Cytospins stained with May-Grunwald and Giemsa stains (**b**) Flow cytometry plot of HUDEP-2 cells stained with erythrocyte cell surface markers CD71 (transferrin receptor) and CD235a (Glycophorin A) shows that the cells are double positive for CD71 and CD235a

differentiation required as an experimental endpoint. We routinely limit this phase to 2 days because the majority of HUDEP-2 cells are at the orthochromatic erythroblast stage by this time point (*see* **Note 2**).

3.3 Cytospin Preparation to Assess the Stage of Maturation of HUDEP Cells

1. Harvest 100,000 cells. Centrifuge at 300 × *g* for 5 min in a microcentrifuge tube.
2. Discard the culture medium and resuspend the cells in 100 μL PBS.
3. Load cells into the chamber of the cytospin funnel.
4. Spin the cells at 350 rpm (15 × *g*) for 3 min in a cytocentrifuge.
5. Air-dry the slides before staining with May-Grünwald and Giemsa stains.

3.4 Flow Cytometry Analysis of Erythroid Cell Surface Markers and Percentage of Enucleation to Assess Stage of Erythroid Differentiation

1. Harvest 100,000 cells. Centrifuge at 300 × *g* for 5 min.
2. Discard the culture medium and resuspend cells in 100 mcl cell staining buffer.
3. Add 5 mcl of PECy7-conjugated anti-CD71 antibody, 5 mcl of APC-conjugated anti-CD235a antibody and 0.02 mcl of Hoechst 33342 (1 mcl of a 1:50 dilution of the stock reagent).
4. Incubate for 15 min in the dark at room temperature.
5. Add 400 mcl of cell staining buffer.
6. Centrifuge for 5 min at 300 × *g*.
7. Resuspend in 300 mcl cell staining buffer.
8. Analyze by flow cytometry.

3.5 Lentiviral Transduction of HUDEP Cells

1. Prepare lentivirus using a suitable packaging system. We use psPAX2 and pCMV-VSV-G as our packaging and envelope vectors, respectively [12]. Two lentiviruses are prepared separately, one with a Cas9-carrying transfer vector and another with the sgRNA-carrying transfer vector (*see* **Note 3**).
2. Transduce 10,000 cells (or as required) first with the Cas9 lentivirus at an MOI = 1.
3. Select for successfully transduced cells either by antibiotic selection or by sorting the cells for a fluorescence marker. We have had success with both blasticidin selection (10 μg/mL final concentration) and puromycin selection (1 μg/mL final concentration). If using a fluorescence marker for selection, take into consideration the expression of endogenous Kusabira Orange in the HUDEP cells, while choosing a suitable fluorophore (*see* **Note 4**).
4. Transduce Cas9-transduced cells with sgRNA-carrying transfer vector and select for successfully transduced cells.

3.6 Plating HUDEP Cells to Obtain a Clonal Cell Population

Cells that have successfully been transduced with both Cas9 and a sgRNA targeting the gene of interest have the following three potential outcomes at the targeted site as a result of DNA cleavage and nonhomologous end joining (NHEJ): alleles carrying an insertion/deletion (indel) resulting in a frameshift, alleles carrying an indel resulting in an in-frame mutation, and unedited alleles. Bulk populations consisting of gene-edited cells with various editing outcomes can be screened for phenotypes of interest. Cell clones that carry biallelic mutations, making both alleles of the gene of interest nonfunctional, may be characterized using the following steps.

1. Culture the cells for 1 week after transduction before limiting dilution to avoid mixed clones in which gene editing occurs after single-cell plating.
2. Plate the cells in the HUDEP expansion medium at a limiting dilution in 96-well round-bottom plates. To enhance the likelihood that clones are derived from single cells, plate 30 cells per 96-well plate, by diluting the cells to 30 cells per 10 mL of medium and distributing 100 μL in each well. The number of 96-well plates to be prepared is determined by the calculated probability of the desired gene editing outcome.
3. Grow for 10–14 days to expand cells for experiments. The level of medium in the wells should be carefully monitored over this course of time due to evaporation from wells especially at the edges of the plate. Once cell clones have expanded to cover the bottom of the well, they can be consolidated to a single 96-well flat-bottom plate or expanded to larger volumes.
4. Our experience with cell clones indicates that there may be nongenetic clonal variation. Therefore, for each transduction, 3–10 clones of interest should be characterized along with unedited clones and non-targeted control clones plated simultaneously.

4 Notes

1. As few as 5000–10,000 HUDEP-2 cells can be used to start cultures in a 12- or 24-well plate; fewer in a 96-well plate. If growing in T25 culture flasks, we typically plate at 50,000–100,000 cells/mL. The cells double every day and must always be maintained below 800,000 cells/mL. Once they reach 1 million cells/mL, there will be cell death. The HPV16-E6/E7 transgene used for immortalization is a tetracycline-inducible cassette and hence doxycycline must be added to culture media to maintain the cells in a proliferative state. The HUDEP cells are dependent on doxycycline for continuous proliferation. However, the natural process of terminal erythroid differentiation necessitates cell cycle exit [13],

suggesting that the continuous signal to proliferate provided by doxycycline-induced E6/E7 expression may be unfavorable to terminal erythroid differentiation. To model/mimic this pattern of decreased proliferation, we withdraw doxycycline during the final stage of HUDEP differentiation. Of note, the withdrawal of doxycycline is accompanied by a loss of cell viability.

2. The rate of enucleation is minimal. Experiments are ongoing to improve enucleation of the HUDEP-2 cells. We have had some success by co-culturing the HUDEP-2 cells with MS-5 cells, a murine stromal cell line, during the EDM3-dox phase. HUPEP-2 cells in EDM3 (+dox) were resuspended in EDM (-dox) at a concentration of 1 million cells/mL and plated on 80% confluent MS5-cells in a T75 tissue culture flask. After 6 days of coculture with MS-5 cells, we observed that the cells begin to show enucleation and 6% of the cells were enucleated (Fig. 3a–c). After 11 days of coculture with MS-5 cells we observed an increase in enucleation with 38% enucleated cells (Fig. 3d–f). There is a constant decline in cell number during this phase. On Day 6 of coculture there were 0.26 million cells/mL remaining and by Day 11 of coculture there were only 20,000 cells/mL, a 50-fold reduction in cell number from the 1 million cells/mL initially plated on MS-5 cells. The remaining cells appear viable by trypan blue staining. Further optimization of the culture protocol would be required to achieve enucleation with minimal loss of cells.
3. We typically use lentiCas9-Blast (blasticidin resistance, Addgene plasmid #52962) to introduce Cas9 and lentiGuide-Puro (puromycin resistance, Addgene plasmid #52963) to introduce the sgRNA into HUDEP cells.
4. The lentiviral vector used for the immortalization of the cord blood cells contains the humanized Kusabira Orange fluorescent protein gene in addition to the HPV16-E6/E7 coding sequence. Kusabira Orange emits at 561 nm upon excitation at 500 nm [14].

Acknowledgments

D.E.B. is supported by NIDDK (K08DK093705, R03DK109232), Burroughs Wellcome Fund, American Society of Hematology, and the Doris Duke Charitable, Charles H. Hood, and Cooley's Anemia Foundations.

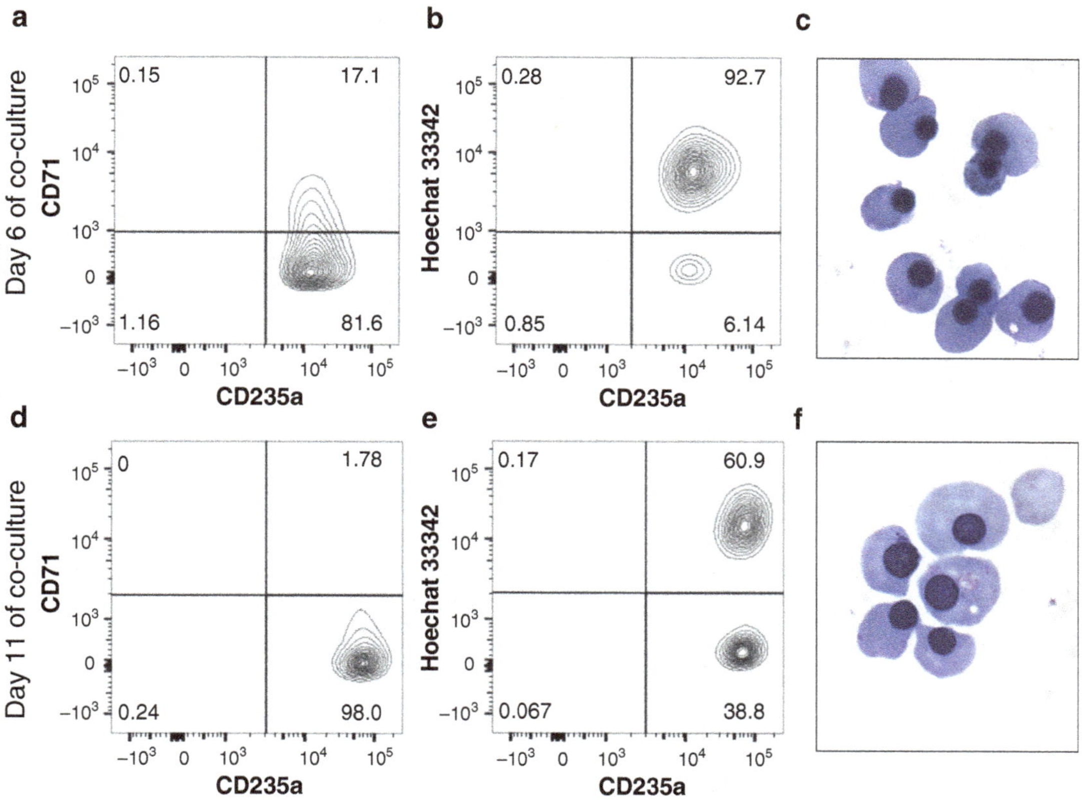

Fig. 3 Terminal erythroid differentiation of HUDEP-2 cells. On Day 6 of coculture with MS-5 cells, HUDEP-2 cells show (**a**) decreased CD71 staining, (**b**) approximately 6% of the cells are enucleated, and (**c**) most cells are at the orthochromatic erythroblast stage by May-Grunwald and Giemsa staining. On day 11 of coculture with MS-5 cells, the majority of HUDEP-2 cells are (**d**) $CD71^{-}CD235a^{+}$, (**e**) 38% of the cells are enucleated, and (**f**) enucleated cells are observed in May-Grunwald and Giemsa-stained cytospins

References

1. Tsiftsoglou AS, Vizirianakis IS, Strouboulis J (2009) Erythropoiesis: model systems, molecular regulators, and developmental programs. IUBMB Life 61(8):800–830. doi:10.1002/iub.226. Review. PubMed PMID: 19621348
2. Olivier E, Qiu C, Bouhassira EE (2012) Novel, high-yield red blood cell production methods from CD34-positive cells derived from human embryonic stem, yolk sac, fetal liver, cord blood, and peripheral blood. Stem Cells Transl Med 1(8):604–614. doi:10.5966/sctm.2012-0059. PubMed PMID: 23197866; PubMed Central PMCID: PMC3659727
3. Lapillonne H, Kobari L, Mazurier C, Tropel P, Giarratana MC, Zanella-Cleon I, Kiger L, Wattenhofer-Donzé M, Puccio H, Hebert N, Francina A, Andreu G, Viville S, Douay L (2010) Red blood cell generation from human induced pluripotent stem cells: perspectives for transfusion medicine. Haematologica 95(10):1651–1659. doi:10.3324/haematol.2010.023556. PubMed PMID: 20494935; PubMed Central PMCID: PMC2948089
4. Migliaccio G, Di Pietro R, di Giacomo V, Di Baldassarre A, Migliaccio AR, Maccioni L, Galanello R, Papayannopoulou T (2002) In vitro mass production of human erythroid cells from the blood of normal donors and of thalassemic patients. Blood Cells Mol Dis 28 (2):169–180. PubMed PMID: 12064913
5. Giarratana MC, Rouard H, Dumont A, Kiger L, Safeukui I, Le Pennec PY, François S, Trugnan G, Peyrard T, Marie T, Jolly S, Hebert N, Mazurier C, Mario N, Harmand L, Lapillonne H, Devaux JY, Douay L (2011) Proof of principle for transfusion of in vitro-generated red blood cells. Blood 118(19):5071–5079.

doi:10.1182/blood-2011-06-362038. PubMed PMID: 21885599; PubMed Central PMCID: PMC3217398

6. Kurita R, Suda N, Sudo K, Miharada K, Hiroyama T, Miyoshi H, Tani K, Nakamura Y (2013) Establishment of immortalized human erythroid progenitor cell lines able to produce enucleated red blood cells. PLoS One 8(3): e59890. doi:10.1371/journal.pone.0059890. PubMed PMID: 23533656; PubMed Central PMCID: PMC3606290
7. Münger K, Howley PM (2002) Human papillomavirus immortalization and transformation functions. Virus Res 89(2):213–228. Review. PubMed PMID: 12445661
8. Canver MC, Smith EC, Sher F, Pinello L, Sanjana NE, Shalem O, Chen DD, Schupp PG, Vinjamur DS, Garcia SP, Luc S, Kurita R, Nakamura Y, Fujiwara Y, Maeda T, Yuan GC, Zhang F, Orkin SH, Bauer DE (2015) BCL11A enhancer dissection by Cas9-mediated in situ saturating mutagenesis. Nature 527(7577):192–197. doi:10.1038/nature15521. PubMed PMID: 26375006; PubMed Central PMCID: PMC4644101
9. Masuda T, Wang X, Maeda M, Canver MC, Sher F, Funnell AP, Fisher C, Suciu M, Martyn GE, Norton LJ, Zhu C, Kurita R, Nakamura Y, Xu J, Higgs DR, Crossley M, Bauer DE, Orkin SH, Kharchenko PV, Maeda T (2016) Transcription factors LRF and BCL11A independently repress expression of fetal hemoglobin. Science 351(6270):285–289. doi:10.1126/science.aad3312. PubMed PMID: 26816381; PubMedCentral PMCID: PMC4778394
10. Traxler EA, Yao Y, Wang YD, Woodard KJ, Kurita R, Nakamura Y, Hughes JR, Hardison RC, Blobel GA, Li C, Weiss MJ (2016) A genome-editing strategy to treat β-hemoglobinopathies that recapitulates a mutation associated with a benign genetic condition. Nat Med 22(9):987–990. doi:10.1038/nm.4170. PubMed PMID: 27525524
11. Addgene. https://www.addgene.org/crispr/cut/. Accessed Nov 2016
12. Addgene. https://www.addgene.org/viral-vectors/lentivirus/. Accessed Nov 2016
13. Dzierzak E, Philipsen S (2013) Erythropoiesis: development and differentiation. Cold Spring Harb Perspect Med 3(4):a011601. doi:10.1101/cshperspect.a011601. Review. PubMed PMID: 23545573; PubMed Central PMCID: PMC3684002
14. Karasawa S, Araki T, Nagai T, Mizuno H, Miyawaki A (2004) Cyan-emitting and orange-emitting fluorescent proteins as a donor/acceptor pair for fluorescence resonance energy transfer. Biochem J 381(Pt 1):307–312. PubMed PMID: 15065984; PubMed Central PMCID: PMC1133789

Chapter 18

Good Manufacturing Practice (GMP) Translation of Advanced Cellular Therapeutics: Lessons for the Manufacture of Erythrocytes as Medicinal Products

Neil W.A. McGowan, John D.M. Campbell, and Joanne C. Mountford

Abstract

Blood transfusion is a mainstay of modern medical practice. In many parts of the world the use of this life-saving therapy is hampered by issues of supply and the potential for transfusion transmitted infections. Accordingly, there are many studies seeking to find an alternative to donated red blood cells (RBCs) for transfusion, including large-scale production from adult and pluripotent stem cells, or erythroid cell lines. Translating basic studies, using any cell lineage, into protocols that are suitable for the generation of cellular therapies requires a wide range of biological and regulatory procedures to be put in place. Additionally, there are specific challenges for the production of RBCs caused by the number of cells needed for a single dose (approx. 1–2 $\times$ 10^{12}). In this chapter, we will review critical areas in the development and good manufacturing practice (GMP) translation of cellular therapeutics through to early phase clinical trials and how this learning can be applied to in vitro RBC therapies.

Key words Good manufacturing practice (GMP), Cellular therapy, Translation, Red blood cells, Regulatory requirements

1 Introduction

Basic studies of erythropoiesis and RBC function present great potential for clinical translation. Blood transfusion is a mainstay of modern medical practice and in many parts of the world the use of this vital therapy is restricted by supply issues, compatibility (matching blood groups to patients), and the potential for transfusion transmitted infections [1, 2]. There are many studies seeking to find an alternative to donated RBCs for transfusion including large-scale production from adult and pluripotent stem cells, or erythroid cell lines [3] some of which are described in the previous chapters. However, there are many considerations before basic studies of any cell lineage can be translated into protocols that are suitable for the

Joyce A. Lloyd (ed.), *Erythropoiesis: Methods and Protocols*, Methods in Molecular Biology, vol. 1698,
DOI 10.1007/978-1-4939-7428-3_18, © Springer Science+Business Media LLC 2018

generation of cellular therapies; these include the use of raw materials of sufficient grade, preparation of highly defined and validated standard operating procedures (SOPS), separation technology for recovery of the final product, and availability of licensed manufacturing facilities. There are specific challenges for the production of RBCs. The number of cells needed for a single dose is very high (approx. $1–2 \times 10^{12}$ RBCs in a standard unit of red cell concentrate), while other somatic cell therapies are often used in $10^{6}–10^{8}$ scale, at least in early trials [4]. Other areas include: the increased risk of nucleated, undifferentiated cells surviving the production and purification processes; the cost of goods; the very large-scale manufacturing facilities required and the need for fully validated assays for in-process and final product characterization (including functional markers). Some of these issues are somewhat mitigated by the lack of a nucleus in the final reticulocyte/RBC product. There are also very well-described and established standards for filtration and storage of RBC concentrate for transfusion that can be used to formulate in vitro RBC [5].

2 What Is a Cellular Therapy?

Cellular therapy is defined as '*a product in which human cells are administered to the body, for the purpose of replacing, repairing, regenerating, or enhancing function of tissues*' [6].

The main stages and challenges in the development of a novel cellular therapy and how this could apply to the manufacture of an in vitro RBC product [7] are summarized in Fig. 1. The manufacture of cellular therapies for clinical use is subjected to increasingly stringent regulation, which is largely dependent on the level of manipulation or potential use in the recipient. In general terms, cell therapies are classified as medicinal products or biologics, and in this regard the regulation of these therapies belongs to two main areas of legislation, public health legislation and pharmaceutical legislation.

The regulatory framework for cellular therapies differs between the EU, US, and other countries [8, 9] and the detailed legal requirements for each stage remain extensive. Reference should be made to the appropriate regulatory bodies and/or competent authorities in the country of origin to determine the appropriate route, although the vast majority of countries active in cellular therapy manufacture broadly follow the process outlined in Fig. 1.

3 Research & Development (R&D)

The first step toward a cellular therapy involves R&D familiar to basic and clinical researchers; this usually incorporates the transition from in vitro cell line models, often of animal origin, to a system

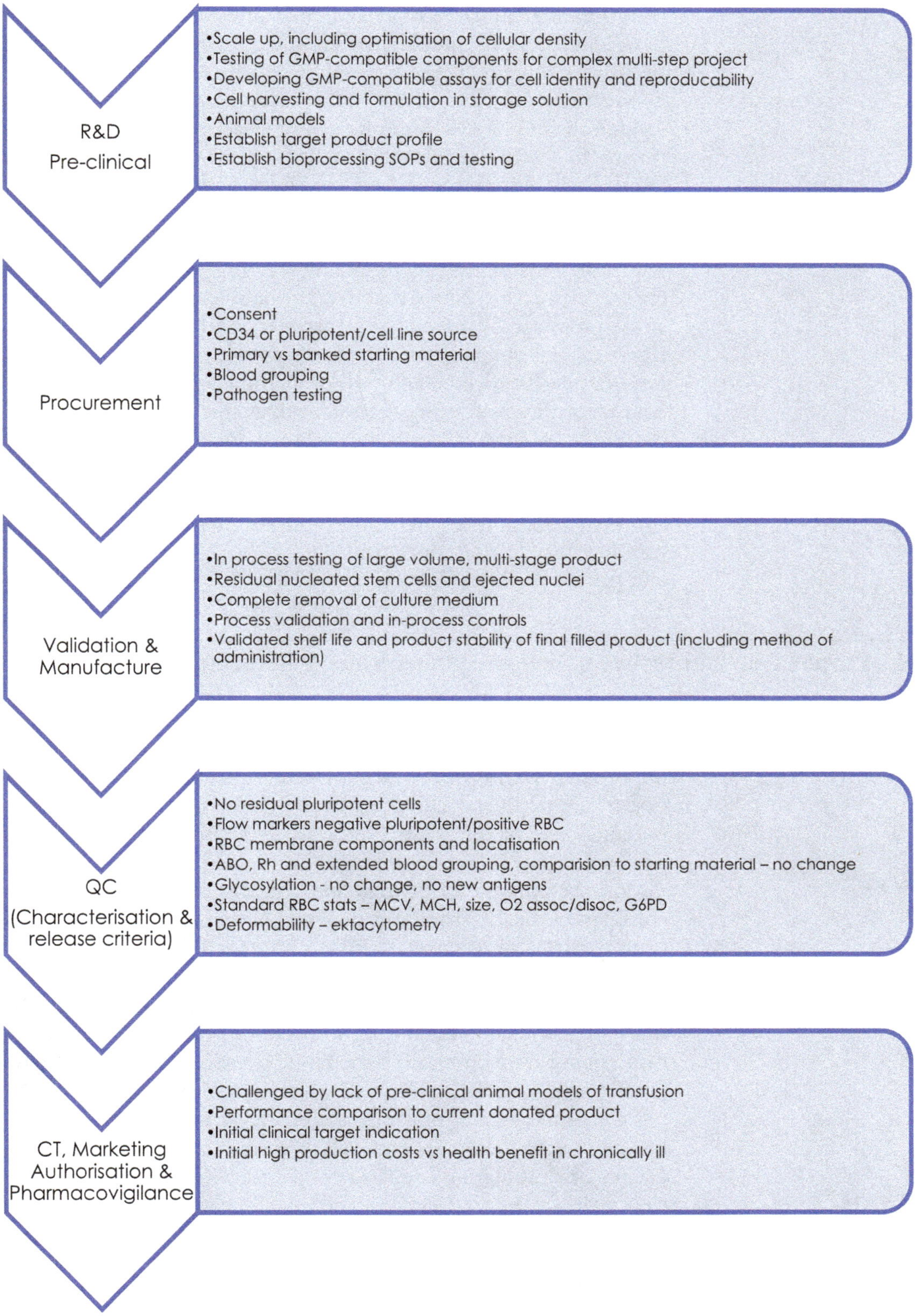

Fig. 1 Stages in the development of an RBC medicinal product

incorporating human tissues and cells and/or demonstration of desired effects in an animal model. While this stage is not covered by the stringent regulatory requirements observed further downstream in the process, it is essential to consider future validation and Good Manufacturing Practice (GMP) requirements to avoid significant changes to protocols or reagents later. Early engagement between development, manufacture, and Quality Control departments (QC) will reduce the chances of often costly delays. In an ideal scenario, research and development in consultation with manufacturing and QC staff will ensure a process that is readily translatable to GMP, with the subsequent establishment of in-process controls and release criteria. In this regard, consideration should be given to Good Laboratory Practice (GLP) [9], or at the very least high-quality data recording as documentation even at this early stage is often required for submission within regulatory product dossiers as part of the authorization process. The aim of this stage is to have developed a robust process that can be used to generate the final product, key to this is the establishment of a target product profile and optimization of the bio-processing steps and protocols that will be applied [10].

4 Procurement and Testing of Human Starting Material for Cellular Therapy

For all cellular therapies manufacture begins with the procurement and testing of the donor and/or starting material. Of paramount importance in this process is informed consent with the donor or donor family (in the case of cadaveric tissues) being aware of the potential use of any tissues or cells and impact of any testing performed. Good Clinical Practice (GCP) [11] should be followed during the consenting process and controls must be in place to consider medical history and possible contraindications to donation, particularly when materials are intended for allogeneic use. With procurement begins GMP in that various controls are introduced at this stage to mediate this process, for example controlled transport, storage, and labeling to preserve product quality, while maintaining anonymity and traceability. Consideration must also be given to the appropriate disposal of any waste material due to the often sensitive nature of the procurement process. The minimum testing criteria for tissues and cells, regardless of whether for allogeneic or autologous use, usually include human immunodeficiency viruses 1 and 2 (HIV-1, HIV-2), hepatitis B and C viruses (HBV, HCV), syphilis, and human T-lymphotrophic virus-1 (HTLV-1). Additional testing requirements can be included and while only serological testing is often required virological testing is preferred to minimize any risk of transmission.

5 Manufacture of Cellular Therapies

Manufacture of cellular therapies as medicinal products must take place in appropriately licensed facilities that conform to GMP requirements [12, 13] and operate under appropriate Quality Management Systems (QMS). As described earlier, effective communication and engagement between R&D and manufacturing will ensure that GMP translation of protocols remains a relatively uncomplicated transition with the main outcome being to fully document the manufacturing process, define in-process controls and acceptance criteria of the cell therapy product that can then be manufactured in a reproducible manner as defined in a Standard Operating Procedure (SOP). The main attributes of a QMS include document control, incident reporting, change control, risk assessment, and validation, all of which will need to be established and adhered to as part of the GMP manufacturing process.

Development runs using GMP should be performed where possible and it is acceptable (and indeed often beneficial) to perform these in nonclassified work areas in the first instance before applying full GMP in a cleanroom environment. This will allow early issues to be identified and compared against the R&D process to ensure consistency of product, while mitigating the expense associated with working to full GMP. In addition, this will hopefully avoid the need for changes that will require additional validation prior to implementation. It is therefore vital at this stage to utilize the reagents and consumables of the appropriate grade to fully assess the impact of moving to GMP-grade materials.

Given the nature of many cellular therapies, it is often impossible to incorporate the terminal sterilization measures used for chemical or biological agents without detrimental effect on product quality [12, 13]. In this regard, all raw materials particularly those contacting product or excipients must be subjected to thorough risk assessment with GMP-grade materials being employed where available. Materials of animal origin should be avoided to eliminate the risk of transmissible spongiform encephalopathy (TSE) and other adventitious agents and it is essential to develop good working relationships with suppliers to gain access to formulation data and information on material manufacture. Supply from dedicated manufacturing equipment and production lines for GMP raw materials is always preferable, and declaration of any material of animal origin used in the preparation of the raw materials for risk assessment is required. Any synthetic raw materials should confirm to the appropriate Pharmacopoeial monograph for the country of origin for example in the UK [14] and USA [15].

Following development GMP runs, the next stage is process validation in which the process should be conducted to full GMP within the appropriate classified environment (Grade A/B/C cleanroom suite) using final versions of all SOPs, with all equipment and tests/assays in a fully qualified state. Defined in-process controls and acceptance criteria should be agreed upon prior to commencing process validation and this should include all tests, even those that would normally be retrospective following product release. The ultimate aim of process validation is to demonstrate the consistent production of a fully defined, sterile efficacious product [12, 13]. This should include the demonstration that intermediates also satisfy in-process acceptance criteria, particularly when manufacturing involves stages, e.g., in master cell banking, where acceptance criteria pre- and post-cryopreservation must be applied. Stability studies should also be incorporated into process validation to define the cellular product shelf-life under controlled conditions. Process simulations must also be established to replicate the process using microbial nutrient broth to demonstrate that the process can be performed from end-to-end in an aseptic manner and to identify any intervention steps that present a risk, and therefore where implementation of a specific microbial control strategy is needed.

Product release criteria can be defined upon the outcomes of acceptance criteria set for process validation, which as a minimum should include assessment of identity, purity, viability, potency, and sterility. Upon the basis of evidence gathered during process validation, it is often acceptable to reduce the acceptance criteria applied, to release criteria in order that products can be released in a timely manner, given the often short shelf-lives associated with cellular therapies. While not being appropriate for clinical use it is often suitable to utilize products prepared during process validation in additional nonclinical safety studies, including animal models, all of which would be included in subsequent regulatory submissions.

6 Clinical Trials of Cellular Therapies

All cellular therapy clinical trials are conducted in compliance with GCP [7] and follow the conventional Phase I through IV process, with initial trials designed to confirm preclinical studies and safety, followed by Phase II/III assessing efficacy in increasing larger patient cohorts. Initial studies will often incorporate a dose escalation phase and special consideration needs to be given to the nature of the product given the shelf-life restrictions and method of administration [16]. Process validation should incorporate as much of the product administration process as possible, e.g., if administration sets with filters, storage at certain temperatures, or reconditioning is required prior to infusion/transplantation the impact of this on product quality should be assessed.

7 Marketing Authorization and Pharmacovigilance

All cell therapies require authorization prior to marketing by the EMA or FDA in the EU or US, respectively (and other relevant regulatory bodies in other countries). Post-authorization considerations such as pharmacovigilance are still in relative infancy for cellular therapies given the novel nature of these agents. Nevertheless, given the therapeutic potential and range of these therapies in clinical trial this is anticipated to become a considerable area of expertise that will influence the future direction of cellular therapy development and manufacture [17].

8 Developing an In Vitro RBC Medicinal Product

These general cellular therapy regulations and best practices should be applied to the development of in vitro derived red blood cells for transfusion use as illustrated in Fig. 1. The use of donated RBC is perhaps the most established and well-characterized cellular therapy, having been an important part of clinical practice since the establishment of blood banks in the early twentieth century [18], resulting in extensive guidance on the specification of products and processes involved in their production. If in vitro RBC are to be considered a potential alternative to donated RBC, it will be necessary to demonstrate equivalent, or improved, functional and safety characteristics in the new product; thus, in vitro RBC should be assessed using the existing criteria for donated RBC. The pre-existing guidelines and product specification may be seen as beneficial for the development of this product, as it may be more challenging to demonstrate suitability of other stem cell-derived cellular therapies with no current clinical equivalent, e.g., dopaminergic neurons, hepatocytes, cardiomyocytes.

References

1. Mountford JC, Turner M (2011) In vitro production of red blood cells. Transfus Apher Sci 45(1):85–89. doi:10.1016/j.transci.2011.06.007. Review. PubMed PMID: 21723197
2. Watkins NA, Dobra S, Bennett P, Cairns J, Turner ML (2012) The management of blood safety in the presence of uncertain risk: a United kingdom perspective. Transfus Med Rev 26(3):238–251. doi:10.1016/j.tmrv.2011.09.003. Review. PubMed PMID: 22126710
3. Olivier EN, Marenah L, McCahill A, Condie A, Cowan S, Mountford JC (2016) High-efficiency serum-free feeder-free Erythroid differentiation of human pluripotent stem cells using small molecules. Stem Cells Transl Med 5(10):1394–1405. PubMed PMID: 27400796; PubMed Central PMCID: PMC5031182
4. https://ct.catapult.org.uk/resources/cell-therapy-catapult-clinical-trials-database
5. Norfolk D (ed) (2013) Handbook of transfusion medicine, 5th edn. TSO, Norwich. isbn:9780117068469
6. PAS 83:2012. Developing human cells for clinical applications in the European Union and the United States of America
7. Mittra J, Tait J, Mastroeni M, Turner ML, Mountford JC, Bruce K (2015) Identifying viable regulatory and innovation pathways for

regenerative medicine: a case study of cultured red blood cells. New Biotechnol 32 (1):180–190. doi:10.1016/j.nbt.2014.07.008. PubMed PMID: 25094050

8. https://www.pactgroup.net/content/clinical-research-roadmap-0
9. http://www.oecd.org/env/ehs/testing/good laboratorypracticeglp.htm
10. Campbell A, Brieva T, Raviv L, Rowley J, Niss K, Brandwein H, Oh S, Karnieli O (2015) Concise review: process development considerations for cell therapy. Stem Cells Transl Med 4(10):1155–1163. doi:10.5966/sctm.2014-0294. Review. PubMed PMID: 26315572; PubMed Central PMCID: PMC4572896
11. http://www.ich.org/products/guidelines/efficacy/efficacy-single/article/good-clinical-practice.html
12. http://ec.europa.eu/health/documents/eudralex/vol-4/index_en.htm
13. http://www.fda.gov/Drugs/GuidanceComplianceRegulatoryInformation/Guidances/ucm064971.htm
14. https://www.edqm.eu/en/european-pharmacopoeia-8th-edition-1563.html
15. US FDA (2015) Considerations for the Design of Early-Phase Clinical Trials of Cellular and Gene Therapy Products; Guidance for Industry. http://www.fda.gov/downloads/BiologicsBloodVaccines/GuidanceComplianceRegulatoryInformation/Guidances/CellularandGene Therapy/UCM359073.pdf
16. http://www.usp.org/
17. https://ct.catapult.org.uk/case-study/creating-eu-market-access-roadmaps-cell-and-gene-therapies
18. Giangrande PL (2000) The history of blood transfusion. Br J Haematol 110(4):758–767. PubMed PMID: 11054057

Index

Joyce A. Lloyd (ed.), *Erythropoiesis: Methods and Protocols*, Methods in Molecular Biology, vol. 1698, DOI 10.1007/978-1-4939-7428-3, © Springer Science+Business Media LLC 2018

D

E

F

G

H

I

www.ingramcontent.com/pod-product-compliance
Lightning Source LLC
Chambersburg PA
CBHW082211270126
38381CB00036B/2585
* 9 7 8 1 4 9 3 9 7 4 2 7 6 *